T0338387

Bio Monomers for Green Polymeric Composite Materials

Bio Monomers for Green Polymeric Composite Materials

Edited by

P.M. Visakh
TUSUR University
Russia

Oguz Bayraktar
Ege University
Turkey

Gopalakrishnan Menon
Tomsk State University
Russia

This edition first published 2020
© 2020 John Wiley & Sons Ltd

The right of P.M. Visakh, Oguz Bayraktar, and Gopalakrishnan Menon to be identified as the authors of the editorial material in this work has been asserted in accordance with law.

Registered Offices
John Wiley & Sons, Inc., 111 River Street, Hoboken, NJ 07030, USA
John Wiley & Sons Ltd, The Atrium, Southern Gate, Chichester, West Sussex, PO19 8SQ, UK

Editorial Office
The Atrium, Southern Gate, Chichester, West Sussex, PO19 8SQ, UK

For details of our global editorial offices, customer services, and more information about Wiley products visit us at www.wiley.com.

Wiley also publishes its books in a variety of electronic formats and by print-on-demand. Some content that appears in standard print versions of this book may not be available in other formats.

Limit of Liability/Disclaimer of Warranty
In view of ongoing research, equipment modifications, changes in governmental regulations, and the constant flow of information relating to the use of experimental reagents, equipment, and devices, the reader is urged to review and evaluate the information provided in the package insert or instructions for each chemical, piece of equipment, reagent, or device for, among other things, any changes in the instructions or indication of usage and for added warnings and precautions. While the publisher and authors have used their best efforts in preparing this work, they make no representations or warranties with respect to the accuracy or completeness of the contents of this work and specifically disclaim all warranties, including without limitation any implied warranties of merchantability or fitness for a particular purpose. No warranty may be created or extended by sales representatives, written sales materials or promotional statements for this work. The fact that an organization, website, or product is referred to in this work as a citation and/or potential source of further information does not mean that the publisher and authors endorse the information or services the organization, website, or product may provide or recommendations it may make. This work is sold with the understanding that the publisher is not engaged in rendering professional services. The advice and strategies contained herein may not be suitable for your situation. You should consult with a specialist where appropriate. Further, readers should be aware that websites listed in this work may have changed or disappeared between when this work was written and when it is read. Neither the publisher nor authors shall be liable for any loss of profit or any other commercial damages, including but not limited to special, incidental, consequential, or other damages.

Library of Congress Cataloging-in-Publication Data

Names: P.M., Visakh, editor. | Bayraktar, Oguz, editor. | Menon,
    Gopalakrishnan, 1980- editor.
Title: Bio monomers for green polymeric composite materials / edited by Dr.
    P.M. Visakh, TUSUR University, Russia, Dr. Oguz Bayraktar, Department of
    Chemical Engineering, Ege University, Bornova, Izmir, Turkey, Dr.
    Gopalakrishnan Menon, Tomsk State University, Russian Federation.
Description: First edition. | Hoboken, NJ : John Wiley & Sons, Inc., [2020] |
    Includes bibliographical references and index. |
Identifiers: LCCN 2018061443 (print) | LCCN 2019000950 (ebook) | ISBN
    9781119301691 (Adobe PDF) | ISBN 9781119301707 (ePub) | ISBN 9781119301646
    (hardcover)
Subjects: LCSH: Biomedical materials. | Polymeric composites. | Polymer
    colloids. | Biofilms. | Green chemistry. | Monomers.
Classification: LCC R857.M3 (ebook) | LCC R857.M3 B45 2020 (print) | DDC
    610.28/4—dc23
LC record available at https://lccn.loc.gov/2018061443

Cover Design: Wiley
Cover Image: © T.Thinnapat/Shutterstock

Set in 10/12pt WarnockPro by SPi Global, Chennai, India
Printed and bound in Singapore by Markono Print Media Pte Ltd

10 9 8 7 6 5 4 3 2 1

Contents

List of Contributors

Bilahari Aryat
International and Inter University Center
for Nanoscience and Nanotechnology,
Mahatma Gandhi University,
India

Nor Asikin Awang
Faculty of Chemical and Energy
Engineering, Advanced Membrane
Technology Research Centre (AMTEC),
School of Chemical and Energy
Engineering, Faculty of Engineering,
University Teknologi Malaysia,
Malaysia

Mohamad Azuwa Mohamed
Faculty of Chemical and Energy
Engineering, Advanced Membrane
Technology Research Centre (AMTEC),
School of Chemical and Energy
Engineering, Faculty of Engineering,
University Teknologi Malaysia,
Malaysia

Dipali R. Bagal-Kestwal
Institute of Food Science and Technology,
National Taiwan University,
section 4, Taiwan, ROC

Muhammad Taqi Zahid Butt
Faculty of Engineering and Technology,
Department of Metallurgy and Materials,
University of the Punjab,
Lahore, Pakistan

Been Huang Chiang
Institute of Food Science and Technology,
National Taiwan University,
Taiwan, ROC

Deepu A. Gopakumar
International and Inter-University Center
for Nanoscience and Nanotechnology,
Mahatma Gandhi University,
India

and

School of Industrial Technology,
Universiti Sains Malaysia,
Malaysia

Nafisa Gull
Research Scholar, Department of Polymer
Engineering and Technology,
University of the Punjab,
Lahore, Pakistan

Atif Islam
Research Scholar, Department of Polymer
Engineering and Technology,
University of the Punjab,
Lahore, Pakistan

Ahmad Fauzi Ismail
Faculty of Chemical and Energy
Engineering, Advanced Membrane
Technology Research Centre (AMTEC),
School of Chemical and Energy
Engineering, Faculty of Engineering,
University Teknologi Malaysia,
Malaysia

Athira Johnson
International and Inter-University Center
for Nanoscience and Nanotechnology,
Mahatma Gandhi University,
India

M. Karthika
International and Inter-University Center
for Nanoscience and Nanotechnology,
Mahatma Gandhi University,
India

Shahzad Maqsood Khan
Research Scholar, Department of Polymer
Engineering and Technology,
University of the Punjab,
Lahore, Pakistan

M.H. Pan
Institute of Food Science and Technology,
National Taiwan University, No.1,
Roosevelt Road, section 4, Taiwan, ROC,
106

Muhammad Abdur Rehman
Department of Geology and
Department of Chemistry,
Jalan University,
Kuala Lumpur,
Malaysia

and

Quality Control Laboratory,
Commander Agro,
Multan,
Pakistan

Zia ur Rehman
University of Agriculture Faisalabad,
Toba Tek Singh, Pakistan

Amin Saboktakin
Nanostructured Laboratory, NanoBMat
Company,
GmbH, Hamburg, Germany

Mohammadreza Saboktakin
Nanostructured Laboratory, NanoBMat
Company GmbH,
Hamburg, Germany

Neelakandan M. Santhosh
International and Inter-University Center
for Nanoscience and Nanotechnology,
Mahatma Gandhi University,
India

Nitheesha Shaji
International and Inter-University Center
for Nanoscience and Nanotechnology,
Mahatma Gandhi University,
India

Sabu Thomas
International and Inter-University Center
for Nanoscience and Nanotechnology,
Mahatma Gandhi University,
India

Şükrü Tüzmen
Molecular Biology and Genetics Program,
Department of Biological Sciences,
Eastern Mediterranean University,
North Cyprus, Famagusta, Turkey

P.M. Visakh
Department of Physical Electronics,
TUSUR University,
Tomsk, Russia

Wan Norharyati Wan Salleh
Advanced Membrane Technology
Research Centre (AMTEC),
Faculty of Chemical and Energy
Engineering and Faculty of Engineering,
University Teknologi Malaysia
Malaysia

V.K. YaduNath
International and Inter University Center
for Nanoscience and Nanotechnology,
Mahatma Gandhi University,
India

Zulal Yalinca
Department of Chemistry,
Eastern Mediterranean University,
North Cyprus, Famagusta
Turkey

Preface

This book summarizes many of the recent research accomplishments in the area of biomonomers for green polymeric composites and their nanocomposites. It includes an introduction to biomonomers for green polymers, the current status of these compounds, new challenges, opportunities, and processing methods for bionanocomposites, biopolymeric material-based blends, the preparation, characterization, and applications of biomonomers and their nanocomposites, applications of biopolymeric gels in medical biotechnology, an introduction to green polymeric membranes, properties and applications of gelatins, pectins, and carrageenans gels, the biodegradation of green polymeric composite materials, applications of green polymeric composites materials, hydrogels used for biomedical applications, and natural aerogels as thermal insulations.

This book is a valuable reference source for university and college faculties, professionals, post-doctoral research fellows, senior graduate students, and researchers from R&D laboratories working in the area of biomonomers and green polymeric composites materials. The chapters are contributed by prominent researchers from industry, academia, and government/private research laboratories across the globe and present an up-to-date record of the major findings and observations in the field of biomonomers and green polymeric composite materials.

The first chapter discusses the state of art and new challenges of biomonomers and green polymeric composite materials.

The second chapter covers several topics, including classification of nanobiocomposites, general processing methods, and properties. The author also includes many subtopics, such as polysaccharide nanocomposites, animal protein-based nanocomposites, plant protein-based nanocomposites, metal nanocomposites, and inorganic nanocomposites. General processing methods such as pressure extrusion, solid-state shear pulverization, electrospinning, solution casting, evaporation, melt intercalation, *in situ* polymerization, drying techniques and polymer grafting are also discussed in this chapter.

Various topics on biopolymeric material-based blends, such as preparation, characterization, and applications, are addressed in Chapter 3. This chapter presents a comprehensive study of biopolymeric material-based blends, including general preparative methods, aqueous blending technology, their hydrophilic or hydrophobic nature, degradation problems, thermodynamics of miscibility, and their opportunities or challenges. Biopolymeric blends have attracted academic, research, and industrial scientists' attention as their properties are desirable for various applications. The challenges due to the unique structure, preparative methods, and resultant properties of blends of natural polymeric materials are discussed, and examples are drawn from the scientific literature. The various forms of natural polymers, i.e. polysaccharides, proteins, lipids, natural rubber, chitosan, starch, and silk-based blends, are reviewed with respect to preparative techniques, characterization methods, and various applications.

The fourth chapter of this book discusses the applications of biopolymeric gels in medical biotechnology. The primary objective of this chapter is to review the literature regarding the classification of the properties of hydrogels and their biomedical applications. The composition and structure of hydrogels, especially their use in biological fields, makes them ideal candidates for biopharmaceutical implementation. Innovations in recent manufacturing and world-wide resources of hydrogels are also reported.

The fifth chapter introduces green polymeric membranes, covering types of green polymeric membranes, their physicochemical properties, and their potential applications. The application of these materials in various industries was facilitated by their tremendous and significant physicochemical properties. However, despite these advances, there are still some drawbacks which prevent the wider commercialization of green polymeric membranes in many applications. This chapter reviews the current trend of research involving green polymeric membranes that focuses on the fabrication method, processing, and surface and structure modification. In addition, the long-term stability and durability of green polymeric membranes for specific applications has become a challenge to researchers all around the world. The introduction of nanostructure fillers (e.g. graphene oxide, metal oxides, carbon nanotubes, nano-clay, etc.) and the blending with other polymers, or the making of new copolymers, has significantly improved their overall properties and performance. These improvements are generally attained at low filler content, and this nano-reinforcement is a very attractive route to generate new functional green polymeric membranes for various applications. It should be noted that the development of green polymeric membranes with specific physicochemical properties for specific functionalities is crucial for practical applications in industry. Green polymeric membranes with various physicochemical properties have a promising contribution to make in various applications.

Chapter 6 discusses the properties and applications of gelatins, pectins, and carrageenans gels. For each of these substances the authors cover various subtopics, such as structural units, molecular structure, properties, thickening ability, gelling ability, film-forming properties, microbiological properties, food applications, cosmetics applications, and pharmaceutical applications. In the chapter on biodegradation of green polymeric composite materials, the authors consider a wide range of review studies on this subject, including biomechanical pathways for the degradation of green polymers and green polymer composites. Several studies have been carried out to design polymers with biodegradable properties to help keep the environment safe and clean.

In Chapter 8, on applications of green polymeric composites materials, the authors discuss several different topics, including a series of interesting green polymer composites developed from thermoplastic starch and its blends, poly(lactic acids) and its modifications, cellulose, gelatin, and chitosan. The authors also describe how natural fibers have more environmentally friendly properties than synthetic fibers synthesized from agricultural sources such as jute, banana, bamboo, and coconut coir, etc. There is thus a wide range of possible applications of nanocomposites from agriculture to automobiles. However, problems of poor adhesion of matrix and fiber, difficulty with fiber orientation, achieving nanoscale sizes, and the evolution of truly green polymers that are environmentally friendly and renewable must first be solved.

In Chapter 9, constituents, fabrication, crosslinking and clinical applications of hydrogels are described. Hydrogels are extensively found in everyday products although their potential has so far not been thoroughly investigated. The authors review the fabrication and composition of hydrogels along with their different properties, and the natural and synthetic polymers used for the development of hydrogels in the presence of different crosslinking agents. The major characteristics of hydrogels related to clinical, pharmaceutical, and biomedical applications are also identified, particularly for applications of hydrogels in contact lenses, oral drug delivery, wound healing, tissue engineering matrices, and gene delivery.

The final chapter examines the use of natural aerogels as thermal insulation and in other applications. A further review of natural aerogel-based composites and nanocomposites is also provided.

The editors would like to express their sincere gratitude to all the contributors to this book, whose excellent support led to the successful completion of this venture. We are grateful to them for the commitment and the sincerity they have shown toward their contribution to the book. Without their enthusiasm and support, the compilation of this book would have not been possible. We would like to thank all the reviewers who have given their valuable time to make critical comments on each chapter. We also thank Wiley for recognizing the demand for such a book, and for realizing the increasing importance of the area of bio monomers for green polymeric composites materials and supporting this project.

Tomsk, Russia, March 2019

Dr. P.M. Visakh
Dr. Oguz Bayraktar
Dr. Gopalakrishnan Menon

1

Biomonomers for Green Polymers: Introduction

P. M. Visakh

Department of Physical Electronics, TUSUR University, Tomsk, Russia

1.1 Processing Methods for Bionanocomposites

The new generation of hybrid nanostructured materials has two crucial properties: biocompatibility and biodegradability [1, 2]. Exploitation of various biopolymers such as proteins, nucleic acids, polysaccharides, etc. for preparation of nanocomposites has been done in last few decades [3]. Processing methods for matrix and filler are sometimes the same. However, some matrices are prepared using combinations of techniques to achieve the desired quality of bionanocomposites, therefore we will discuss the processing methods for bionanocomposites with suitable examples. Bionanocomposites of polysaccharide matrices are mainly prepared by solvent intercalation or melt processing and not through in situ polymerization where nature of the polysaccharide directly influences the route of preparation. Some polysaccharides with nanostructure fillers are discussed as examples. Most of the cellulose whiskers-reinforced poly(lactic acid) (PLA) nanocomposites are prepared by melt extrusion to avoid agglomeration and aggregation during drying [4]. Porous networks and thickened cellulose ribbons in gelatin/nanocellulose composites are prepared using an enzymatically modified form of gelatin [5]. Cellulose nanocomposites based on nanoparticles, such as clay [6–14], carbon nanotubes (CNTs) [15], graphene, layered double hydroxide (LDH) [16], and silica [17] have been prepared.

Starch is another abundant, inexpensive, naturally renewable and biodegradable polysaccharide, produced by most green plants as an energy store. It is the most common carbohydrate in human diets and animal feeds. Starch nanocomposites are mixtures of starch-based biopolymers with nanofillers (solid layered clays, synthetic polymer nanofibers, cellulose nanowhiskers, CNTs, and other metal nanostructures). Environmentally friendly starch nanocomposites exhibit significant improvements in mechanical properties, dimensional stability, transparency, improved processability, and solvent or gas resistance. Chitosan (CS)/chitin, the second most abundant natural biopolymer, also can be integrated with clay, graphene, and carbon nanostructures to prepare bionanocomposites [18–21]. Due to its high content of amino ($-NH_2$) and hydroxyl ($-OH$) groups, chitosan and its derivatives are excellent adsorbents for the removal of heavy metal ions, fluoride, and organic dyes. Films of spin-coated chitosan–alginate nanocomposite have potential uses in bioapplications. Lignin-based nanocomposite films have been prepared using CNCs (carbon nanocomposites) and used in various applications such as medical, biological, optical and sensors, and electronic [22]. They are also used as adhesives, stabilizing agents, and precursors for many aromatic chemicals. Modified lignins, such as lignosulfates, kraft lignin, and acetylated lignin, contain CNCs or commercial derivatives or nanocellulosic polysaccharides. Polyethylene terephthalate (PET) film coated with graphene oxide (GO)/pullulan nanocomposite can be used in food/pharmaceutical applications [23]. Bionanocomposites with enriched properties based on two microbial polysaccharides, pullulan and bacterial cellulose (BC), were prepared by Trovatti et al. for possible application in organic electronics, dry food packaging and the biomedical field [24]. Pullulan composites with many materials, including chitosan [25], caseinate [26], starch nanocrystals [27], collagen [28], poly (vinyl alcohol) [29], and hydrogel with methacrylate [30], have excellent compatibility.

Their biodegradability, low cost, and surfaced modification with active functional groups for catching targeting molecules make these matrices feasible candidates for applications in the pharmaceutical industry [31]. Electrospun collagen-chitosan nanofibers were stabilized by glutaraldehyde vapor via crosslinking, which afforded a biomimetic extracellular matrix (ECM) for cell growth [32]. Collagen is regarded as one of the most useful biomaterials, exhibiting a number of biological advantages. The outstanding performance and biomedical application of this protein biomaterial have induced researcher interests in synthetic composite material fabrication. Soy protein isolate (SPI) has been extensively studied for bioderived packaging materials. Several recent studies have investigated the improvement of mechanical and barrier properties of nanocomposite films after incorporating nanoclays such as montmorillonite (MMT) [33–41]. Further, these nanocomposite films have also been reported for decreased water vapor and oxygen permeability, and increased elastic modulus and tensile strength, which makes them suitable for packaging industry. Recent studies have also reported that the SPI-based nanocomposite bioplastics with highly exfoliated MMT have significantly improved mechanical strength and thermal stability [42]. Thus, bio-based polycaprolactone–SPI is not only ecofriendly but intercalated nanocomposites with enhanced tensile and dynamic mechanical properties when produced by the melt compounding method [43].

In the case of biocomposites, the properties of the composites produced are dependent on the inter-phase interaction of the reinforced material and matrix. Filler is also a value-added material, but wise selection of processing methodology, optimum conditions, and compatible phase components is needed. Polymer/metal nanocomposites consisting of polymer as matrix and metal nanoparticles as nanofiller commonly exhibit several attractive advantages, such as electrical, mechanical, and optical characteristics [44]. Metal nanocomposites with protein, nucleic acid, and polysaccharides have shown potential applications in drug delivery, tissue engineering, bioimaging, wound healing, biomedicine, energy production and storage, and electronic devices such as biosensors, affinity materials, etc. [45]. Bottom-up methods are found to be promising for controlling the properties and specific orientation of nanomaterials. Thermal evaporation and sputtering techniques have been considered as facile, simple, low-cost, and high-yield methods for synthesis of high-quality nanomaterials/nanostructures [46, 47]. Various immobilization methods, including entrapment, adsorption, crosslinking, electro-polymerization, and encapsulation, have been used for capturing biological moieties in the matrix. This is one of the main processes employed in the manufacturing of nanobiocomposites (NBCs) [48]. There are two main types of extrusion: reactive extrusion and extrusion cooking. Reactive extrusion uses chemical modification via crosslinking [49]. Generally, extrusion technology used in the food industry is referred to as extrusion cooking and results in different physical and chemical properties of the extrudates depending on the raw materials and extrusion conditions used [50]. Various starch nanocomposite varieties have been prepared and reported by many researchers for biodegradable packaging applications in food industry. Moigne et al. developed a continuous CO_2 assisted extrusion process to prepare poly(3-hydroxybutyrate-co-3-hydroxyvalerate)/clays NBC foams with better homogeneity and high porosity [51]. Inventor Torkelson has successfully produced a well-dispersed graphite–polymer nanocomposite [52]. Taking advantage of near-ambient-temperature processing, solid-state shear pulverization (SSSP) was recently used to produce biodegradable polymer matrix composites with starch [53], rice husk ash [54], and eggshell filler [55, 56]. This technique that has proven to effectively disperse nanoscale structural entities to achieve compatibilized polymer blends and exfoliated polymer nanocomposites.

This physical method uses extrusion of the polymer solution with reinforcement of nanomaterials and biological entity for the preparation of NBCs. Polymers, molten at high temperature, can also be made into nanofibers by electrically charging a suspended droplet of polymer melt or solution [57–63]. Instead of a solution, the polymer melt is introduced into the capillary tube. The major difference is that a compound spinneret with two (or more) components can be fed through different coaxial capillary channels [64]. Wet-dry electrospinning and wet-wet electrospinning techniques are used for volatile and non-volatile solvents respectively. Both techniques offer the possibility of producing nanofibers with controlled fiber diameter to make film or membrane or an oriental controlled fiber. Such fibrous scaffolds are ideal for the purpose of tissue regeneration because their dimensions are similar to the components of the extracellular matrix and mimic its fibrillar structure, providing essential signals for cellular assembly and proliferation. Core–shell structured nanofibers where collagen as the shell and poly(ε-caprolactone) (PCL) as core were prepared by co-axial electrospinning and show the advantage of controlled shell thickness and manipulative mechanical strength and degradation properties of the resulting composite nanofibers, without affecting biocompatibility [65]. Thus, such core–shell structured composite nanofibers have potential uses in drug or growth factor encapsulation and the development of highly sensitive sensors and tissue engineering applications [66].

The solution casting method is based on a solvent system in which the polymer or pre-polymer is soluble. The polymer is usually dissolved in a suitable solvent while the nanoparticles are dispersed in the same or a different solvent before the two are mixed. For example, during the preparation of bionanocomposites based on clays, the solvent is normally used to pre-swell the clays [67]. In case of non-water-soluble polymers, additional steps are required before processing. Solvent exchange, use of surfactant, freeze-drying, and chemical modification can be used for this purpose. Melt intercalation is a mechano-chemical process that is highly preferred in case of clay/silicate biocomposites, and this processing method is compatible with extrusion and injection molding. This method is also fast and clean, ecofriendly, and can alter the lifecycle analysis [68]. In this technique nanomaterials and/or biomaterials are mixed with the polymer in the molten state. The process involves mixing the particles with the polymer and heating the mixture above the softening point of the polymer, statically or under shear.

Like other techniques, proper dispersion of the nanoparticles is always a goal during processing. During melt processing, a number of factors are important to achieve homogenous dispersion of reinforced nanomaterial into the polymer matrix, including enthalpic interaction between the polymer matrix, biocomponent, and nanoparticle. Multi-walled CNT-poly methyl methacrylate (MWNT/PMMA) nanocomposite has been prepared by in situ polymerization of MMA dispersed with MWNTs with fairly good dispersion stability [69]. Grafting of long chains can also be used to transform the nanofiller into a co-continuous material by increasing apolar character through grafting agents bearing a reactive end group and a long "compatibilizing" tail [70].

1.2 Biopolymeric Material-based Blends: Preparation, Characterization, and Applications

Preparation of biopolymeric blends is encouraged due to their biocompatible and biodegradable properties [71, 72]. The reason for the increased research in the preparation of versatile biopolymer blends is their broad application, e.g. biomedical applications [73]. The biopolymer blends in this regard have emerged as promising materials with suitable thermal, biocompatible, and mechanical properties for use in the intended applications [74, 75]. The main biopolymers used in the preparation of blends for various applications include collagen, chitin, chitosan, keratin, silk, and elastin, all natural polymers derived from animals [76, 77]. Property of biopolymers is useful in blends formation with the other soluble polymers. The polymers with little solvent affinity, e.g. elastin, silk, or keratin, have a problem during blend formation [78, 79]. The environmentally friendly nature of the biopolymers due to their biodegradable properties is also advantageous [80]. It is also notable that polymers have dominant hydrophobic properties, degrade after use, have mechanical properties, and behavior toward aqueous environments [81–84]. Biopolymeric blends are non-biodegradable in nature [85] and therefore these do not have the natural advantage of environmental friendly properties.

The melt process in making biopolymeric blends is advantageous to overcome various shortcomings associated with the basic physicochemical properties of biopolymers. The new techniques were developed to make biopolymeric blends, e.g. reaction extrusion technology to make starch–cellulose–acetate blends. In this technology, a number of materials were used during the blend formation process. The biodegradable hydrogels were prepared by biopolymeric blends formed by the combination of starch derived from corn starch and cellulose acetate. The blending reaction proceeds through free radical mechanization after reaction between methacrylate or acrylic acid monomers. The free radical reaction was initiated through a redox system consisting of 4-dimethylaminobenzyl alcohols and benzyl peroxide at ambient temperature. The addition of hydroxyapatite content in these blends provided a biocompatible character in the blended materials along with osteoconductive or oleophilic properties. The effect of polyol on the thermal stability, mechanical strength, and water or gas permeability was monitored to establish the usability of these blends. The study findings suggest that the polyol or water contents have a positive impact on the mechanical properties or stability of the biopolymeric blends [85]. The use of acetic acid promoted the formation of sheet-like structures in the blended materials. The characterization results of the biopolymeric blends showed improved thermal stability and degradation stability. In another study, Lazaridou and Biliaderis studied the effect of blended materials on thermal and mechanical properties [86].

Blends formed by the injection molding method show higher tensile strengths, lower water absorption/adsorption, and longer elongation values. The impacts of the chemical or physical properties of the starch on the biopolymeric blend properties were studied by Park and Im [87]. The starch was gelatinized by the addition of a mixture of water and glycerol in a twin-screw mixer. The microstructural properties of the starch blends with PLA, polyethylene, and vernonia oil were prepared by the melt processing technique and acid hydrolysis. Surface fractures were observed during scanning electron microscopic (SEM) examination of the blends. Polyvinyl alcohols (PVOHs) were also used as compatibilizers in PLA-starch-based biopolymeric blends [88]. This study reported that the use of PVOHs enables preparation of blends with better compatibility and improved mechanical strengths.

1.3 Applications of Biopolymeric Gels in Medical Biotechnology

Based on natural and synthetic polymers, hydrogels can be utilized in research into cell encapsulation and in particular have facilitated the establishment of a novel field in tissue engineering. Hydrogels are a significant class of biomaterials in medical biotechnology. These biomaterials categories may include tissue engineering and regeneration, diagnostics, cellular immobilization, cellular biomolecular separation, and utilization of barrier materials for biological adhesion for regulation. Hydrogels are invaluable in their three-dimensional network capacity to capture and release active compounds and biomolecules. Hydrogels have the sponge-like capacity to absorb water due to their hydrophilic functional groups. The water absorbed into hydrogels permits diffusion of certain molecules while the polymer component of the hydrogel acts as a matrix for holding water molecules together. Hydrogels have the ability to imitate the physical, chemical, biological, and electrical features of many tissues. These features earn them the ability to serve as potential candidates for biomaterials. These novel approaches involve super porous hydrogels [89], comb-like grafted hydrogels [90–92], self-assembling hydrogels [93, 94], and recombinant triblock copolymers [95–97]. To construct hydrogel systems with well defined chemical characteristics, information regarding polymer chemistry and synthesis, features of the materials to be utilized, parameters of mode of interaction, material release capability, and delivery systems need to be taken in to consideration. In order to construct hydrogel systems with well-defined chemical characteristics, information regarding polymer chemistry and synthesis, the features of the materials to be utilized, the parameters of the mode of interaction, material release capability and delivery systems need to be taken into consideration. Crosslinking is the most versatile method to facilitate biopolymeric deficiencies [98]. Mechanical properties and the stability of biomaterials can be ameliorated by crosslinking agents. However, limitations exist due to reduced degradability and lack of functional groups and potential cytotoxicity introduced by crosslinking agents [99, 100]. The chemical and physical nature of hydrogels depends on the concentration of crosslinking agents, the degree of crosslinking, the kind of crosslinker/monomer, and the method of preparation used. Hydrogels are excellent candidates for biotechnological applications, including drug delivery systems, since they possess soft, hydrophilic, high swelling ability, a viscoelastic nature, are biodegradable, and have biocompatible characteristics [101]. Hydrogels can provide protection towards small biological molecules and chemical compounds against stringent environmental conditions.

The mode of response of hydrogels may depend on the chemical composition of the polymeric networks. Hydrogels can be modified to respond to environmentally triggered stimuli, including changes to pH, ionic strength, and temperature. Novel treatments for site-specific drug release utilize specifications such as disease-specific enzymatic activities to prompt the release of drugs from hydrogels [102]. Hence, the use of hydrogels provides feasible drug release due to their chemical composition in response to environmental stimuli [103]. Hydrogel-based products are potential candidates for drug-delivery systems, providing the necessary conditions for drug release [104, 105]. Hydrogels are considered to be one of the best-qualified materials for this purpose. Successful applications of tissue engineering are facilitated by the generation of functional hydrogels [106].

1.4 Introduction to Green Polymeric Membranes

Green polymeric membranes have been extensively commercialized in the field of membrane science and technology, and involve various membrane separation processes, e.g. microfiltration, ultrafiltration, nanofiltration, reverse osmosis, gas separation, pervaporation, and renewable energy [107–116]. There is a wide range of naturally occurring polymers produced from renewable resources that are available for numerous material applications. These renewable resources include cellulose and chitosan, which are widely employed in manufacturing. Chitosan-based membranes can be prepared by an immersion–precipitation process that applies silica particles as porogen [117]. Chitosan powder is first dissolved in 2 wt% HNO_3 before being cast as a film. Chitosan-based ceramic membrane can also be prepared via the dip coating technique [118]. Bierhalz et al. formulated chitosan from white mushroom and shrimp shells [119]. Celluloses and chitosan can form cohesive films since they are flexible and they can be cast into different sizes of films. Moreover, there is a strong interaction between cellulose chain molecules due to intramolecular and intermolecular hydrogen bonding within the cellulose molecule. Cellulose can be considered to possess moderate thermal stability. It has been reported that rapid chemical decomposition of cellulose occurs between 315 and 400 °C [120], while chitosan decomposition occurs between 175 and 400 °C. Modification of the surface chemistry of the cellulose also allows the surface groups to be tailored to specific applications. For example, the surface silylation of cellulose will increase the hydrophobicity. It has been reported that cellulose films with a high water contact angle (117–146°) can be prepared from modified cellulose by solution casting [121]. Technologies such as coagulation and ion exchange that were developed for heavy metal removal work well, but have several disadvantages, such as low efficiency and high cost for energy and materials. Besides cellulose, chitosan is another significant biopolymer that can be obtained from chitin by the deacetylation process. Chitosan is biocompatible, non-toxic, and biodegradable in nature. It differs from cellulose in that it will dissolve in most organic solvents, whereas chitosan is soluble in water in which a small amount of acetic acid is present.

Effluent consisting of pharmaceutical compounds such as antibiotics, vasodilators, β-blockers, organic pollutants (e.g. phenolic compounds), and anti-epileptics has been found in most wastewater, sewage, groundwater, and drinking water [122]. Various kinds of bio-based polymers are used for many applications nowadays since they are environmentally friendly and low in cost. Mohamed et al. had successfully prepared regenerated cellulose membrane with photocatalytic properties in an effort to produce green portable photocatalysts and photocatalytic membranes from the degradation of organic pollutants [123]. Thakuro et al. reviewed an application of water purification which involved the separation of pure water from various mixtures of tetrachloride (CCl_4), chloroform ($CHCl_3$), and dichloromethane (CH_2Cl_2) [124]. The fabricated composite membrane was crosslinked with tetraethyl ortosilicate (TEOS) to minimize the swelling and improve the selectivity of the membrane. Such waste discharge, containing colored substances, normally comes from the textile industries and other dyeing fields such as paper, printing, food, and plastics. The approximate volume of discharged wastewater, especially from the textile process, is between 40 and 65 l kg^{-1} of the product [125]. The hybrid CS/OSR (Oxidized starch)/silica membrane produced showed that there was an improvement in terms of thermal stability as well as the ability to lower the degree of swelling in the water.

The applications of biopolymers in dye removal focus on chitosan, but also involved the use of cellulose. As we know, cellulose can be abundantly modified and is a renewable biopolymer in nature. The structure of the cellulose, with the presence of a hydroxyl group, allows it to undergo chemical modification that can improve its efficiency for dye removal from water. Tsurumi et al. reported similar research on the capability of cuprammonium regenerated cellulose hollow fiber (BMM Hollow Fiber) for virus removal applications. In addition, graphene oxide-modified chitosan/polyvinyl pyrrolidone nanocomposite membranes [126] and polycaprolactone membrane coated with chitosan-silver nanoparticles [127] prepared by the electrospinning technique demonstrated remarkable potential applicability in wound-healing tissue engineering applications.

1.5 Properties and Applications of Gelatin, Pectin, and Carrageenan Gels

Gels are defined as non-fluid colloidal networks or polymer networks that expanded throughout whole volume by a fluid [128]. They are liquid by weight but due to a three-dimensional crosslinked network within the liquid they behave like solids. Crosslinking within the fluid results in a gel structure because of the contribution of hydrogen bonds, helix formation, and complexation at the network junction points. A hydrogel is a network of polymer chains that are hydrophilic, and sometimes exhibits as a colloidal gel in which water is the dispersion medium. Hydrogels are highly absorbent and they can contain over 90% water. Hydrogels also possess a degree of flexibility very similar to natural tissue due to their significant water content. The first appearance of the term "hydrogel" in the literature was in 1894 [129]. Hydrogels can be classified by various methods based on source (natural or synthetic) [130], configuration (crystalline, semi-crystalline, or non-crystalline), crosslinking type (chemical or physical), and polymeric composition (homopolymeric, copolymeric, or multipolymer interpenetrating polymeric hydrogel (IPN)), and they are an important and integral part of tissue engineering, cosmetics, dentistry, and the food industry. Readers can get more detailed information from review articles [131–134]. In this chapter we will focus on biopolymeric hydrogels, especially gelatin, pectin, and carrageenan, which are widely used in the food industry.

Gelatin is obtained from the acid, alkaline, or enzymatic hydrolysis of collagen, the chief protein component of the skin, bones, and connective tissue of animals, including fish and poultry [135]. Fish gelatin has similar functional characteristics to mammalian gelatin and has received considerable attention in recent years [136, 137]. Molecular weight (MW) distribution and amino acid composition influence the physical and structural properties of gelatins from various sources [138–140]. Because of their similar amino acid composition and, to some extent, structure, collagen and gelatin show similarities in their properties [141–144]. However, gelatin as a polymeric product exhibits characteristic properties that are important in various industrial applications. The hydrolytic conversion of collagen to gelatin yields different mass peptide chains. Thus, gelatin is not a single chemical entity, but a mixture of fractions composed entirely of amino acids joined by peptide linkages to form polymers varying in molecular mass from 15 to 400 kD [145–152].

The rate of the formation of a helical structure of gelatin depends on numerous factors such as the presence of covalent bonds [153], gelatin molecular weight [154], the presence of imino acids, and the gelatin concentration in the solution. Gelatin is insoluble in less polar organic solvents such as benzene, acetone, primary alcohols, and dimethyl-formamide, but soluble in aqueous solutions of polyhydric alcohols such as glycerol, propylene glycol, etc. It is also soluble in highly polar organic solvents such as acetic acid, trifluoroethanol, and formamide, and partially soluble in benzene, acetone, primary alcohols, and dimethylformamide [155]. Gelatin is hygroscopic, meaning that it absorbs water depending on the relative humidity at which it is stored. Gelatin has positive and negative charges along with uncharged hydrophilic and hydrophobic groups, which make it a polyampholyte [156]. Together with water gelatin forms a semi-solid colloidal gel which is thermoreversible. This is certainly its most interesting property. One part of gelatin can trap 99 parts of water. The two most important properties of gelatin are its melt-in-the-mouth characteristics and its ability to form thermoreversible gels. Gelatin is also stable at a wide range of pH and remains unaffected by ionic strength. It is preferred in many applications because of its clarity and bland flavor. In most cases, except for the food industry, gelatin is used in the solid state. Gelatin is sometimes used in the baking sector. Other than as a food ingredient, gelatin is also used as a food additive and acts as a thickening agent, gelling agent, stabilizer, and emulsifier.

A study by Gomez-Guillén et al. [140] revealed antimicrobial activity associated with gelatin. Gelatin films have commercial application as food packaging films [157, 158] because of their edible, biodegradable and antibacterial, heat sealing, moisture, and oxygen barrier properties. Gelatin is used as a binder in tablet formulations and as a coating to ease swallowing or mask unpleasant tastes. Cosmetics companies are also important users. The properties of gelatin are particularly well suited to the encapsulation of bath oils, and for use in moisturizing lotions and skin creams, making it an important contributor to these products. Neutral sugars such as L-rhamnose, xylose, galactose, and arabinose are also present in gelatin side chains. However, the total content of neutral sugars varies with the source, the extraction conditions, and subsequent treatments

[159]. Gelatin also carries non-sugar substituents, usually methanol, acetic acid, phenolic acids, and occasionally amide groups.

The junction zones resulting from polymer molecule interactions must be of limited size. As a result of large chain size, precipitation may happen instead of gel formation. Pectin use in edible films, paper substitute, foams, and plasticizers, etc. In addition to pectolytic degradation, pectins are susceptible to heat degradation during processing, and this degradation is influenced by the nature of the ions and salts present in the system [160]. Pectin is used in sulfuric in lead accumulators. Carrageenan is a mixture of water-soluble, linear, sulfated galactans. It is an anionic linear sulfated polygalactan polymer with 15–40% ester-sulfate content. The molecular weight of carrageenan is usually high but depends on many factors, such as type of seaweed species, age of seaweed, harvesting season, extraction method, and condition. Carrageenan recovery can be achieved by several methods. Alcohol precipitation or potassium chloride followed by steam heating is commonly used to get "refined carrageenan" (RC) [161]. An alternative process involves immersion and boiling in hot aqueous potassium hydroxide.

1.6 Biodegradation of Green Polymeric Composite Materials

The term "biodegradability" means the decomposition of materials by microorganisms into methane, carbon dioxide, inorganic compounds, water, and biomass [162]. A huge number of biopolymers (collagen, cellulose. starch, chitin, etc.) have been identified and extracted from various biological sources. Based on the source of the material, biopolymers can be classified into biomass products, microorganism-derived products, biotechnologically obtained products, and oil products. Biopolymers like starch, cellulose, and chitin are the major examples of agro polymers and these have a wide range of applications because of their high availability and biodegradability, and low toxicity. The degradation of polymeric materials follows various stages, including biodeterioration, depolymerization, assimilation, and mineralization. One major factor is that the degradation can stop at each stage. Agro-based polymers can be polysaccharides like cellulose that consist of glycosidic bonds and proteins obtained from amino acids. Polymers based on natural products are both biocompatible and biodegradable, but they still need to be technologically acceptable. The major advantages of biodegradable polymers are that they can be composted with organic wastes and returned to enrich the soil, their use will not only reduce environmental pollution but will also lessen the labor cost for the removal of plastic waste in the environment because they are degraded naturally, and their decomposition will help increase the longevity and stability of landfills by reducing the volume of garbage [163]. Apart from biotic environmental factors like microorganisms, abiotic factors like photodegradation, hydrolysis, and oxidation add to the biodegradation process [164–166]. Conversion of biodegradable materials or biomass to gases, water, salts, minerals, and residual biomass is called mineralization [167].The degradation process can be either aerobic or anaerobic [168]. Biodegradation of polymers includes a change in the properties, such as tensile strength, color, shape, molar mass, etc., of a polymer or polymer-based product under the action of environmental agents like heat, light, or chemicals [169].

A polymer's complexity, structure, and composition are the most important aspects that govern its biodegradability. Another structural characteristic of polymers is the possible branching of chains or the formation of networks (crosslinked polymers). The microbes digest the starch, yielding a porous, sponge-like structure with a high surface area and low structural strength. Once the starch element has been depleted, the compound matrix begins to be degraded by an enzymatic attack. Several other polymers are degraded by exposure to chemicals and broken to small particles, which are further degraded upon microbial degradation. It has been reported that soil microbes are able to start the depolymerization of many natural polymers such as cellulose and hemicelluloses [170]. Maria Ratajska et al. (1998) investigated the biodegradation of new polymeric materials of natural origin and their mixtures with other natural and synthetic polymers [171]. Biodegradable composites have great importance because of the problem of the solid waste generated by plastic materials after their use. By blending natural fibers with the polymers, such as PLA, starch blends, cellulose acetates, and polyhydroxyalkanoates (PHA), a fully biodegradable material can be manufactured. Materials of this kind are known to decay under defined conditions, which are usually different from the ambient conditions under which they are used. To reduce their harmful impact on the environment and allow them to be altered during organic waste recycling processes, recently various materials have been added to plastics to improve their biodegradability. Natural fiber reinforced composites are renewable, environmentally friendly, low cost, lightweight, and have high specific performance. The chemical structure and constitution of the composites determine the biodegradability of plastics. Biodegradation is brought about by biological activity predominantly by the enzymatic action of microorganisms and can be measured by standard tests for a specified period of time.

Molecular degradation in aerobic and anaerobic conditions is triggered by enzymes, leading to complete or partial removal of the residue from the environment. The rate of biodegradation of composites of natural polymers has been studied in various environments, such as soil, compost, and weather. The degradability of thermoplastic starch and PVOH blends under anaerobic conditions to simulate the most common disposal environment for household wastes. Also mostly PVOH remained at the end of the digestion and that starch was almost entirely degraded. The degradability of thermoplastic starch and PVOH blends under anaerobic conditions to simulate the most common disposal environment for house hold wastes. However, the PVOH content significantly impacted the rate of starch solubilization [172].

1.7 Applications of Green Polymeric Composite Materials

Green polymeric composite materials are formed by the combination of a biodegradable polymer as a matrix and natural fibers as reinforced materials [173]. These ecofriendly composites have good mechanical (e.g. strength and elastic modulus), thermal, electrical, and chemical properties [174]. Based on their renewability, green polymeric composites can be divided into two categories: renewable composites and partially renewable composites. Renewable composites contain both matrix and reinforced materials that are renewable, but in partially renewable composites either the matrix or the reinforced material is renewable [175]. Biodegradable polymers have huge potential for applications in clinical, packing, and agricultural industry. Thin films of biodegradable polymers are used for early cropping and interrupt early weed formation [176]. Green polymers like cellulose have an excellent barrier properties and keep materials under airtight [177]. The polyethylene glycol (PEG)-derived detergent Triton X-114 can be dispersed in a buffer, stirred with a crude protein mixture, and heated above the cloud point to form a two-phase system and partition the proteins. A chromatographic form of partitioning can also be derived by immobilizing the dextran–water phase on a chromatographic support and eluting with the PEG–water phase.

Examples of medically useful PEG-altered proteins include PEG–asparaginase for analysis of acute leukemias [178], PEG–adenosine deaminase for therapy of acute combined immunodeficiency disease [179], and PEG–superoxide dismutase and PEG–catalase for reducing tissue damage emanating from reactive oxygen species correlated with ischemia and associated pathological circumstances [180]. PEG displays attractive characteristics when applied as a tether or linker to couple an active molecule to a surface. In this treatment, the PEG operates to suppress non-specific protein adsorption on the surface. Additionally, the tethered molecules have been demonstrated to be remarkably active, performing virtually as loose molecules in solution. PLA is a biodegradable polymer that has a collection of applications. It has been extensively employed in the biomedical and pharmaceutical fields for several decades due to its biocompatibility and biodegradability.

Green polymer-based nanocomposites are ideal for making packaging materials and product casings such as cell phone covers and biodegradable bags and wraps. Their biocompatibility along with the antimicrobial properties and UV absorption properties of the fillers gives these composites an advantage over traditional packaging and casing materials. Biodegradable scaffolds can be used to support and guide the in-growth of cells. Tissue scaffolds must be biodegradable, biocompatible, and also sterilizable. An ideal scaffold should have enough mechanical strength for withstanding physiological strains and must provide a suitable environment for cellular growth. Biocompatible composites, by solution mixing and freeze-drying processes, can be utilized in scaffold preparation. Hydrogels can be developed from polymers and can be used in tissue engineering. Thermoresponsive polymeric nanocomposite gels can impart large elongation at break and high moduli and strength. Connective tissue membranes for wound healing can be developed using a lithium chloride/dimethylacetamide mixture through solvent casting as they possess adequate swelling and moisture transmission abilities. They also have excellent antibacterial properties [181].

Recent developments in this PCL based composites area exploring antibacterial, antioxidant properties of the biomaterials filler have been found ideal for packaging materials [182]. Three types of technology are commonly used for food packaging materials: nanoreinforcement packaging, nanocomposite active packaging, and nanocomposite smart packaging. Nanoreinforcement can enhance polymer flexibility, temperature/moisture stability, and gas barrier properties.

1.8 Constituents, Fabrication, Crosslinking, and Clinical Applications of Hydrogels

Biodegradable and biocompatible natural polymers demonstrate a renewable and versatile substitute for synthetic polymers. Physical properties such as permeation, swelling, mechanical strength, and surface characteristics can be tailored via structural modifications [183]. Moreover, hydrogels can be prepared in different physical forms, such as films, coatings, nanoparticles, microparticles, and slabs. Natural polymer-based hydrogels are suitable for drug delivery and tissue engineering of bioactive molecules [184]. In medicine, numerous hydrogel products based on natural or synthetic polymers have had a huge impact on patient care in the modern era. Hydrogels are used in a broad range of experimental medicine and clinical practice applications, e.g. regenerative medicine and tissue engineering [185], diagnostics [186], cellular immobilization [187], as barrier materials to regulate biological adhesions, separation of biomolecules or cells, etc. [188]. Wichterle and Lim illustrated the polymerization of HEMA and a crosslinker using water and other solvents. Instead of hard and brittle, they found it to be elastic, water swollen, clear, and a soft gel. Formulations of hydrogels have progressively developed over the years [89] and they can now be fabricated by different chemical methods. Natural polymer-based hydrogels are normally biocompatible and have nominal stimulation to the immunological or inflammatory receptiveness of the host tissues. Numerous natural polymers have been extensively explored as biomaterials for reparative medicine and tissue engineering.

Physical or chemical compositing and amendments are used to induce specific interactions (electrostatic interactions and hydrogen bonding), functionalities and/or well-defined micro- or nanostructures in implants to enhance the toughness, strength, and bioactivity of the implants. Synthetic and natural polymers are combined to form a strengthened hydrogel, known as a hybrid hydrogel. Collagen strands can self-assemble to form strong fibers [189]. Additionally, fibers and scaffolds can be formed from collagen and their mechanical strength can be improved by using different crosslinkers (i.e. carbodiimide, formaldehyde, glutaraldehyde) [190, 191], by crosslinking with physical treatments (i.e. heating, freeze-drying, ultraviolet (UV) irradiation) [192], and by blending with other natural and synthetic polymers, i.e. polyethylene oxide (PEO), chitosan (CS), poly(lactic-co-glycolic acid) (PLGA), poly(glycolic acid), PLA, and HA [193–195].

Protein-based hydrogels can be prepared by a thermal gelation process and their mechanical characteristics can be improved via chemical crosslinking agents like glutaraldehyde [196]. The growth factors in matrigel impart cell migration through the activation of a G-protein, modulation of cell attachment, and remodeling of the cytoskeleton [197, 198], which modifies the morphology of cells through the contribution of actin dynamics [199]. The polysaccharides vary in molecular weight (MW), substituent of saccharide units, and linkage types and sites. These divergences in chemical structure result in diverse physical characteristics like gelling behavior, viscosity, properties at interface and surface, electrostatic behavior, solubility, and mechanical strength [200]. The polysaccharides are transparent, soft, and have excellent permeability so oxygen, in particular, can easily permeate through them. One example is polyalkylimide hydrogel, which is 96% water. It is a polyacrylate gel with alkyl amide and imide groups in its structure. It is normally used in the form of injectable endoprosthesis for the simple non-invasive rectification of small defects [201, 202].

Merrill et al. developed a hydrophilic biomaterial based on PEO and PEG [203], and presented adsorption of PEO on a glass surface which prevented the adsorption of protein. Since then, different surface-modified PEG materials have been used to promote surface biocompatibility and surface protein resistance. Hydrogels with physical crosslinking are not homogeneous, as bundles of polymer chain entanglements and ionic and hydrophobic interactions can cause inhomogeneities.

1.9 Natural Aerogels as Thermal Insulation

It is generally recognized that various non-covalent forces (including hydrogen bonding, ionic, amphiphilic, and $\pi-\pi$ interactions) exist in graphene-based hydrogels. Traditional hydrogels and aerogels have drawbacks such as poor mechanical properties and limited functional properties. In order to integrate nanostructural materials into macroscopic devices, a great effort has been made to make graphene nanosheet (GNS) self-assemble and to improve traditional hydrogels. The extraordinary low thermal conductivity of aerogels as well as their optical transparency allow their use in insulating building facades and window panes. Two different types of silica aerogel-based insulating materials are used in the building sector: opaque silica aerogel-based materials and translucent insulation materials. The pore structure of aerogels is comparable to that of large-pore mesostructures, therefore aerogels that have an open-pore structure can be readily adapted to polymer nanocomposites as reinforcing agents.

There are several strategies to overcoming the drawbacks associated with the weakness and brittleness of silica aerogels [204–208]. Aerogels are typically fabricated by subjecting a sol-gel precursor to critical-point drying in order to remove background liquid without collapsing the network. Supercritical drying is difficult to apply on an industrial scale because of its expensive and potentially dangerous character. Aerogels are composed of tenuous networks of clustered nanoparticles, and these materials often have unique properties, including very high strength-to-weight and surface-area-to-volume ratios. To date most aerogels have been fabricated from silica [209] or pyrolyzed organic polymers [210, 211]. CNT aerogel's electrical and structural properties are expected to be similar to those of pure SWNT (single walled carbon nanocomposites) samples because the electrical [212] and tensile [213] properties of bulk SWNTs and DWNTs (Double walled carbon Nano tube) are comparable. Aerogels are typically inorganic materials composed primarily of air, with inorganic skeletal structures. Silica is the material most commonly converted to aerogels – these materials have been known since the 1930s, when Kistler coined the term "aerogel" [214, 215].

The aerogel blanket is a new material that has very low thermal conductivity, which makes this material a good candidate for insulating walls. In fact, aerogel blankets are used to improve the energy performance of existing walls. Using aerogel blankets as insulation material could be an interesting alternative wherever space is an important factor. The performance of nanogel aerogel is based on the unique nanostructure of the particles, which is not impacted by size.

References

1 Darder, M., Aranda, P., and Ruiz-Hitzky, E. (2007). *Adv. Mater.* 19: 1309. http://onlinelibrary.wiley.com/doi/10.1002/adma.200602328/pdf.

2 Bagal-Kestwal, D.R., Kestwal, R.M., and Chiang, B.H. (2016). Bio-based nanomaterials and their bio-nanocomposites. In: *Book of Nanomaterials and Nanocomposites, Zero- to Three-Dimensional Materials and their Composites* (ed. M. Visakh), 255. Germany: Wiley-VCH Verlag GmbH and Co. ISBN: 978-3-527-33780-4.

3 Gross, R.A. and Kalra, B. (2002). *Science* 297: 803.

4 Zhang, X., Liu, X., Zheng, W., and Zhu, J. (2012). *Carbohydr. Polym.* 88: 26.

5 Dufresne, A. (2006). *J. Nanosci. Nanotechnol.* 6: 322.

6 Park, H.M., Mohanty, A., Drzal, L. et al. (2006). *J. Polym. Environ.* 14: 27.

7 Wibowo, A.C., Misra, M., Park, H.M. et al. (2006). *Compos. Part A. Appl. Sci. Manuf.* 37: 33.

8 Park, H.M., Misra, M., Drzal, L.T., and Mohanty, A.K. (2004). *Biomacromolecules* 5: 2281.

9 Misra, M., Park, H.M., Mohanty, A.K., Drzal, L.T., (2004) Proceedings of the NSTI Nanotechnology Conference, 3, 316. http://www.nsti.org/Nanotech2004/500.pdf.

10 Park, H.M., Liang, X., Mohanty, A. et al. (2004). *Macromolecules* 37: 9076.

11 Delhom, C.D., White-Ghoorahoo, L.A., and Pang, S.S. (2010). *Compos. Part B Eng.* 41: 475.

12 White, L.A. (2004). *J. Appl. Polym. Sci.* 92: 2125.

13 Romero, R.B., Leite, C.A.P., and Gonc, a.M.C. (2009). *Polymer* 50: 161.

14 Rodríguez, F.J., Galotto, M.J., Guarda, A., and Bruna, J.E. (2012). *J. Food Eng.* 10: 262.

15 Li, M., Kim, I.H., and Jeong, Y.G. (2010). *Appl. Polym. Sci.* 118: 2475.

16 Kang, H., Huang, G., Ma, S. et al. (2009). *J. Phys. Chem. C* 113: 9157.

17 Yano, S., Maeda, H., Nakajima, M. et al. (2008). *Cellulose* 15: 111.

18 Corre Le, D. and Angellier-Coussy, H. (2014). *React. Funct. Polym.* 85: 97.

19 Haniffa, M.A.C.M., Ching, Y.C., Abdullah, L.C. et al. (2016). *Polymers* 8: 246.

20 Ojijo, V. and Ray, S.S. (2013). *Prog. Polym. Sci.* 38: 1543.

21 Wang, L., Sun, Y., and Yang, X. (2014). *Ceram. Int.* 40: 4869.

22 Hambardzumyan, A., Foulon, L., Chabbert, B., and Aguie-Beghin, V. (2012). *Biomacromolecules* 13: 4081.

23 Saha, B.C. and Bothast, R.J. (1993). *J. Ind. Microbiol.* 12: 413.

24 Trovatti, E., Fernandes, S.C.M., Rubatat, L., and Freire, C.S.R. (2012). *Cellulose* 19: 729.

25 Biliaderis, C.G. and Lazaridou, A. (2002). *Carbohydr. Polym.* 48: 179.

26 Biliaderis, C.G., Kristo, E., and Zampraka, A. (2007). *Food Chem.* 101: 753.

27 Biliaderis, C.G. and Kristo, E. (2007). *Carbohydr. Polym.* 68: 146.

28 Gurtner, G.C., Wong, V.W., Rustad, K.C. et al. (2011). *Tissue Eng.* 17: 631.

29 Teramoto, N., Saitoh, M., Kuroiwa, J. et al. (2001). *J. Appl. Polym. Sci.* 82: 2273.

30 Khademhosseini, A., Bae, H., Ahari, A.F. et al. (2011). *Soft Matter* 7: 1903.

31 Lee, C.H., Singla, A., and Lee, Y. (2001). Biomedical applications of collagen. *Int. J. Pharm.* 221: 1–22.

32 Chen, Z.G., Wang, P.W., Wei, B. et al. (2010). *Acta Biomater.* 6: 372.

33 Arora, A. and Padua, G.W. (2010). Review: nanocomposites in food packaging. *J. Food Sci.* 75 (1): R43–R49.

34 Chang, P.R., Yang, Y., Huang, J. et al. (2009). *J. Appl. Polym. Sci.* 113 (2): 1247.

35 Chen, P. and Zhang, L. (2006). *Biomacromolecules* 7 (6): 1700.

36 Guilherme, M.R., Mattoso, L.H.C., Gontard, N. et al. (2010). *Compos. Part A-Appl. Sci. Manuf.* 41 (3): 375–382.

37 Kumar, P. (2009). Development of bio-nanocomposite films with enhanced mechanical and barrier properties using extrusion processing. PhD dissertation, North Carolina State University.

38 Kumar, P., Sandeep, K.P., Alavi, S. et al. (2010). *J. Food Sci.* 75 (5): N46.

39 Kumar, P., Sandeep, K.P., Alavi, S. et al. (2010). *J. Food Eng.* 100 (3): 480.

40 Lee, J.E. and Kim, K.M. (2010). *J. Appl. Polym. Sci.* 118 (4): 2257–2263.

41 Martucci, J.F. and Ruseckaite, R.A. (2010). *J. Food Eng.* 99 (3): 377.

42 Alig, I., Pötschke, P., Lellinger, D. et al. (2012). *Polymer* 53: e28.

43 Capek, I. (2006). *Nanocomposite Structures and Dispersions: Science and Nanotechnology – Fundamental Principles and Colloidal Particles*. Nanocomposite Structures and Dispersions, vol. 23, 312. eBook ISBN: 9780080479590, Hardcover ISBN: 9780444527165.

44 Liu, Z., Jiao, Y., Wang, Y. et al. (2008). *Adv. Drug Delivery Rev.* 60: 1650.

45 Vaia, R.A. and Giannelis, E.P. (1997). *Macromolecules* 30: 7990.

46 Bagal-Kestwal, D.R., Kestwal, R., and Chiang, B.H. (2013). Biosensors based on nanomaterials and their applications. In: *Applications of Nanomaterials* (ed. R.S. Chaughule and S.C. Watawe), 1–52. USA: American Scientific Publishers. ISBN: 1-58883-181-7.

47 Kestwal, R., Bagal-Kestwal, D.R., and Chiang, B.H. (2012). *Plant Food Hum. Nutr.* 67: 136.

48 Moad, G. (2011). *Prog. Polym. Sci.* 36: 218.

49 García, N.L., D, Accorso, F.N.B., and Goyanes, S. (2015). Biodegradable starch nanocomposites. In: *Eco-Friendly Polymer Nanocomposites, Advanced Structured Materials* (ed. V.K. Thakur and M.K. Thakur), 75. India: Springer https://doi.org/10.1007/978-81-322-2470-9_2.

50 Galicia-García, T., Martínez-Bustos, F., Jiménez-Arévalo, O.A. et al. (2012). *J. Appl. Polym. Sci.* 126: 327.

51 Moigne, N., Sauceau, M., Benyakhlef, M. et al. (2014). *ECCM16 – 16th European Conference on Composite Materials*, 1. Spain: Seville.

52 Torkelson, J., (2014) Patent number- 8734696, https://www.scholars.northwestern.edu/en/publications/polymer-graphite-nanocomposites-via-solid-state-shear-pulverizati-3.

53 Iwamoto, S., Yamamoto, S., Lee, S.H., and Endo, T. (2014). *Cellulose* 21: 1573.

54 Walker, A., Tao, Y., and Torkelson, J. (2007). *Polymer* 48: 1066.

55 Iyer, K.A. and Torkelson, J.M. (2013). *Polym. Compos.* 34: 1211.

56 Iyer, K.A. and Torkelson, J.M. (2014). *Compos. Sci. Technol.* 102: 152.

57 Sahay, R., Kumar, S.P., Sridhar, R. et al. (2012). *J. Mater. Chem.* 22: 12953.

58 Reneker, D.H., Yarin, A.L., Fong, H., and Koombhangse, S. (2010). *J. Appl. Phys.* 87: 4531.

59 Reneker, D.H., Yarin, A.L., Zussman, E., and Xu, H. (2007). Electrospinning of nanofibers from polymer solutions and melts. *Adv. Appl. Mech.* 41: 43.

60 Frenot, A. and Chronakis, I.S. (2003). *Curr. Opin. Colloid Interface Sci.* 8: 64.

61 Huang, Z.M., Zhang, Y.Z., Kotaki, M., and Ramakrishna, S. (2003). *Compos. Sci. Technol.* 63: 2223.

62 Dzenis, Y. (2004). *Science* 304: 1917.

63 Aleksander, G., Rahul, S., Velmurugan, T., and Ramakrishna, S. (2011). *Polym. Rev.* 51: 265.

64 Zhang, Y.Z., Venugopal, J., Huang, Z.M. et al. (2005). *Biomacromolecules* 6: 2583.

65 Khajavi, R. and Abbasipour, M. (2012). *Scientia Iranica* 19 (6): 2029.

66 Ye, S.H., Zhang, D., Liu, H.Q., and Zhou, J.P. (2011). *J. Appl. Polym. Sci.* 121: 1757.

67 Pushparaj, V.L., Shaijumon, M.M., Kumar, A. et al. (2007). *Proc. Nat. Acad. Sci. USA* 104: 13574.

68 Sulak, M.T., Erhan, E., and Keskinler, B. (2012). *Sens. Mater.* 24: 141.

69 Degner, B.M., Chung, C., Schlegel, V. et al. (2014). *Compr. Rev. Food Sci. Food Saf.* 13 (2): 98.

70 Dufresne, A. (2010). *Molecules* 15: 4111.

71 Wu, R.L., Wang, X.L., Li, F., Li, H.Z., and Wang Y.Z. (2009). Green composite films prepared from cellulose, starch and lignin in room-temperature ionic liquid. *Bioresour. Technol.* 100: 2569–2574.

72 Sionkowska, A. (2011). Current research on the blends of natural and synthetic polymers as new biomaterials. *Prog. Polym. Sci.* 36: 1254–1276.

73 Costa-Júnior, E.S., Barbosa-Stancioli, E.F., Mansur, A.A.P. et al. (2009). Preparation and characterization of chitosan/poly (vinyl alcohol) chemically crosslinked blends for biomedical applications. *Carbohydr. Polym.* 76: 472–481.

74 Jawad, H., Lyon, A.R., Harding, S.E. et al. (2008). Myocardial tissue engineering. *Br. Med. Bull.* 87: 31–47.

75 Triplett, R.G. and Budinskaya, O. (2017). New frontiers in biomaterials. *Oral Maxillofac. Surg. Clin.* 29: 105–115.

76 Chicatun, F., Griffanti, G., McKee, M.D., and Nazhat, S.N. (2017). Collagen/chitosan composite scaffolds for bone and cartilage tissue engineering. In: *Biomedical Composites*, 2e. Elsevier. 163–198.

77 Liu, M., Min, L., Zhu, C. et al. (2017). Preparation, characterization and antioxidant activity of silk peptides grafted carboxymethyl chitosan. *Int. J. Biol. Macromol.* 732–738.

78 Chen, L., Zhou, M.-L., Qian, Z.-G. et al. (2017). Fabrication of protein films from genetically engineered silk-elastin-like proteins by controlled cross-Linking. *ACS Biomater. Sci. Eng.* 3: 335–431.

79 Shavandi, A., Silva, T.H., Bekhit, A.A., and Bekhit, A.E.-D. (2017). Dissolution, extraction and biomedical application of keratin: methods and factors affecting the extraction and physicochemical properties of keratin. *Biomater. Sci.* 5 (9): 1699–1735.

80 Hernández, N., Williams, R.C., and Cochran, E.W. (2014). The battle for the "green" polymer. Different approaches for biopolymer synthesis: bioadvantaged vs. bioreplacement. *Org. Biomol. Chem.* 12: 2834–2849.

81 Averous, L., Fauconnier, N., Moro, L., and Fringant, C. (2000). Blends of thermoplastic starch and polyesteramide: processing and properties. *J. Appl. Polym. Sci.* 76: 1117–1128.

82 The, D.P., Debeaufort, F., Voilley, A., and Luu, D. (2009). Biopolymer interactions affect the functional properties of edible films based on agar, cassava starch and arabinoxylan blends. *J. Food Eng.* 90: 548–558.

83 Rhim, J.-W. and Wang, L.-F. (2013). Mechanical and water barrier properties of agar/κ-carrageenan/konjac glucomannan ternary blend biohydrogel films. *Carbohydr. Polym.* 96: 71–81.

84 Yu, L., Dean, K., and Li, L. (2006). Polymer blends and composites from renewable resources. *Prog. Polym. Sci.* 31: 576–602.

85 Parulekar, Y. and Mohanty, A.K. (2006). Biodegradable toughened polymers from renewable resources: blends of polyhydroxybutyrate with epoxidized natural rubber and maleated polybutadiene. *Green Chem.* 8: 206–213.

86 Biliaderis, C.G., Lazaridou, A., and Arvanitoyannis, I. (1999). Glass transition and physical properties of polyol-plasticised pullulan–starch blends at low moisture. *Carbohydr. Polym.* 40: 29–47.

87 Park, J.W. and Im, S.S. (2002). Phase behavior and morphology in blends of poly (L-lactic acid) and poly (butylene succinate). *J. Appl. Polym. Sci.* 86: 647–655.

88 Mungara, P., Chang, T., Zhu, J., and Jane, J. (2002). Processing and physical properties of plastics made from soy protein polyester blends. *J. Polym. Environ.* 10: 31–37.

89 Ullah, F., Othman, M.B.H., Javed, F. et al. (2015). Classification, processing and application of hydrogels: a review. *Mater. Sci. Eng. C-Mater. Biol. Appl.* 57: 414–433:https://doi.org/10.1016/j.msec.2015.07.053.

90 Chen, S.Q., Li, J.M., Pan, T.T. et al. (2016). Comb-type grafted hydrogels of PNIPAM and PDMAEMA with reversed network-graft architectures from controlled radical polymerizations. *Polymers* 8 (2): 38: https://doi.org/10.3390/polym8020038.

91 Gonzalez-Gomez, R., Ortega, A., Lazo, L.M., and Burillo, G. (2014). Retention of heavy metal ions on comb-type hydrogels based on acrylic acid and 4-vinylpyridine, synthesized by gamma radiation. *Radiat. Phys. Chem.* 102: 117–123: https://doi.org/10.1016/j.radphyschem.2014.04.026.

92 Miao, L., Lu, M.G., Yang, C.L. et al. (2013). Preparation and microstructural analysis of poly(ethylene oxide) comb-type grafted poly(N-isopropyl acrylamide) hydrogels crosslinked by poly(epsilon-caprolactone). *J. Appl. Polym. Sci.* 128 (1): 275–282: https://doi.org/10.1002/app.38172.

93 Ryan, D.M. and Nilsson, B.L. (2012). Self-assembled amino acids and dipeptides as noncovalent hydrogels for tissue engineering. *Polym. Chem.* 3 (1): 18–33: https://doi.org/10.1039/c1py00335f.

94 Sun, T.L., Wu, Z.L., and Gong, J.P. (2012). Self-assembled structures of a semi-rigid polyanion in aqueous solutions and hydrogels. *Sci. China-Chem.* 55 (5): 735–742: https://doi.org/10.1007/s11426-012-4497-x.

95 Alexander, A., Ajazuddin, Khan, J., and Saraf, S. (2013). Poly(ethylene glycol)-poly(lactic-co-glycolic acid) based thermosensitive injectable hydrogels for biomedical applications. *J. Controlled Release* 172 (3): 715–729: https://doi.org/10.1016/j.jconrel.2013.10.006.

96 Khodaverdi, E., Tekie, F.S.M., Hadizadeh, F. et al. (2014). Hydrogels composed of cyclodextrin inclusion complexes with PLGA-PEG-PLGA triblock copolymers as drug delivery systems. *Aaps Pharmscitech* 15 (1): 177–188: https://doi.org/10.1208/s12249-013-0051-1.

97 Tabassi, S.A.S., Tekie, F.S.M., Hadizadeh, F. et al. (2014). Sustained release drug delivery using supramolecular hydrogels of the triblock copolymer PCL-PEG-PCL and alpha-cyclodextrin. *J. Sol-Gel Sci. Technol.* 69 (1): 166–171: https://doi.org/10.1007/s10971-013-3200-9.

98 Akhtar, M.F., Hanif, M., and Ranjha, N.M. (2016). Methods of synthesis of hydrogels ... a review. *Saudi Pharm. J.* 24 (5): 554–559: https://doi.org/10.1016/j.jsps.2015.03.022.

99 Hoare, T.R. and Kohane, D.S. (2008). Hydrogels in drug delivery: progress and challenges. *Polymer* 49 (8): 1993–2007: https://doi.org/10.1016/j.polymer.2008.01.027.

100 Reddy, N., Reddy, R., and Jiang, Q.R. (2015). Crosslinking biopolymers for biomedical applications. *Trends Biotechnol.* 33 (6): 362–369: https://doi.org/10.1016/j.tibtech.2015.03.008.

101 Rizwan, M., Yahya, R., Hassan, A. et al. (2017). pH Sensitive hydrogels in drug delivery: brief history, properties, swelling, and release mechanism, material selection and applications. *Polymers* 9 (4): 137: https://doi.org/10.3390/polym9040137.

102 Ulijn, R.V., Bibi, N., Jayawarna, V. et al. (2007). Bioresponsive hydrogels. *Mater. Today* 10 (4): 40–48: https://doi.org/10.1016/s1369-7021(07)70049-4.

103 Lee, P.I. and Kim, C.J. (1991). Probing the mechanisms of drug release from hydrogels. *J. Controlled Release* 16 (1–2): 229–236: https://doi.org/10.1016/0168-3659(91)90046-g.

104 Calo, E. and Khutoryanskiy, V.V. (2015). Biomedical applications of hydrogels: a review of patents and commercial products. *Eur. Polym. J.* 65: 252–267. https://doi.org/10.1016/j.eurpolymj.2014.11.024.

105 Hersel, U., Dahmen, C., and Kessler, H. (2003). RGD modified polymers: biomaterials for stimulated cell adhesion and beyond. *Biomaterials* 24 (24): 4385–4415. https://doi.org/10.1016/s0142-9612(03)00343-0.

106 Perale, G., Rossi, F., Sundstrom, E. et al. (2011). Hydrogels in spinal cord injury repair strategies. *ACS Chem. Neurosci.* 2 (7): 336–345. https://doi.org/10.1021/cn200030w.

107 Ramesh Babu, P. and Gaikar, V.G. (2001). Membrane characteristics as determinant in fouling of UF membranes. *Sep. Purif. Technol.* 24: 23–34. https://doi.org/10.1016/S1383-5866(00)00207-0.

108 Dogan, H. and Hilmioglu, N.D. (2010). Zeolite-filled regenerated cellulose membranes for pervaporative dehydration of glycerol. *Vacuum* 84: 1123–1132. https://doi.org/10.1016/j.vacuum.2010.01.043.

109 Xiong, X., Duan, J., Zou, W. et al. (2010). A pH-sensitive regenerated cellulose membrane. *J. Membr. Sci.* 363: 96–102. https://doi.org/10.1016/j.memsci.2010.07.031.

110 Yang, Q., Fukuzumi, H., Saito, T. et al. (2011). Transparent cellulose films with high gas barrier properties fabricated from aqueous alkali/urea solutions. *Biomacromolecules* 12: 2766–2771. https://doi.org/10.1021/bm200766v.

111 Singh, N., Chen, Z., Tomer, N. et al. (2008). Modification of regenerated cellulose ultrafiltration membranes by surface-initiated atom transfer radical polymerization. *J. Membr. Sci.* 311: 225–234. https://doi.org/10.1016/j.memsci.2007.12.036.

112 Fukuzumi, H., Saito, T., Iwata, T. et al. (2009). Transparent and high gas barrier films of cellulose nanofibers prepared by TEMPO-mediated oxidation. *Biomacromolecules* 10: 162–165. https://doi.org/10.1021/bm801065u.

113 Ma, H., Burger, C., Hsiao, B.S., and Chu, B. (2012). Nanofibrous microfiltration membrane based on cellulose nanowhiskers. *Biomacromolecules* 13: 180–186. https://doi.org/10.1021/bm201421g.

114 Zhu, T., Lin, Y., Luo, Y. et al. (2012). Preparation and characterization of TiO2-regenerated cellulose inorganic–polymer hybrid membranes for dehydration of caprolactam. *Carbohydr. Polym.* 87: 901–909. https://doi.org/10.1016/j.carbpol.2011.08.088.

115 Mohamed, M.A., Salleh, W.N.W., Jaafar, J. et al. (2016). Physicochemical characteristic of regenerated cellulose/N-doped TiO2 nanocomposite membrane fabricated from recycled newspaper with photocatalytic activity under UV and visible light irradiation. *Chem. Eng. J.* 284: 202–215. https://doi.org/10.1016/j.cej.2015.08.128.

116 Purwanto, M., Atmaja, L., Mohamed, M.A. et al. (2016). Biopolymer-based electrolyte membranes from chitosan incorporated with montmorillonite-crosslinked GPTMS for direct methanol fuel cells. *RSC Adv.* 6: 2314–2322. https://doi.org/10.1039/C5RA22420A.

117 Zhang, X., Yu, H., Yang, H. et al. (2015). Graphene oxide caged in cellulose microbeads for removal of malachite green dye from aqueous solution. *J. Colloid Interface Sci.* 437: 277–282. https://doi.org/10.1016/j.jcis.2014.09.048.

118 Jana, S., Saikia, A., Purkait, M.K., and Mohanty, K. (2011). Chitosan based ceramic ultrafiltration membrane: preparation, characterization and application to remove Hg(II) and As(III) using polymer enhanced ultrafiltration. *Chem. Eng. J.* 170: 209–219. https://doi.org/10.1016/j.cej.2011.03.056.

119 Bierhalz, A.C.K., Westin, C.B., and Moraes, Â.M. (2016). Comparison of the properties of membranes produced with alginate and chitosan from mushroom and from shrimp. *Int. J. Biol. Macromol.* 91: 496–504. https://doi.org/10.1016/j.ijbiomac.2016.05.095.

120 Yang, H., Yan, R., Chen, H. et al. (2007). Characteristics of hemicellulose, cellulose and lignin pyrolysis. *Fuel* 86: 1781–1788. https://doi.org/10.1016/j.fuel.2006.12.013.

121 Andresen, M., Johansson, L.S., Tanem, B.S., and Stenius, P. (2006). Properties and characterization of hydrophobized microfibrillated cellulose. *Cellulose* 13: 665–677. https://doi.org/10.1007/s10570-006-9072-1.

122 Altintas, Z., Chianella, I., Da Ponte, G. et al. (2016). Development of functionalized nanostructured polymeric membranes for water purification. *Chem. Eng. J.* 300: 358–366. https://doi.org/10.1016/j.cej.2016.04.121.

123 Mohamed, M.A., Salleh, W.N.W., Jaafar, J. et al. (2015). Incorporation of N-doped TiO_2 nanorods in regenerated cellulose thin films fabricated from recycled newspaper as a green portable photocatalyst. *Carbohydr. Polym.* 133: 429–437. https://doi.org/10.1016/j.carbpol.2015.07.057.

124 Thakur, V.K. and Voicu, S.I. (2016). Recent advances in cellulose and chitosan based membranes for water purification: a concise review. *Carbohydr. Polym.* 146: 148–165. https://doi.org/10.1016/j.carbpol.2016.03.030.

125 Vakili, M., Rafatullah, M., Salamatinia, B. et al. (2014). Application of chitosan and its derivatives as adsorbents for dye removal from water and wastewater: a review. *Carbohydr. Polym.* 113: 115–130. https://doi.org/10.1016/j.carbpol.2014.07.007.

126 Mahmoudi, N. and Simchi, A. (2017). On the biological performance of graphene oxide-modified chitosan/polyvinyl pyrrolidone nanocomposite membranes: In vitro and in vivo effects of graphene oxide. *Mater. Sci. Eng., C* 70: 121–131. https://doi .org/10.1016/j.msec.2016.08.063.

127 Nhi, T.T., Khon, H.C., Hoai, N.T.T. et al. (2016). Fabrication of electrospun poly-caprolactone coated with chitosan-silver nanoparticles membranes for wound dressing applications. *J. Mater. Sci. – Mater. Med.* 27: 156. https://doi.org/10.1007/s10856-016-5768-4.

128 Ferry, J.D. (1980). *Viscoelastic Properties of Polymers*, 3e. New York: Wiley. ISBN: 978-0-471-04894-7.

129 Bemmelen, J.M. (1907). *Colloid. Polym. Sci.* 1 (7): 213.

130 Wen, Z., Xing, J., Yang, C. et al. (2013). *J. Chem. Technol. Biotechnol.* 88: 327.

131 Ahmed, E.M. (2015). *J. Adv. Res.* (6): 105.

132 Calo, E. and Khutoryanskiy, V.V. (2015). *Eur. Polym. J.* 65: 252.

133 Pal, K., Bhatinda, A.K., and Mujumdar, D.K. (2009). Polymeric hydrogels: characterization and biomedical applications. *Des. Monomers Polym.* 12 (3): 197–220.

134 Kozlov, P.V. (1983). *Polymer* 24: 651.

135 US Pharmacopoeia 34/National Formulary 29 (2011). *Food Chemicals Codex 7*. Rockville, MD: United States Pharmacopeial Convention, Inc.

136 Nur Hanani, Z.A., Roos, Y.H., and Kerry, J.P. (2014). *Int. J. Biol. Macromol.* 71: 94.

137 Chiou, B.S., Avena-Bustillos, R.J., Bechtel, P.J. et al. (2008). *Eur. Polym. J.* 44: 3748.

138 Nur Hanani, Z.A., Roos, Y.H., and Kerry, J.P. (2012). *Food Hydrocolloids* 29: 144.

139 Mhd Sarbon, N., Badii, F., and Howell, N.K. (2013). *Food Hydrocolloids* 30: 143.

140 Gómez-Guillén, M.C., Giménez, B., López-Caballero, M.E., and Montero, M.P. (2011). *Food Hydrocolloids* 25: 1813.

141 Sokolov, S.I., (1937) Physical chemistry of collagen and its derivatives, Gizlegprom, Moscow-Leningrad.

142 Zaides, A.L., (1961). The structure of collagen and its changes in 1960. *Biochem. J.* 81, 356.

143 Mikhailov, A.N. (1971). *Collagen of Skin and Principles of Its Processing*. Moscow: Legkaya Insustriya.

144 Raikh, G. (1969). *Kollagen (Collagen)*. Moscow: Legkaya Industriya.

145 Aoyagi, S. (1985). Photographic gelatin. In: *Proceedings of the Fourth IAG Conference (1983) Internatonale Arbeitsgemeinschaft Fur Photogelatine*, (ed. H. Ammann-Brass and J. Pouradier), 79–94.

146 Larry, D. and Vedrines, M. (1985). Photographic gelatin. In: *Proceedings of the Fourth IAG Conference (1983) Internatonale Arbeitsgemeinschaft Fur Photogelatine*, (ed. H. Ammann-Brass and J. Pouradier), 35–54.

147 Chen, X. and Peng, B. (1985). Photographic gelatin. In: *Proceedings of the Fourth IAG Conference (1983) Internatonale Arbeitsgemeinschaft Fur Photogelatine*, (ed. H. Ammann-Brass and J. Pouradier), 55–64.

148 Beutel, J. (1985). Photographic Gelatin. In: *Proceedings of the Fourth IAG Conference (1983) Internatonale Arbeitsgemeinschaft Fur Photogelatine*, (ed. H. Ammann-Brass and J. Pouradier), 65–78.

149 Koepff, P., (1984) Photographic gelatin reports 1970–1982, (eds. H. Ammann-Brass and J. Pouradier), pp. 197–209.

150 Tomka, I., (1984) Photographic gelatin reports 1970–1982, *Internatonale Arbeitsgemeinschaft Fur Photogelatine*, (eds. H. Ammann-Brass and J. Pouradier), pp. 210.

151 Bohonek, J., Spuhler, A., Ribeaud, M., and Tomka, I. (1967). *In Photographic Gelatin II Chapter-2*, (ed. R.J. Cox), 37–55. London: Academic Press.

152 Courts, A. and Stainsby, C. (1958). *Recent Advances in Gelatin and Glue Research*, 100. New York, USA: Pergamon Press.

153 Papon, P., Leblon, J., and Meijer, P.H.E. (2007). *The Physics of Phase Transitions*, 22. Berlin: Springer.

154 Ross, P.I. (1987). *Encyclopedia of Polymer Science and Engineering, In Fibers, optical to Hydrogenation*, vol. 7 (ed. H.F. Mark, N.M. Bikales, C.G. Overberger, et al.), 488. New York: Wiley-Interscience.

155 Vold, R.D. and Vold, M.J. (1983). *Colloid and Interface Chemistry*. Reading, MA: Addison-Wesley.

156 Howe, A.M. (2000). *Curr. Opin. Colloid Interface Sci.* 5: 288.

157 Villegas, R., O, Connor, T.P., Kerry, J.P., and Buckley, D.J. (1999). *Int. J. Food Sci. Technol.* 34: 385.

158 Hoque, M.S., Benjakul, S., and Prodpran, T. (2011). *Int. Aquat. Res*, 3: 165.

159 BeMiller, J.N., (1986) An introduction to pectins: structure and properties, ACS Symposium Series; American Chemical Society: Washington, DC.

160 Thiele, H., German Patent 1,249,517, (1967) Chem. Abstr. 67, 109390m.

161 Bourgade, G., 1871. Improvement in treating marine plants to obtain gelatin. US Patent 112, 535.

162 Avérous, L. and Pollet, E. (2012). Biodegradable polymers. In: *Environmental Silicate Nano-Biocomposites*, 13–39. Springer.

163 Hernandez, N., Williams, R.C., and Cochran, E.W. (2014). The battle for the "green" polymer. Different approaches for biopolymer synthesis: bio advantaged vs. bio-replacement. *Org. Biomol. Chem.* 12: 2834–2849.

164 Engineer, C., Parikh, J., and Raval, A. (2011). Review on hydrolytic degradation behavior of biodegradable polymers from controlled drug delivery system. *Trends Biomater. Artif. Organs* 25: 79–85.

165 Hawkins, W.L. (1964). Thermal and oxidative degradation of polymers. *Polym. Eng. Sci.* 4: 187–192.

166 Yousif, E. and Haddad, R. (2013). Photodegradation and photostabilization of polymers, especially polystyrene: review. *SpringerPlus* 2: 398.

167 Mohan, S.K. and Srivastava, T. (2010). Microbial deterioration and degradation of polymeric materials. *J. Biochem. Technol.* 2 (4): 210–215.

168 Doble, M. Biodegradation of polymers, 2005.

169 Porter, R.S., Cantow, M.J.R., and Johnson, J.F. (1967). Polymer degradation. V. Changes in molecular weight distributions during sonic irradiation of polyisobutene. *J. Appl. Polym. Sci.* 11: 335–340.

170 López-Mondéjar, R., Zühlke, D., Becher, D. et al. (2016). Cellulose and hemicellulose decomposition by forest soil bacteria proceeds by the action of structurally variable enzymatic systems. *Sci. Rep.* 6: 25279.

171 Ratajska, M. and Boryniec, S. (1998). Physical and chemical aspects of biodegradation of natural polymers. *React. Funct. Polym.* 38: 35–49.

172 Russo, M.A., O, Sullivan, C., Rounsefell, B. et al. (2009). The anaerobic degradability of thermoplastic starch: polyvinyl alcohol blends: potential biodegradable food packaging materials. *Bioresour. Technol.* 100: 1705–1710.

173 Castleton, H.F., Stovin, V., Beck, S.B.M., and Davison, J.B. (2010). *Energy Build.* 42: 1582.

174 S. Adeosun, G. Lawal, S. Balogun, and E. Akpan, Scirp.Org 11, 483 (2012).

175 Thakur, V.K., Thakur, M.K., and Gupta, R.K. (2014). *Int. J. Polym. Anal. Charact.* 19: 256.

176 N.C. Billingham, *Degradable Polymers: Principles and Applications* (1996).

177 Tang, X. and Alavi, S. (2012). Structure and Physical Properties of Starch/Poly Vinyl Alcohol/Laponite RD Nanocomposite Films. *J. Agric. Food. Chem.* 60: 1954.

178 Abuchowski, A., Kazo, G.M., Verhoest, C.R.J. et al. (1984). *Cancer Biochem. Biophys.* 7: 175.

179 Hershfield, M.S., Buckley, R.H., Greenberg, M.L. et al. (1987). Treatment of adenosine deaminase deficiency with polyethylene glycol-modified adenosine deaminase. *Engl. J. Med.* 316: 589.

180 Beckman, J.S., Minor, R.L. Jr, White, C.W. Repine J.E., Rosen G.M, Freeman B.A. (1988). Superoxide dismutase and catalase conjugated to polyethylene glycolincreases endothelial enzyme activity and oxidant resistance. *J. Biol. Chem.* 263: 6884–6892.

181 Anitha, A., Sowmya, S., Kumar, P.T.S. et al. (2014). Chitin and Chitosan in Selected Biomedical Applications. *Prog. Polym. Sci.* 39: 1644.

182 Chen, G., Lu, J., Lam, C., and Yu, Y. (2014). A novel green synthesis approach for polymer nanocomposites decorated with silver nanoparticles and their antibacterial activity. *Analyst* 139: 5793.

183 Wichterle, O. and Lim, D. (1960). Hydrophilic gels for biological use. *Nature* 185: 117–118.

184 Giri, T.K., Thakur, A., Alexander, A. et al. (2012). Modified chitosan hydrogels as drug delivery and tissue engineering systems: present status and applications. *Acta Pharm. Sin. B* 2: 439–449.

185 Lee, K.Y. and Mooney, D.J. (2001). Hydrogels for tissue engineering. *Chem. Rev.* 101: 1869–1880.

186 Van der Linden, H.J., Herber, S., Olthuis, W., and Bergveld, P. (2003). Stimulus-sensitive hydrogels and their applications in chemical (micro)analysis. *Analyst* 128: 325–331.

187 Jen, A.C., Wake, M.C., and Mikos, A.G. (1996). Review: hydrogels for cell immobilization. *Biotechnol. Bioeng.* 50: 357–364.

188 Bennett, S.L., Melanson, D.A., Torchiana, D.F. et al. (2003). Next-generation hydrogel films as tissue sealants and adhesion barriers. *J. Cardiac Surg.* 18: 494–499.

189 Lee, C.H., Singla, A., and Lee, Y. (2001). Biomedical applications of collagen. *Int. J. Pharm.* 221: 1–22.

190 Lee, C.R., Grodzinsky, A.J., and Spector, M. (2001). The effects of cross-linking of collagen-glycosaminoglycan scaffolds on compressive stiffness, chondrocyte-mediated contraction, proliferation, and biosynthesis. *Biomaterials* 22: 3145–3154.

191 Park, S.N., Park, J.C., Kim, H.O. et al. (2002). Characterization of porous collage/hyaluronic acid scaffold modified by 1-ethyl-3-(3-dimethylaminopropyl) carbodiimide cross-linking. *Biomaterials* 23: 1205–1212.

192 Schoof, H., Apel, J., Heschel, I., and Rau, G. (2001). Control of pore structure and size in freeze-dried collagen sponges. *J. Biomed. Mater. Res.* 58: 352–357.

193 Tan, W., Krishnaraj, R., and Desai, T.A. (2001). Evaluation of nanostructured composite collagen–chitosan matricies for tissue engineering. *Tissue Eng.* 7: 203–210.

194 Chen, G., Ushida, T., and Tateishi, T. (2001). Development of biodegradable porous scaffolds for tissue engineering. *Mater. Sci. Eng., C* 17: 63–69.

195 Huang, L., Naqapudi, K., Apkarian, R.P., and Chaikof, E.L. (2001). Engineered collagen—PEO nanofibers and fabrics. *J. Biomater. Sci., Polym. Ed.* 12: 979–993.

196 Zhu, J. and Marchant, R.E. (2011). Design properties of hydrogel tissue-engineering scaffolds. *Expert Rev. Med. Devices* 8: 607–626.

197 Santarius, M., Lee, C.H., and Anderson, R.A. (2006). Supervised membrane swimming: Small G-protein lifeguards regulate PIPK signalling and monitor intracellular Ptdins(4,5)P2 pools. *Biochem. J.* 398 (1): 13.

198 Schmitz, A.A., Govek, E.E., Böttner, B., and Van Aelst, L. (2000). Rho GTpases: Signaling, migration, and invasion. *Exp. Cell. Res.* 261: 1–12.

199 Tamura, M., Yanagawa, F., Sugiura, S. et al. (2015). Click-crosslinkable and photodegradable gelatin hydrogels for cytocompatible optical cell manipulation in natural environment. *Sci. Rep.* 5: 1–12.

200 Zhao, W., Jin, X., Cong, Y. et al. (2013). Degradable natural polymer hydrogels for articular cartilage tissue engineering. *J. Chem. Technol. Biotechnol.* 88: 327–339.

201 Gibas, I. and Janik, H. (2010). Review: synthetic polymer hydrogels for biomedical applications. *Chem. Chem. Technol.* 4: 297–304.

202 Ramires, P.A., Miccoli, M.A., Panzarini, E. et al. (2005). In vitro and in vivo biocompatibility evaluation of a polyalkylimide hydrogel for soft tissue augmentation. *J. Biomed. Mater. Res. Part B: Appl. Biomater.* 72: 230–238.

203 Merrill, E.W., Salzman, E.W., Wan, S. et al. (1982). Platelet-compatible hydrophilic segmented polyurethanes from polyethylene glycols and cyclohexane diisocyanate. *Trans.– Am. Soc. Artif. Intern. Organs* 28: 482–487.

204 Jin, W. and Brennan, J.D. (2002). Properties and applications of proteins capsulated within sol-gel derived materials. *Anal. Chim. Acta* 461: 1–36.

205 Zheng, J.K., Pang, J.B., Qiu, K.Y., and Wei, Y. (2000). Synthesis of mesoporous silica materials with hydroxyacetic acid derivateds as templates via a sol-gel process. *J. Inorg. Organomet. Polym.* 10: 103–113.

206 Smirnova, I., Suttruengwong, S., and Arlt, W. (2004). Feasibility study of hydrophilic and hydrophobic silica aerogels as drug delivery systems. *J. Non-Cryst. Solids* 350: 54–60.

207 Soleimani, D. and Abbasi, M.H. (2008). Silica aerogel; synthesis, properties and characterization. *J. Mater. Proc. Technol.* I9 (9): I0–I26.

208 Graham, A.L., Carison, C.A., and Edmiston, P.L. (2002). Development and characterization of molecularly imprinted sol-gel materials for the selective detection of DDT. *Anal. Chem.* 74: 458–467.

209 Hrubesh, L.W. (1998). *J. Non-Cryst. Solids* 225: 335.

210 Pekala, R.W., Farmer, J.C., Alviso, C.T. et al. (1998). Carbon aerogels for electrochemical applications *J. Non-Cryst. Solids* 225: 74.

211 Pekala, R.W. (1989). Organic aerogels from the polycondensation of resorcinol with formaldehyde *J. Mater. Sci.* 24: 3221.

212 Zhang, M., Fang, S.L., Zakhidov, A.A. et al. (2005). Strong, transparent, multifunctional, carbonnanotube sheets. *Science* 309: 1215.

213 Islam, M.F., Rojas, E., Bergey, D.M. et al. (2003). High weight fraction surfactant solubilization of single-wall carbon nanotubes in water. *Nano Lett.* 3: 269.

214 Kistler, S.S. (1932). Colloidal silica: fundamentals and applications. *J. Phys. Chem.* 36: 52–64.

215 Kistler, S.S., Sol-Gel Processing and Applications, US Patent 2,093,454, September 21, 1937.

2

Processing Methods for Bionanocomposites

Dipali R. Bagal-Kestwal, M.H. Pan and Been-Huang Chiang

Institute of Food Science and Technology, National Taiwan University, No.1, Roosevelt Road, section 4, Taipei, Taiwan, ROC

2.1 Introduction

In the last few decades nanobiocomposites (NBC) have formed a new research field with an excellent and innovative future. Their fast development has inspired many fields, including materials science, life science, and nanotechnology. NBCs are a combination of composite materials with multiphase solid materials in which one of the phases has one-, two- or three-dimensional nanoscale structure, and biological materials with nanoscale value-added structures [1–4]. Like general composites, NBCs are composed of a main continuous phase and a discontinuous phase. The continuous phase is also known as the matrix or polymer base while the discontinuous phase is called the reinforcement or reinforcing material or fillers [5–7]. The discontinuous phase consists of one or more components: a biological entity and one or more nano-scaled material(s). Either the biological or the nanomaterial component plays a crucial role in influencing the system and contributing to the quality improvement of manufactured composite materials.

This new generation of hybrid nanostructured materials has two crucial properties: biocompatibility and biodegradability [8, 9]. The exploitation of various biopolymers such as proteins, nucleic acids, polysaccharides, etc. for the preparation of nanocomposites has been done in last few decades [9, 10]. NBCs consist of inorganic nanoparticles dispersed in a biopolymer matrix, which may improve the mechanical, barrier, and thermo-resistance properties of natural biopolymer-based packaging films [11–14]. These properties currently form the technological challenges of developing biopackaging to replace the conventional petro-plastics that are currently used [15]. Moreover, NBCs are high-performance materials that combine unusual properties and unique architecture. NBCs also mimic their precursors to strengthen the properties of the new matrix to provide high biocompatibility and mechanical strength during in the synthesis process. They pervade almost all application fields required by the society, starting from the basic human needs of habitat to food packaging, transportation, entertainment, etc. [16].

In this chapter we will discuss the commonly used physical and chemical processes for preparation of NBC matrices and reinforcement/filler materials. The processing methods for matrix and filler are sometimes the same, but some matrices are prepared using

Bio Monomers for Green Polymeric Composite Materials, First Edition.
Edited by P.M. Visakh, Oguz Bayraktar and Gopalakrishnan Menon.
© 2020 John Wiley & Sons Ltd. Published 2020 by John Wiley & Sons Ltd.

combinations of one or multi-techniques to achieve the desire quality of NBC. We will therefore discuss the processing methods for NBCs with suitable examples.

2.2 Classification of NBCs

Still in their infancy, NBCs have not yet had any foolproof classification. Based on phase, either nanocrystalline or matrix, NBCs have been classified in the same way as nanocomposites and biocomposites. Matrix-based NBCs can be categorized into biodegradable organic NBCs and non-biodegradable inorganic NBCs. The sub-categorization of organic NBCs may include clay or ceramic-based NBCs, metal/metal oxide-based NBCs, biopolymer-based NBCs, hybrid NBCs, etc. Filler reinforcement-based NBCs can be divided into biodegradable and non-biodegradable, and both categories can be divided into particle, fiber, structure, and hybrid reinforced NBCs. Matrix-based organic NBCs can be further classified into silica, ceramic/clay, graphene, hydroxyapatite, carbon nanotube (CNT), cellulose, chitosan/chitin, starch, polylactic acid (PLA), bacterial-based polyester, and polyhydroxyalkanoates (PHA), etc. (see Figure 2.1).

2.2.1 Matrix-based NBCs

2.2.1.1 Polysaccharide Nanocomposites

Polysaccharides are the most readily available macromolecules in the biosphere [17]. This group of polymers consists of molecules such as cellulose, chitin/chitosan, starch, alginates, etc. [10]. NBCs of polysaccharide matrices are mainly prepared by solvent intercalation or melt processing and not through in situ polymerization where nature of

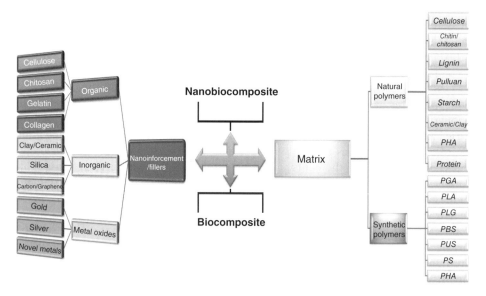

Figure 2.1 Schematic presentation of nanoreinforcement material and matrix used for the preparation of biocomposite and nanobiocomposite materials.

the polysaccharide directly influences the route of preparation. Some polysaccharides with nanostructure fillers are discussed as examples.

Cellulose is the most common organic compound and biopolymer on earth. About 33% of all plant matter is cellulose. In composite, the matrix's role is to support, protect the filler materials, and transmit and distribute the applied load to them [10]. Cellulose is an ideal supporting matrix because it allows the creation and design of new composite materials with high flexibility and improved properties [18–24]. It is the stronger and stiffer component as a reinforcing filler as well as the matrix.

Most cellulose whiskers-reinforced PLA NBCs are prepared by melt extrusion to avoid agglomeration and aggregation during drying [25]. Porous networks and thickened cellulose ribbons in gelatin/nanocellulose composites are prepared using an enzymatically modified form of gelatin [26]. Cellulose nanocomposites based on nanoparticles, such as clay [27–39], CNTs [24, 36], graphene [25], layered double hydroxide (LDH) [37], and silica [38], have been prepared. Injection-molded nanocomposites have been successfully fabricated from cellulose acetate (CA), ecofriendly triethyl citrate (TEC) plasticizer, and organically modified clay with and without maleic anhydride grafted cellulose acetate butyrate (CAB-g-MA) as a compatibilizer [29].

Starch is another abundant, inexpensive, naturally renewable, and biodegradable polysaccharide, produced by most green plants as an energy store. It is the most common carbohydrate in human diets and animal feeds [38–40]. Starch nanocomposites are mixtures of starch-based biopolymers with nanofillers (solid layered clays, synthetic polymer nanofibers, cellulose nanowhiskers, CNTs, and other metal nanostructures). Environmentally friendly starch NBCs exhibit significant improvements in mechanical properties, dimensional stability, transparency, processability, and solvent or gas resistance [10]. Nanocomposites of starch are new hybrid materials with enhanced properties due to a high aspect ratio and easy phase-to-phase energy transfer with a low amount of filler [39, 40]. Processing of starch/clay nanocomposites is dependent on the melting temperature of starch. Melt extrusion can be used with glycerol [40, 41] and sorbitol [42], or nitrogen-based plasticizers [43]. However, solvent processing, solution casting, melt-mixing/blending/intercalation [44–52], and melt extrusion [53–60] are also used to obtain proper dispersion of the filler into the starch matrix for paper coating, tissue complexing or even as a replacement for polyvinyl acetate in medical applications [61].

Chitosan/chitin, the second most abundant natural biopolymer, can also be integrated with clay, graphene, and carbon nanostructures to prepare NBC [62–64]. Due to its high content of amino ($-NH_2$) and hydroxyl ($-OH$) groups, chitosan and its derivatives are excellent adsorbents for the removal of heavy metal ions, fluoride, and organic dyes [9]. Films of spin-coated chitosan–alginate NBCs have potential uses in bioapplications [65]. An alkaline precipitation method and ultrasonication techniques were used to produce nanocrystals of magnetite (Fe_3O_4) coated with chitosan [66]. These magnetic chitosan nanoparticles showed low toxicity and proved to be a good candidate for hyperthermia therapy applications. Multi-tissue engineering applications of chitosan-based NBCs have been reported by many researchers using different nanotubes including CNTs due to the high aspect ratio and easy phase-to-phase energy transfer [67–77]. Such NBCs showed promising properties even at very low filler concentration, provided the fillers have been uniformly and completely dispersed in the host matrix [15, 16].

Lignin is also a plant-derived natural renewable biopolymer used as a continuous matrix in NBCs [63, 78]. The term "lignin" from the Latin word lignum, meaning "wood," was first used by the Swiss botanist Candolle [79]. The immense potential of lignin as a sustainable resource has made them an attractive candidate for the synthesis of NBCs. Lignin-based nanocomposite films are prepared using CNCs and have been used in various applications such as medical, biological, optical and sensors, and electronic [80]. They are also used as adhesives [19], stabilizing agents [30], and precursors for many aromatic chemicals [28]. Modified lignin, such as lignosulfates, kraft lignin, and acetylated lignin, contain CNCs or commercial derivatives or nanocellulosic polysaccharides. Incorporation of lignin not only improves hydrophobicity and mechanical resistance, but also the oxygen barrier properties of the hybrid materials [63]. Natural rubber/lignin NBCs from lignin-cationic polyelectrolyte colloids and rubber latex were prepared by the co-precipitation method [80, 81]. Nanocomposites with PLA [82, 83], polyaniline [84, 85], poly (N-methylaniline) [86], poly (N-butylaniline) [87, 88], polyethylene (PE) [89], polystyrene (PS) poly (methyl methacrylate) (PMMA) [90], and poly (vinyl alcohol) (PVA) [91, 92] also have tremendous potential in industrial applications.

Pullulan is a water-soluble biopolymer produced by a polymorphic fungus, *Aureobasidium pullulans* [93]. Nanocomposites of pullulan are prepared by electrospinning and used as filters and in protective clothing [94]. Adriamycin-loaded pullulan acetate (PA) and sulfonamide conjugate nanoparticles as well as PA and oligo-sulfadimethoxine conjugated self-assembled pH-sensitive hydrogel nanoparticles are used for treating tumors, ischemia, and inflammation [95]. Cast films of the pullulan composited with nanofibrillated cellulose show improved thermal and mechanical properties. Their malleability and mechanical properties could be further improved by adding glycerol [96]. Polyethylene terephthalate (PET) film coated with a graphene oxide (GO)/pullulan NBC is used in food/pharmaceutical applications [93]. NBCs with enriched properties based on two microbial polysaccharides, pullulan and bacterial cellulose (BC), were prepared by Trovatti et al. for possible application in organic electronics, dry food packaging, and the biomedical field [97]. Pullulan composites with many materials, including chitosan [98], caseinate [99], starch nanocrystals [100], collagen [101], PVA [102], and hydrogel with methacrylate [103], have proven excellent compatibility.

2.2.1.2 Animal Protein-based Nanocomposites

Protein is a natural polymer derived from animals and plants. Collagen, gelatin, heparin, albumin, milk protein, keratin, etc. are animal-derived proteins that are used in many biomedical applications due to their biocompatibility [10]. Their biodegradability, low cost and surfaced modification with active functional groups for catching targeting molecules make these matrices useful compounds in pharmaceutical industry [104]. Electrospun collagen–chitosan nanofibers were stabilized by glutaraldehyde vapor via crosslinking, which afforded a biomimetic extracellular matrix for cell growth [105]. Gelatin also can be used as matrix, for instance gelatin linked with antibodies [106], gelatin entrapped with heparin and chitosan nanoparticles [107], gelatin–clay nanocomposite film [108], and gelatin–collagen in artificial skin [109]. Gelatin-based montmorillonite (MMT) composites can be prepared by the solution casting technique. This clay–protein structure has an exfoliated and intercalated structure with excellent barrier properties and reduced water sensitivity [110, 111].

Collagen is regarded as one of the most useful biomaterials, exhibiting a number of biological advantages [104]. The outstanding performance and biomedical application of this protein biomaterial have induced research interest in synthetic composite material fabrication. As the composite matrix, collagen can be combined with a recombinant human bone morphogenetic protein-2 (rhBMP-2) for implant-bone formation [112]. Collagen-HA composite has been used in implant applications due to its mechanical properties and biocompatibility [112][113].

Heparin-coated superparamagnetic iron oxide nanoparticles (NPs) were also tested as potential image-guided anticancer drug delivery vehicles using co-precipitation [114]. Heparin-conjugated fibrin has been used as an injectable system for sustained delivery of BMP-2 [17]. BMP-2 released from the heparin–fibrin gel is bioactive and thus this composite can be used for bone regeneration.

2.2.1.3   Plant Protein-based Nanocomposites

Plant proteins such as soybean protein, zein (corn protein), and wheat gluten are mainly used in the preparation of NBCs and their commercial applications [113–115]. Nanocomposite films of whey protein with TiO_2 nanoparticles were found to be suitable for food and cosmetics packaging materials [116]. These transparent fills can improve the antibacterial properties of the film [117]. Soy protein isolate (SPI) has been extensively studied for bioderived packaging materials [115–118]. Several recent studies are investigating the improvement in the mechanical and barrier properties of nanocomposite films after incorporating nanoclays such as MMT [118–126]. Further, these NBC films have also been reported to decrease water vapor and oxygen permeability and increase elastic modulus and tensile strength, which makes them suitable for the packaging industry [125]. Shear mixing and sonication/mixing techniques were used for homogeneous dispersion of CNTs in polymer matrices. NBCs preparation using CNTs as nanofillers in SPI matrices can help to improved bond strength and water resistance in the development of environmental friendly adhesives [127, 128]. Recent studies have reported that SPI-based nanocomposite bioplastics with highly exfoliated MMT have significantly improved mechanical strengths and thermal stabilities [120, 129]. Thus, bio-based polycaprolactone-SPI is an ecofriendly and intercalated nanocomposite with enhanced tensile and dynamic mechanical properties when produced by the melt compounding method [130].

2.2.2   Reinforcement-based NBCs

Filler is incorporated into a polymer to improve its physical properties, regulate the viscosity, modify the surface properties or simply reduce cost [9]. Commonly used fillers such as talc, calcium carbonate fibers, and wood flour are used in large amounts in polymers [9]. The composition, dimension, and shape of the nanofillers as well as their spatial distribution in the matrix are important factors that impact the properties of nanocomposite products [131]. The reinforcement materials employed in the production of layered and non-layered matrices can be classified on the basis of their dimensions as follows [132]:

- Zero dimensional nanomaterials with three dimensions in the nanometer range, for example spherical nanoparticles, semiconductor nanoclusters, etc. [132].

- One-dimensional nanomaterials with two dimensions in the nanometer range and one dimension in the micrometer range with elongated structure, for example nanotubes (CNTs), whiskers (cellulose whiskers), etc. [133–135].
- Two-dimensional nanomaterials with only one dimension in the nanometer range, for example sheet/membrane-like structures where thickness is in the nanometer range and length is in the hundreds to thousands of nanometers [40, 136–138].

As for NBCs, at least one dimension of the incorporated fillers is in the nanometer range [139]. The structure-property relationship illustrates that the final structure of nanocomposite materials greatly impacts the properties and performance of the final products. Therefore, achieving either intercalation or exfoliation of the layered nanomaterial is critical to improve the properties of NBC materials [139–141]. Clays, CNTs, graphene, silica, hydroxyapatite, silver, and organic nanofillers are routinely used to process NBCs [140].

NBCs with nanoscale reinforcements exhibit excellent mechanical properties with low volume fraction of the reinforcement. Crosslinking/chemical modification of filler material/matrix promotes the formation of strong hydrogen, electrostatic, and covalent bonding between the reinforced material and the continuous phase. These interactions also enhance and strengthen the intrinsic properties of the biopolymer matrix. In case of biocomposites, the properties of the composites are dependent on the inter-phase interaction of the reinforced material and the matrix. Filler is also a value-added material, but wise selection of processing methodology, optimum conditions, and compatible phase components is needed. Properties such as high aspect ratio and easy phase-to-phase energy transfer are greatly improved if the filler is uniformly and completely dispersed in the host matrix [142, 143]. The most commonly used techniques are chemical modification and physical modification [144]. Chemical modification methods mostly include impregnation of fibers with matrix compatible polymer [145], graft copolymerization [146], acetylation, and mercerization [147–149], while techniques like corona discharge, and thermal and plasma treatment [150] are physical modifications. Increased tensile properties and improved processability of resin NBCs in soy protein concentrate were also obtained by crosslinking with glutaraldehyde. These processed compounds can further be used to manufacture the flax fabric-reinforced composites [13].

2.2.2.1 Metal Nanocomposites

Polymer/metal nanocomposites consisting of polymer as matrix and metal nanoparticles as nanofiller commonly show several attractive advantages such as electrical, mechanical, and optical characteristics [151]. Metal NBCs with protein, nucleic acids, and polysaccharides have showed potential applications in drug delivery, tissue engineering, bio-imaging, wound healing, biomedicine, energy production and storage, electronic devices such as biosensors, affinity materials, etc. [10, 112, 119, 152]. Metal nanoparticles such as silver (Ag), gold (Au), copper (Cu), palladium (Pd), platinum (Pt) and titanium (Ti) were most often used in this area because they exhibit significant physical, chemical, and biological properties. The sensing properties of polymer/metal nanocomposites depend on the structural and chemical changes produced by metal nanoparticles in the polymer matrix as well as the different electronic structures of the nanocomposite [153].

2.2.2.2 Inorganic Nanocomposites

Polymer-clay nanocomposites (PCNs) is a group of nanocomposites in which nanoclays are the nanoscale fillers dispersed in a polymer matrix. The inorganic nanoclays have at least one dimension in the nanometer range.

The high aspect ratio and high surface area of incorporated nanoclays and their interfacial interactions within the matrix effectively enhance the physical properties of PCNs. This includes mechanical properties such as tensile strength, tensile modulus, and percent elongation at break, and barrier properties for gas, moisture, and volatiles for biodegradable and renewable natural raw materials [17]. As a naturally existing clay, MMT is one of the most widely studied layered silicates in developing biopolymer nanocomposites due to its high active surface, low cost, abundant supply, large specific aspect ratio, and very high elastic modulus [17]. MMT silicates are active fillers which can improve the mechanical strength of hydrogel networks but inhibit the effect of ionic strength on protein–protein aggregations [118, 119]. Highly intercalated or exfoliated MMT has repeatedly been shown to improve the mechanical and barrier properties of NBC materials [122–124]. Well-dispersed MMT in SPI matrices resulted in significant improvements in thermal-stability and mechanical strength. Many methods have been described for the preparation of ceramic matrix nanocomposites. The most common methodologies, as used for microcomposites, are the conventional powder method[122], the polymer precursor route[123], spray pyrolysis[124], vapor techniques (chemical vapor deposition [CVD] and PVD) and chemical methods, which include the sol-gel process, colloidal precipitation approaches, and template synthesis[122–125].

Carbon Nanocomposites CNTs have been routinely employed as a filler or nano-reinforcement materials and employed for detection in electrochemical immune-sensor fabrication as they possess metallic or semi-conducting properties [117]. Their large specific surface area may increase immobilization of the primary antibody onto the assay surface and thus increase sensitivity. Foodborne pathogen detection using biomolecules and nanomaterials has led to platforms for rapid and simple electronic biosensing[118]. Single-walled carbon nanotubes (SWCNTs) and immobilized anti-bodies were incorporated into a disposable bio-nano combinatorial junction sensor and used for detection of Escherichia coli K-12 [118–122].

2.3 General Processing Methods for NBCs

A combination of different techniques is used for the preparation of various types of NBCs. Combination of manufacturing single or multiple steps with different composite phases can be useful in various fields of science and technology especially new biomedical technologies [154]. For example, matrix-based NBCs can be prepared by physical and/or chemical methods, including the conventional powder method, the polymer precursor route, the sol-gel process, CVD and/or PVD vapor deposition, and electrodeposition [5]. Top-down and bottom-up methods are used for the synthesis of nanomaterials. Bottom-up methods are found to be more promising for controlled properties and specific orientation of nanomaterials. Thermal evaporation and sputtering techniques have been considered as facile, simple, low-cost, and high-yield methods for the synthesis of high-quality nanomaterials/nanostructures [155, 156].

Biological components are highly sensitive to environmental and processing conditions such as pH, temperature, and processing time. The aforementioned techniques such as sol-gel process, CVD and/or PVD vapor deposition, and electrode-position can be used for the preparation of polymer nanocomposites first. Then the biological entity can be included at the final step. However, as the most important part is the incorporation of the biological entity, one should adopt the processing procedures according to the biological entity. Generally, this can be performed by direct mixing (physical method) or a chemical method. Various immobilization methods, including entrapment, adsorption, crosslinking, electro-polymerization, and encapsulation, have been used for the imprisonment of biological moieties in the matrix [141, 157–160]. Similarly, nanomaterials can be used to improve the stability of biocomponents as they provide large specific surface area, increased biocompatibility, and long shelf life [158]. In this chapter we will focus on the processing of NBCs based on nanomaterials as rein-forcement agents. There are several different routes for composite manufacturing, each with its own specialty and suitability. Some of the most popular physical/mechanical and chemical manufacturing routes are discussed in the following sections [161].

2.3.1 Pressure Extrusion

This is one of the main processes employed in the manufacturing of NBCs [162, 163]. There are two main types of extrusion: reactive extrusion and extrusion cooking. Reactive extrusion uses chemical modification via crosslinking [163]. Generally, extrusion technology in the food industry is referred to as extrusion cooking and results in different physical and chemical properties of the extrudates depending on the raw materials and extrusion conditions used [164]. In the case of starch NBCs, extrusion is an energy-efficient method of breaking down the starch granule structure through a combination of high shear, temperature, and pressure [77]. Various starch nanocomposite varieties have been prepared and reported by many researchers for biodegradable packaging applications in food industry [140, 163–167]. Moigne et al. developed a continuous CO_2-assisted extrusion process to prepare poly(3-hydroxybutyrate-co-3-hydroxyvalerate)/clay NBC foams with good homogene-ity and high porosity [168]. Attempts have also been made to make NBC films using a twin-screw extruder. Plasticized PLA/chitin nanocomposite was prepared by extrusion followed by compounding of the final compound and film blowing of the bags for packaging applications [169].

2.3.2 Solid-state Shear Pulverization

Solid-state shear pulverization (SSSP) is a novel mechanical process that uses a modified twin-screw extruder with cooling zones to maintain the polymer in the solid state during processing. This is a single-step, solventless, continuous, and industrially scalable process that is a modification of a twin-screw melt extruder equipped with a barrel for cooling. Features like high shear and compressive forces help in fusion and repeated fragmentation of matrix in the solid state, producing excellent mixing and dispersion of polymer blends and nanocomposites [170]. The SSSP process has been used to yield nanostructured polymer blends and nanocomposites that cannot be pre-pared using conventional processing techniques because of kinetic and thermodynamic barriers. In this solid-state processing method, mechanical milling techniques (ball and pan milling) have been used for large scale production of the polymer composites and nanocomposites with low-cost, easy-operating machines, for high-efficiency performance [171–173]. Polyetherimide NBCs reinforced with exfoliated graphite (graphene) nanoplatelets (GNP) were fabricated by various processing methods to achieve good dispersion. These methods include melt-extrusion, pre-coating, and solid state ball milling as well as combinations of these methods [174, 175]. Inventor Torkelson has successfully produced a well-dispersed graphite-polymer nanocomposite [176]. Taking advantage of near-ambient-temperature processing, SSSP was recently used to produce biodegradable polymer matrix composites with starch [177], rice husk ash [178], and eggshell fillers [179, 180]. Thus, this technique that has proven to effectively disperse nanoscale structural entities to achieve compatibilized polymer blends and exfoliated polymer nanocomposites.

2.3.3 Electrospinning and Co-axial Electrospinning

Electrospinning has been recognized as an efficient technique for the fabrication of polymer nanofibers [181]. This physical method uses extrusion of the polymer solution with reinforcement of nanomaterials and the biological entity for the preparation of NBCs. Polymers, molten at high temperature, can also be made into nanofibers by electrically charging a suspended droplet of polymer melt or solution [182–188]. Instead of a solution, the polymer melt is introduced into the capillary tube. However, different from the case of polymer solution, the electrospinning process for a polymer melt has to be performed in vacuum conditions [181, 186, 189]. Nanomaterials and/or biomolecules such as proteins/enzymes can be introduced using a syringe needle in the presence of a high-voltage field. Co-axial electrospinning is a modified method that is gaining popularity due to its ease of fabrication and multi-component core-sheath fiber structure of final NBC product. The major difference to traditional electrospinning is that a compound spinneret with two (or more) components can be fed through different coaxial capillary channels [190]. Wet–dry and wet–wet electrospinning techniques are used for volatile and non-volatile solvents, respectively. Both techniques offer the possibility of producing nanofibers with a controlled fiber diameter to make a film or a membrane desired configuration [184, 185]. Such fibrous scaffolds are ideal for tissue regeneration because their dimensions are similar to the components of the extracellular matrix and mimic its fibrillar structure, providing essential signals for cellular assembly and proliferation [183].

  NBCs with water-soluble hydrocolloids such as CA, gelatin, and starch are prepared by wet–dry electrospinning [10]. Numerous synthetic polymers such as poly(ε-caprolactone) (PCL) [191], PLA [192], poly(glycolic acid) (PGA), poly(lactic-co-glycolic acid) (PLGA), polystyrene, polyurethane (PU), PET, poly(L-lactic acid)-co-poly (ε-caprolactone) (PLACL), and biological materials such as collagen, gelatin, and chitosan have been successfully electrospun to obtain fibers [183, 191–199]. Core–shell structured nanofibers with collagen as the shell and PCL as the core have been prepared by co-axial electrospinning and have the advantage of controlled shell thickness and manipulative mechanical strength and degradation properties in the resulting composite nanofibers, without affecting biocompatibility [194]. Thus, such core–shell structured composite nanofibers have potential uses in drug or growth factor encapsulation and the development of highly sensitive sensors and tissue engineering applications [195].

2.3.4 Solution Casting and Evaporation

The solution casting and evaporation method is the oldest and easiest technology for manufacturing composite structures like films. This is the best option for producing NBCs in the form of films. It allows large-scale production with ease of handling and cost-effective way. In the solution casting method, a matrix/polymer solution is heated with nanofiller material first, as temperature affects the gel and/or network architecture of the polymer and eases intercalation. The main and only requisite condition for this technique is that the matrix or polymer must be soluble in either a volatile solvent or water. Another advantage of this method is that solution/solvent drying or evaporation does not need mechanical or thermal stress. Therefore, the biological entity in the NBC remains catalytically active with negligible or no loss. The solution casting method is based on a solvent system in which the polymer or pre-polymer is soluble. The polymer is usually dissolved in a suitable solvent while the nanoparticles are dispersed in the same or a different solvent before the two are mixed. For example, during the preparation of NBC based on clays, the solvent is normally used to pre-swell the clays [200]. In case of non-water-soluble polymers, additional steps are required before processing, for example solvent exchange, use of surfactant, freeze-drying or chemical modification [177].

Film or membrane casting or dip-coating can also be used for the preparation of NBCs. Polysaccharide integrated NBCs are used for electrode modification and subsequently biosensor fabrication [155]. Solution casting not only allows homogeneous dispersion of nano/bio components but also improves their thermal and mechanical properties and their malleability and mechanical properties [96, 201–206]. Bactericidal films with AgNP-lactose-modified chitosan [206], starch/zinc oxide (ZnO) nanocomposites [207], chitosan/nanocellulose [208], cellulose nanocrystals/ZnO [209, 210], chitosan/ZnO [211], chitosan/Au nanoseeds [212], and chitosan/nano-CdS [213] showed not only enhanced antimicrobial properties but also improved dielectric and mechanical properties.

2.3.5 Melt Intercalation Method

Melt intercalation is a mechano-chemical process that is used for clay/silicate biocomposites and is compatible with extrusion and injection molding. This method is also fast and clean, ecofriendly, and can alter the lifecycle analysis [214]. In this technique nanomaterials and/or biomaterials are mixed with the polymer in the molten state. The process involves mixing the nanoparticles with the polymer and heating the mixture above the softening point of the polymer, statically or under shear [141, 215]. This method is used for biopolymers that are not suitable for an in situ polymerization technique. The formation of NBCs via polymer melt intercalation depends upon the thermodynamic interaction between the polymer matrix and the nanoparticle[216]. This technique is especially suitable for clay-containing NBCs where interactions depend upon the transport/diffusion of polymer chains from the bulk melt into the silicate interlayers [153, 217]. The matrix needs to be sufficiently compatible with the nanoparticle surface to ensure proper dispersion. As in other techniques, proper dispersion of the nanoparticles is always a goal during processing. During melt processing, a number of factors are important to achieve homogenous dispersion of the reinforced nanomaterial into the polymer matrix, including enthalpic interaction between the polymer matrix, biocomponent, and the nanoparticles [63]. The optimization of the processing temperature and pressure is also important to avoid the degradation of the biopolymer matrices. If processing conditions are not well optimized, then the resultant interactions may lead to the formation of microcomposites, therefore there is a need to optimize each and every step of the melt processing to avoid formation of tactoids/agglomerations, especially in case of clay-nanocomposites [200, 218].

Mostgraphite-based nanocomposites are prepared by the melt intercalation technique [219–225]. For this purpose, neat graphite (NG) is first converted to intercalated or expandable graphite (EG) through chemical oxidation in the presence of concentrated H_2SO_4 or HNO_3. Expandable graphite is then obtained by expansion and exfoliation by rapid heating in a furnace above 600 °C [225]. An intercalation method for the direct exfoliation of NG in the presence of chitosan solution for the preparation of NG–chitosan NBC has been developed. In this method, graphite is dispersed in a water solution of chitosan with the aid of ultrasonication. This method is solvent-free and results in large-scale exfoliation to give a large amount of multilayer and monolayer graphene. Another nanobiodegradable composite based on corn starch/high density polyethylene (HDPE)/linear low-density polyethylene (LLDPE) and Nanoclay cloisite 30B were prepared by melt mixing. Ethylene-vinyl alcohol (EVOH), ethylene vinyl acetate (EVA), and polyethylene grafted maleic anhydride (PE-gMA) were used as compatibilizers in these composites [226]. Another attempt showed that glycerol-plasticized and citric acid-modified starch/carboxymethyl cellulose (CMC)/MMT bionanocomposite films formed an intercalated structure in the NBCs. Citric acid, CMC, and MMT not only improved the mechanical properties of starch films but also the hydrophilic character [227, 228].

2.3.6 In Situ Polymerization

In situ polymerization involves chemical reaction to form a thermodynamically stable reinforcing step within the matrix. During in situ polymerization, the nanoparticles are pre-mixed with the liquid monomer or monomer solution. Polymerization is then initiated by heat, radiation or suitable initiators to create strong matrix dispersion bonds. This technique is commonly used to prepare CNT-based biocomposites because CNTs have high electrical and thermal conductivity. To improve the processability and electrical, magnetic, and optical properties of CNTs, some conjugated or conducting polymers are attached to their surfaces by in situ polymerization. CNTs can form composites with purely amorphous polymers, like polystyrene and PMMA, and semi-crystalline polymers, like polyethylene and polypropylene (PP) [229]. Their industrial applications include but are not limited to the transportation, automotive, aerospace, defense, sporting goods, energy, infrastructure sectors, etc. Multi-walled carbon nanotube-poly methyl methacrylate (MWNT/PMMA) NBC has been prepared by in situ polymerization of MMA dispersed with MWNTs with fairly good dispersion stability [230].

2.3.7 Drying Techniques (Freeze-drying and Hot Pressing)

Dehydration is a traditional technique that can also be used to prepare nanocomposites and NBCs of thermoplastic polymers/biopolymers [231, 232]. In this technique matrix and reinforcement material is prepared in aqueous medium, mostly water. This product is further processed by hot pressing. Nair and Dufresne reported preparation of nanocomposite materials from a colloidal suspension of chitin whiskers as the reinforcing phase and latex of both unvulcanized and prevulcanized natural rubber as the matrix, and then used freeze-drying and hot-pressing to prepare the chitin whiskers/rubber nanocomposites [233]. In most cases, freeze drying is used to produce porous materials for a heat-sensitive matrix [234, 235]. Biomaterials such as collagen and chitosan and their porous nanocomposites prepared by these techniques are often used in biomedical applications. Cellulose particles such as whiskers and microfibrillated cellulose (MFC) and nanocomposites have been prepared by this technique[234–236]. Drying in combination with electrospinning has been used to prepare chitosan/PLGA nanocomposite nanofibers [236, 237]. Sehaqui et al. reported the manufacture of nanofibrillated cellulose-reinforced hydroxyethylcellulose (HEC) film in a polystyrene Petri dish under air atmosphere at room temperature for application in renewable materials in food packaging [238]. Many scientists have reported various examples of NBCs processed by these drying techniques, including starch [159, 180, 239–249], silk fibroin [250], poly(oxyethylene) (POE) [251–254], PVA [255–259], hydroxypropyl cellulose (HPC) [255, 257], CMC [258], and SPI [259, 260]. There are many review articles available which discuss polysaccharide-based nanocomposites and their applications [151, 261–268].

2.3.8 Polymer Grafting

Graft co-polymerization is an attractive method to impart a variety of functional groups to a polymer. Graft co-polymerization is initiated by chemical treatment, photo-irradiation, and high-energy radiation techniques [269]. Grafting of long chains can also be used to transform the nanofiller into a co-continuous material by increasing its apolar character through grafting agents bearing a reactive end group and a long "compatibilizing" tail [270]. After grafting, the nanocomposites can be further processed by classical methods such as hot-pressing, injection molding or thermoforming. More recently, researchers have developed a new way of processing that allows chemical modification of polysaccharide nanocrystals with larger chains. This process transforms polysaccharide nanocrystals into a co-continuous material to improve the interfacial adhesion between the filler and the matrix. Depending on the molecular weight, entanglements between grafted and ungrafted polymer chains can be obtained. NBCs can be prepared by this procedure, followed by classical methods such as solvent casting, hot-pressing, extrusion, injection molding, and thermoforming [271]. Habibi et al. grafted ramie nanocrystals with PCL by a ring-opening polymerization technique. In this work, the aqueous suspension of ramie nanocrystals was solvent-exchanged to dry toluene by centrifugation and redispersion [272]. The ring-opening polymerization procedure was also used to graft polycaprolactone (CW-*g*-PCL) onto linter nanocrystals [273]. Cellulose-based NBCs with hydrophobic characters have been envisaged and prepared by surface grafting for various applications[274]. The modification of cellulose nanocrystals and MFC with organic compounds via reaction of the surface hydroxyl groups has been carried out with different grafting agents, such as isocyanates [234], anhydrides [234], chlorosilanes [275, 276] or silanes [275–278]. Polyethylene glycol-grafted whiskers [279] are also used widely in the food industry.

2.4 Properties of NBCs

Once biocomposites have been incorporated or modified with nanomaterials, their fundamental properties and material energy changes into new and advanced materials. Nanocomposite materials are mechanically different from conventional composites due to exceptionally high surface-to-volume ratio [6]. NBCs exhibit superior mechanical, barrier, and functional properties with improved stability and durability [62]. A number of research articles have reported that the physical shape, including spheres, rods, triangles, palates, particles, etc., of the nanostructures intensively affects the resultant hybrid material [280–283]. Once NBCs have been synthesized, the final properties of the nanocomposite are usually investigated using electrical impedance spectroscopy, UV-vis spectroscopy, transmission electron microscopy (TEM) or potentiometric measurements [9, 62, 63, 172]. Phase identification and structural analysis are studied by x-ray diffractometry (XRD)[284–288]. Morphology and size are determined using scanning electron microscopy (SEM) and TEM, UV-vis spectroscopy, photoluminescence (PL) spectroscopy, and atomic force microscopy (AFM) [289–297].

2.5 Future and Applications of NBCs

The development of these biopolymeric nanocomposite materials and their applications in different industrial fields offer promising opportunities. Modern bioelectronic nanocomposite-based devices such as nano- and microbiosensor systems, biofuel elements, and biobatteries can be used to improve human lifestyles. The future development of novel NBCs with improved properties and multifunctionality can be envisaged as an emerging and open field for research. There are plenty of possibilities due to the abundance and diversity of biopolymers in nature and their synergistic combination with variety of nanosize materials. However, problems in processing these NBCs will present new challenges to researchers to develop more, ecofriendly, cost-effective technology in the future. New NBCs with nanomaterials form renewable sources will lead to a better environment.

Acknowledgments

We would like to acknowledge financial assistance from the Ministry of Science and Technology (MOST), Taipei, Taiwan (NSC-100-2221-E-002-032-MY3, MOST-105-2311-B-002-037, MOST-106-2311-B-002-036) and National Taiwan University (Project No. 104R4000).

References

1 Ozin, G., Arsenault, A., (2005) Nanochemistry: A Chemical Approach to Nanomaterials, Royal Society of Chemistry, Cambridge, Ch. 10.

2 Dujardin, E., Mann, S., (2002) *Bio-inspired Materials Chemistry*, Adv. Mater. 14, 775.

3 Ruiz-Hitzky, E., Aranda, P., Darder, M., (2009) In Bottom-Up Nanofabrication: Supramolecules, Self-Assemblies, and Organized Films (Eds. K. Ariga, H.S. Nalwa) American Scientific Publishers, 13: 978-1588830791

4 Ruiz-Hitzky, E., Darder, M., Aranda, P., (2005) *Functional biopolymer nanocomposites based on layered solids, J. Mater. Chem.* 15, 3650.

5 Bagal-Kestwal, D.R., Chiang, B.H., (2019) One dimensional nanostructures and their potential application in biosensor fabrication. In *Applications of One-dimensional Nanomaterials* (Eds. R. Chaughule, R. Devan) American Scientific Publishers, Valencia, CA, USA, pp. 1-38, ISBN: 1-58883-263-5.

6 Goetz, L., Mathew, A., Oksman, K., Gatenholm, P., Ragauskas, A.J., (2009) *A Novel Nanocomposite Film Prepared from Crosslinked Cellulosic Whiskers, Carbohyd. Polym.* 75, 85.

7 Zheng, Y., Monty, J., Linhardt, R.J., (2015) *Polysaccharide-based nanocomposites and their applications,Carbohydr. Res.* 405, 23.

8 Darder, M., Aranda, P., Ruiz-Hitzky, E., (2007) *Bionanocomposites: A New Concept of Ecological, Bioinspired, and Functional Hybrid Materials, Adv. Mater.* 19, 1309. http://onlinelibrary.wiley.com/doi/10.1002/adma.200602328/pdf.

9 Bagal-Kestwal, D.R., Kestwal, R.M., Chiang, B.H., (2016) Bio based nanomaterials and their bio-nanocomposites, in *Book of Nanomaterials and Nanocomposites, Zero- to Three-Dimensional Materials and their Composites*, Eds. Visakh, Morlanes, Wiley-VCH Verlag GmbH and Co. Germany. pp. 255. ISBN: 978-3-527-33780-4.

10 Gross, R.A., Kalra, B., (2002) *Biodegradable polymers for the environment. Science,* 297, 803.

11 Chabba, S., Netravali, A.N., (2005) *Green composites Part 1: Characterization of flax fabric and glutaraldehyde modified soy protein concentrate composites, J. Mater. Sci.* 40 (23), 6263.

12 Rubentheren, V., Thomas, W., Ching, Y.C., Praveena, N., (2016) *Effects of heat treatment on chitosan nanocomposite film reinforced with nanocrystalline cellulose and tannic acid, Carbohydr. Polym.* 140, 202.

13 Svagan, A.J., Jensen, P., Dvinskikh, S.V., Furo, I., Berglund, L.A., (2010) *Towards tailored hierarchical structures in cellulose nanocomposite biofoams prepared by freezing/freeze-drying, J. Mater. Chem.* 20, 6646.

14 Lin, N., Huang, J., Chang, P.R., Feng, L., Yu, J., (2011) *Effect of polysaccharide nanocrystals on structure, properties, and drug release kinetics of alginate-based microspheres, Colloids Surf., B*, 85, 270.

15 Jin, M., (2012) Plant Protein-Based Nanocomposite Materials: Modification of layered nanoclay by surface coating and enhanced interactions by enzymatic and chemical cross-linking, PhD Thesis, University of Tennessee – Knoxville. http://trace.tennessee.edu/cgi/viewcontent.cgi?article=2515&context=utk_graddiss.

16 Satyanarayana, K.G., (2015) *Preparation and characterization of nano silica from equisetum arvenses, J. Bioproces. Biotech.* 5, 205.

17 Singh, S.K., Gross, R.A., (2001) Overview: Introduction to Polysaccharides, Agroproteins, and Poly(amino acids), in Biopolymers from Polysaccharides and Agroproteins, Eds. Gross, R.A.; Scholz, C. ACS Symposium Series; American Chemical Society. Washington, DC. pp. 40.

18 Bagal-Kestwal, D.R., Kestwal, R., Chiang, B.H., (2011) *Development of dip-strip sucrose sensors: Application of plant invertase immobilized in chitosan–guar gum, gelatin and poly-acrylamide films, Sens. Actuators, A.*, 2011, 160, 1026.

19 Helbert, W., Cavaille, J.Y., Dufresne, A., (1996) *Thermoplastic nanocomposites filled with wheat straw cellulose whiskers. Part I: Processing and mechanical behavior, Polym. Compos.*, 17, 604.

20 Bondeson, D., Oksman K., (2007) *Polylactic Acid/Cellulose whisker nanocomposites modified by polyvinyl alcohol, Composites Part A*, 38, 12, 2486.

21 De Menezes, A.J., Siqueira, G., Curvelo, A.A.S., Dufresne, A., (2009) *Extrusion and characterization of functionalized cellulose whiskers reinforced polyethylene nanocomposites, Polymer*, 50, 4552.

22 Goffin, A.L., Raquez, J.M., Duquesne, E., Habibi, Y., Dufresne, A., Dubois, P., (2011) *From interfacial ring-opening polymerization to melt processing of cellulose nanowhisker-filled polylactide-based nanocomposites, Polymer*, 52, 1532.

23 Azizi, S.M.A., Alloin, F., and Dufresne, A., (2005) *Review of recent research into cellulosic whiskers, their properties and their application in nanocomposite field, Biomacromolecules*, 6, 612.

24 Adnan, S., Biak D.R.A., (2012) *The effect of acetylation on the crystallinity of BC/CNTs nanocomposite, J. Chem. Technol. Biotechnol.* 86, 431.

25 Zhang, X., Liu, X., Zheng, W., Zhu, J., (2012) *Regenerated cellulose/graphene nanocomposite films prepared in DMAC/LiCl solution, Carbohydr. Polym.* 88, 26.

26 Dufresne, A., (2006) *Comparing the mechanical properties of high performances polymer nanocomposites from biological sources, J. Nanosci. Nanotechnol.* 6, 322.

27 Park, H.M., Mohanty, A., Drzal, L., Lee, E., Mielewski, D., Misra, M., (2006) *Effect of sequential mixing and compounding conditions on cellulose acetate/layered silicate nanocomposite, J. Polym. Environ.*14, 27.

28 Wibowo, A.C., Misra, M., Park, H.M., Drzal, L.T., Schalek, R., Mohanty A.K., (2006) *Biodegradable nanocomposites from cellulose acetate: Mechanical, morphological, and thermal properties, Composites Part A* 37, 33.

29 Park, H.M., Misra, M., Drzal, L.T., Mohanty, A.K., (2004) *"Green" nanocomposites from cellulose acetate bioplastic and clay: effect of eco-friendly triethyl citrate plasticizer, Biomacromolecules*, 5, 2281.

30 Misra, M., Park, H.M., Mohanty, A.K., Drzal, L.T., (2004) *Injection molded 'Green' nanocomposite materials from renewable resources, Proceedings of the NSTI Nanotechnology Conference*, 3, 316. http://www.nsti.org/Nanotech2004/500.pdf.

31 Park, H.M., Liang, X., Mohanty, A., Misra, M., Drzal, L.T., Lee, E., Mielewski, D., (2004) *Effect of compatibilizer on nanostructure of the biodegradable cellulose acetate/organoclay nanocomposites, Macromolecules*, 37, 9076.

32 Delhom, C.D., White-Ghoorahoo, L.A., Pang, S.S. (2010) *Development and characterization of cellulose/clay nanocomposite, Composites Part B* 41, 475.

33 White, L.A., (2004) *Preparation and thermal analysis of cotton–clay nanocomposites, J. Appl. Polym. Sci.* 92, 2125.

34 Romero, R.B., Leite, C.A.P., Gonçalves MdC., (2009) *The effect of the solvent on the morphology of cellulose acetate/montmorillonite nanocomposites, Polymer*, 50, 161.

35 Rodríguez, F.J., Galotto, M.J., Guarda, A., Bruna, J.E., (2012) *Modification of cellulose acetate films using nanofillers based on organoclays, J. Food Eng.* 10, 262.

36 Li, M., Kim, I.H., Jeong, Y.G., (2010) *Cellulose acetate/multiwalled carbon nanotube nanocomposites with improved mechanical, thermal, and electrical properties, Appl. Polym. Sci.* 118, 2475.

37 Kang, H., Huang, G., Ma, S., Bai, Y, Ma, H., Li, Y., (2009) *Coassembly of inorganic macromolecule of exfoliated ldh nanosheets with cellulose, J. Phys. Chem. C*, 113, 9157.

38 Yano, S., Maeda H., Nakajima, M., Hagiwara, T., Sawaguchi, T., (2008) *Preparation and mechanical properties of bacterial cellulose nanocomposites loaded with silica nanoparticles, Cellulose*, 15, 111.

39 Chivrac, F., Pollet, E., Avérous, L., (2009) *Progress in nano-biocomposites based on polysaccharides and nanoclays, Mater. Sci. Eng. C*, 67, 1.

40 LeCorre, D., Bras, J., Dufresne, A., (2010) *Starch Nanoparticles: A Review, Biomacromolecules* 11, 1139.

41 Pandey, J.K., Singh, R.P., (2005) *Green Nanocomposites from Renewable Resources: Effect of Plasticizer on the Structure and Material Properties of Clay-filled Starch, Starch-Starke*, 57, 8.

42 Wilhelm, H.M., Sierakowski, M.R., Souza, G.P., Wypych, F., (2003) *Starch films reinforced with mineral clay, Carbohydr. Polym.* 52, 101.

43 Ren, P., Shen, T., Wang, F., Wang, X., Zhang, Z., (2009) *Study on biodegradable starch/OMMT nanocomposites for packaging applications, J. Polym. Environ.* 17, 203.

44 Schlemmer, D., Angélica, R.S., Sales, M.J., (2010) *Morphological and thermomechanical characterization of thermoplastic starch/montmorillonite nanocomposites, Compos. Struct.* 92, 2066.

45 Chen, B., Evans, J.R.G., (2005) *Thermoplastic Starch-Clay nanocomposites and their characteristics, Carbohydr. Polym.* 61, 455.

46 Park, H.M., Li, X., Jin, CZ., Park, C.Y., Cho, W.J., Ha, C.S., (2002) *Preparation and properties of biodegradable thermoplastic starch/clay hybrids, Macromol. Mater. Eng.* 287, 553.

47 Chaudhary, D.S., (2008) *Understanding amylose crystallinity in starch–clay nanocomposites, J. Polym. Sci., Part B Polym. Phys.* 46, 979.

48 Dean, K., Yu, L., Wu, D.Y., (2007) *Properties of extruded xanthan-starch-clay nanocomposite films, Compos. Sci. Technol.* 67, 413.

49 Park, H.M., Lee, W.K., Park, C.Y., Cho, W.J., Ha, C.S., (2003) *Environmentally friendly polymer hybrids. Part I Mechanical, thermal and barrier properties of thermoplastic starch/clay nanocomposites, J. Mater. Sci.* 38, 909.

50 Tang, X., Alavi, S., Herald, T.J., (2008) *Effects of plasticizers on the structure and properties of starch–clay nanocomposite films, Carbohydr. Polym.* 74, 552.

51 Wang, N., Zhang, X., Han, N., Bai, S., (2009) *Effect of citric acid and processing on the performance of thermoplastic starch/montmorillonite nanocomposites, Carbohydr. Polym.* 76, 68.

52 Chiou, B.S., Wood, D., Yee, E., Imam, S.H., Glenn, G.M., Orts, W.J., (2007) *Extruded starch-nanoclay nancomposites: effects* of *glycerol and nanoclay concentration, Polym. Eng. Sci.* 47, 1898.

53 Qiao, X., Jiang, W., Sun, K., (2005) *Reinforced thermoplastic acetylated starch with layered silicates, Starch Stärke*, 57, 581.

54 Mondragón, M., Hernández, E.M., Rivera-Armenta, J.L., RodríguezGonzález, F.J., (2009) *Injection molded thermoplastic starch/natural rubber/clay nanocomposites: Morphology and mechanical properties, Carbohydr. Polym.* 77, 80.

55 Lee, S.Y., Chen, H., Hanna, M.A., (2008) *Preparation and characterization of tapioca starch–poly(lactic acid) nanocomposite foams by melt intercalation based on clay type, Ind. Crop. Prod.* 28, 95.

56 Lee, S.Y., Hanna, M.A., (2009) *Tapioca starch-poly(lactic acid)-Cloisite* 30B *nanocomposite foams, Polym. Compos.* 30, 665.

57 Lee, S.Y., Hanna, M.A., (2008) *Preparation and characterization of tapioca starch-poly(lactic acid)±Cloisite Na+ nanocomposite foams, J. Appl. Polym. Sci.* 110, 2337.

58 Arroyo, O.H., Huneault, M.A., Favis, B.D., Bureau, M.N., (2010) *Processing and properties of PLA/thermoplastic starch/montmorillonite nanocomposites, Polym. Compos.* 31, 114.

59 Dean, K.M., Do, M.D., Petinakis, E., Yu, L., (2008) *Key interactions in biodegradable thermoplastic starch/poly (vinyl alcohol)/montmorillonite micro- and nanocomposites, Compos. Sci. Technol.* 68, 1453.

60 Majdzadeh-Ardakani, K., Nazari, B., (2010) *Improvingnthe mechanical properties of thermoplastic starch/poly (vinyl alcohol)/clay nanocomposites, Compos. Sci. Technol.* 70, 1557.

61 Liu X., Xiao S., Liu B., Liu J., Tang D., (2003) CaplusAn 2005:148270 (Patent).

62 Le Corre, D., Angellier-Coussy, H., (2014) *Preparation and application of starch nanoparticles for nanocomposites: A review, React. Funct. Polym.* 85, 97.

63 Haniffa, M.A.C.M., Ching, Y.C., Abdullah, L.C., Poh, S.C., Hock Chuah, C.H., (2016) *Review of bionanocomposite coating films and their applications, Polymers*, 8, 246.

64 Ojijo, V., Ray, S.S., (2013) *Processing strategies in bionanocomposites, Prog. Polym. Sci.* 38, 1543.

65 Wang, L., Sun, Y., Yang, X., (2014) *Fabrication and characterization of ZnxCd1−xS nanoparticles in chitosan alginate nanocomposite films, Ceram. Int.* 40, 4869.

66 Wang, B., Ji, X., Zhao, H., Wang, N., Li, X., Ni, R., Liu, Y., (2014) *An amperometric β-glucan biosensor based on the immobilization of bi-enzyme on Prussian blue–chitosan and gold nanoparticles–chitosan nanocomposite films, Biosens. Bioelectron.* 55, 113.

67 Shete, P.B., Patil, R.M., Thorat, N.D., Prasad, A., Ningthoujam, R.S., Ghosh, S.J., Pawar, S.H., (2014) *Magnetic chitosan nanocomposite for hyperthermia therapy application: Preparation, characterization and in vitro experiments, Appl. Surf. Sci.* 288, 149.

68 Liu, M., Zhang, Y., Wu, C., Xiong, S., Zhou, C., (2012) *Chitosan/halloysite nanotubes bionanocomposites: Structure, mechanical properties and biocompatibility, Int. J. Biol. Macromol.* 51, 566.

69 Kim, S.K., Sudha, P., Aisverya, S., Rose, M.H., Venkatesan, J., (2013) Bionanocomposites of chitosan for multitissue engineering applications. In *Chitin and Chitosan Derivatives: Advances in drug discovery and developments*, Eds. S.K. Kim, CRC Press: Boca Raton, FL, USA, pp. 451.

70 Hajiali, H., Karbasi, S., Hosseinalipour, M., Rezaie, H.R., (2010) *Preparation of a novel biodegradable nanocomposite scaffold based on poly (3-hydroxybutyrate)/bioglass nanoparticles for bone tissue engineering, J. Mater. Sci. Mater. Med.* 21, 2125.

71 Okamoto, M., John, B., (2013) *Synthetic biopolymer nanocomposites for tissue engineering scaffolds, Prog. Polym. Sci.*, 38, 1487.

72 Cao, X., Dong, H., Li, C.M., Lucia, L.A., (2009) *The enhanced mechanical properties of a covalently bound chitosan-multiwalled carbon nanotube nanocomposite, J. Appl. Polym. Sci.* 113, 466.

73 Liu, Y.L., Chen, W.H., Chang, Y.H., (2009) *Preparation and properties of chitosan/carbon nanotube nanocomposites using poly(styrene sulfonic acid)-modified CNTs, Carbohydr. Polym.* 76, 232.

74 Zhang, J.P., Wang, A.Q., (2009) *Synergistic effects of Na+-montmorillonite and multi-walled carbon nanotubes on mechanical properties of chitosan film, eXPRESS Polym. Lett.* 3, 302.

75 Shieh, Y.T., Wu, H.M., Twu, Y.K., Chung, Y.C., (2010) *An investigation on dispersion of carbon nanotubes in chitosan aqueous solutions, Coll. Polym. Sci.* 288, 377.

76 Wang, S.F., Shen, L., Zhang, W.D., Tong, Y.J., (2005) *Preparation and mechanical properties of chitosan/carbon nanotubes composites, Biomacromolecules* 6, 3067.

77 Saad, R., Hawari, J., (2013) *Grafting of lignin onto nanostructured silica SBA-15: preparation and characterization, J. Porous. Mater.* 20, 227.

78 Ruiz, H.V., Martinez, E.S.M., Mendez, M.A.A., (2011) *Biodegradability of polyethylene–starch blends prepared by extrusion and molded by injection: Evaluated by response surface methodology, Starch-Stärke*, 63, 42.

79 Calvo-Flores, G.F., Dobado, J.A., (2010) *Lignin as renewable raw material, ChemSusChem*, 3, 1227.

80 Hambardzumyan, A., Foulon, L., Chabbert, B., Aguie-Beghin, V., (2012) *Natural organic UV-absorbent coatings based on cellulose and lignin: designed effects on spectroscopic properties, Biomacromolecules*, 13, 4081.

81 Wang, X., Han, G., Shen, Z., Sun, R., (2013) Fabrication and application of chitosan-based inorganic nanocomposites. Eds. X. Wang, In *Nanocomposites: Synthesis, characterization and applications, Series: Nanotechnology Science and Technology*, Nova Science Publishers, New York. pp. 1–26.

82 Chung, Y.L., Olsson, J.V., Li, R.J., Frank, C.W., Waymouth, R.M., Billington, S.L., Sattely, E.S., (2013) *A renewable lignin–lactide copolymer and application in biobased composites, ACS Sustainable Chem. Eng.* 1, 1231.

83 Zhu, H., Lin, X., Zhuo, X., Zhang, C., (2011) *Preparation and characterization of spherical lignin/polylactide composite adsorbent, Adv. Mater. Res.* 221, 640.

84 Li, X., Don, Q., Huang, M., (2008) *Highly efficient fluorescence of NdF$_3$/SiO$_2$ core/shell nanoparticles and the applications for in vivo NIR detection, Prog. Chem.* 20, 227.

85 Lü, Q.F., Wang, C., Cheng, X., (2010) *One-step preparation of conductive polyaniline-lignosulfonate composite hollow nanospheres,* Microchim. *Acta* 169, 233.

86 Lü, Q.F., Luo, J.J., Lin, T.T., Zhang, Y.Z., (2014) N*ovel lignin–poly(n-methylaniline) composite sorbent for silver ion removal and recovery, ACS Sustainable Chem. Eng.* 2, 465.

87 Lü, Q.F., Huang, Z.K., Liu, B., Cheng, X., (2012) *Preparation and heavy metal ions biosorption of graft copolymers from enzymatic hydrolysis lignin and amino acids, Bioresour. Technol.* 104, 111.

88 Lü, Q.F., Zhang, J.Y., He, Z.W., (2012) *Controlled preparation and reactive silver-ion sorption of electrically conductive poly(n-butylaniline)–lignosulfonate composite nanospheres, Chem-Eur. J..* 18, 16571.

89 Samal, S.K., Fernandes, E., Corti, A., Chiellini, E., (2014) *Bio-based polyethylene–lignin composites containing a pro-oxidant/pro-degradant additive: preparation and characterization, J. Polym. Environ.* 22, 58.

90 Hilburg, S.L., Elder, A.N., Chung, H., Ferebee, R.L., Bockstaller, M.R., Washburn, N.R., (2014) *A universal route towards thermoplastic lignin composites with improved mechanical properties, Polymer,* 55, 995.

91 Fernandes, D., Winkler Hechenleitner, A., Job, A., Radovanocic, E., Gómez Pineda, E., (2006) *Thermal and photochemical stability of poly(vinyl alcohol)/modified lignin blends, Polym. Degrad. Stab.* 91, 1192.

92 Kubo, S., Kadla, J.F., (2003) *The formation of strong intermolecular interactions in immiscible blends of poly (vinyl alcohol)(PVA) and lignin, Biomacromolecules,* 4, 561.

93 Saha, B.C., Bothast, R.J., (1993) *Starch conversion by amylases from Aureobasidium pullulans, J. Ind. Microbiol.* 12, 413.

94 Mishra, B., Vuppu, S., and Rath, K. (2011) *The role of microbial pullulan, a biopolymer in pharmaceutical approaches: A review, J. Appl. Pharm. Sci.* 1, 45.

95 Na, K., Lee, T.B., Park, K. H., Shin, E.K., Lee, Y.B., Choi, H.K., (2003) *Self-assembled nanoparticles of hydrophobically-modified polysaccharide bearing vitamin H as a targeted anti-cancer drug delivery system, Eur. J. Pharm. Sci.* 18, 165.

96 Trovatti, E., Fernandes, S.C.M., Rubatat, L., Silva Perez, D., Freire, C.S.R., Silvestre, A.J.D., Neto, C.P., (2012) *Pullulan–nanofibrillated cellulose composite films with improved thermal and mechanical properties, Composites Sci. Technol.* 72, 1556.

97 Trovatti, E., Fernandes, S.C.M., Rubatat, L., Freire, C.S.R., (2012) *Sustainable nanocomposite films based on bacterial cellulose and pullulan, Cellulose* 19, 729.

98 Biliaderis, C.G., Lazaridou, A., (2002) *Thermophysical properties of chitosan, chitosan–starch and chitosan–pullulan films near the glass transition, Carbohydr. Polym.* 48, 179.

99 Biliaderis, C.G., Kristo, E., Zampraka, A., (2007) *Water vapour barrier and tensile properties of composite caseinate-pullulan films: Biopolymer composition effects and impact of beeswax lamination, Food Chem.* 101, 753.

100 Biliaderis, C.G., Kristo, E. (2007) *Physical properties of starch nanocrystal-reinforced pullulan films. Carbohydr. Polym.* 68, 146.

101 Gurtner, G.C., Wong, V.W., Rustad, K.C., Galvez, M.G., Neofyotou, E., Glotzbach J.P., Januszyk, M., Major, M.R., Sorkin, M., Longaker, M.T., Rajadas, J., (2011)

Engineered pullulan–collagen composite dermal hydrogels improve early cutaneous wound healing, Tissue Eng. 17, 631.

102 Teramoto, N., Saitoh, M., Kuroiwa, J., Shibata, M., Yosomiya, R., (2001) *Morphology and mechanical properties of pullulan/poly(vinyl alcohol) blends crosslinked with glyoxal, J. Appl. Polym. Sci.* 82, 2273.

103 Khademhosseini, A., Bae, H., Ahari, A.F., Shin, H., Nichol, J.W., Hutson, C.B., Masaeli, M., Kim, S.H., Aubin, H., Yamanlar, S., (2011) *Cell-laden microengineered pullulan methacrylate hydrogels promote cell proliferation and 3D cluster formation, Soft Matter* 7, 1903.

104 Lee, C.H., Singla, A., Lee, Y., (2001) *Biomedical applications of collagen, Int. J. Pharm.* 221, 1.

105 Chen, Z.G., Wang, P.W., Wei, B., Mo, X.M., Cui, F.Z., (2010) *Electrospun collagen–chitosan nanofiber: A biomimetic extracellular matrix for endothelial cell and smooth muscle cell, Acta Biomater.* 6, 372.

106 Balthasar, S., Michaelis, K., Dinauer, N., Briesen, H.V., Kreuter, J., Langer, K., (2005) *Preparation and characterisation of antibody modified gelatin nanoparticles as drug carrier system for uptake in lymphocytes, Biomaterials*, 26, 2723.

107 Tan, Q., Tang, H., Hu, J., Hu, Y., Zhou, X., Tao, Y., Wu, Z., (2011) *Controlled release of chitosan/heparin nanoparticle-delivered VEGF enhances regeneration of decellularized tissue-engineered scaffolds, Int. J. Nanomedicine* 6, 929.

108 Bae, H.J., (2007) Fish Gelatin-Nanoclay Composite Film. Mechanical and physical properties, Effect of enzyme cross-linking, and as a functional film layer, ProQuest: Clemson, SC, USA.

109 Koide, M., Osaki, K., Konishi, J., Oyamada, K., Katakura, T., Takahashi, A., Yoshizato, K. A., (1993) *A new type of biomaterial for artificial skin: Dehydrothermally cross-linked composites of fibrillar and denatured collagens, J. Biomed. Mater. Res.* 27, 79.

110 Martucci, J., Vázquez, A., Ruseckaite, R., (2007) *Nanocomposites based on gelatin and montmorillonite, morphological and thermal studies, J. Therm. Anal. Calorim.* 89, 117.

111 Rao, Y.Q., (2007) *Gelatin–clay nanocomposites of improved properties, Polymer*, 48, 5369.

112 Murata, M., Maki, F., Sato, D., Shibata, T., Arisue, M., (2000) *Bone augmentation by onlay implant using recombinant human BMP-2 and collagen on adult rat skull without periosteum, Clin. Oral Implant. Res.* 11, 289.

113 Teng, Z., Luo, Y., and Wang, Q., (2012) *Nanoparticles synthesized from soy protein: preparation, characterization, and application for nutraceutical encapsulation, J. Agric. Food Chem.* 60, 2712.

114 Javid, A., Ahmadian, S., Saboury, A.A., Kalantar, S.M., Rezaei-Zarchi, S., (2014) *Novel biodegradable heparin-coated nanocomposite system for targeted drug delivery, RSC Adv.* 4, 13719.

115 Lee, S., Lee, M., Song, K., (2005) *Effect of gamma-irradiation on the physicochemical properties of gluten films, Food Chem.* 92, 621.

116 Cuq, B., Gontard, N., Guilbert, S., (1998) *Proteins as agricultural polymers for packaging Production, Cereal Chem.* 75, 1.

117 Sothornvit, R., Krochta, J.M., (2005) Plasticisers in edible films and coatings. In *Innovations in Food Packaging*, Han, J.H., Eds., Elsevier Publishers: New York, pp. 403.

118 Arora, A., Padua, G.W. (2010). Review: nanocomposites in food packaging. *J. Food Sci.*, 75(1), R43-R49.

119 Chang, P.R., Yang, Y., Huang, J., Xia, W. B., Feng, L.D., Wu, J.Y., (2009) *Effects of layered silicate structure on the mechanical properties and structures of protein-based bionanocomposites, J. Appl. Polym. Sci.* 113 (2), 1247.

120 Chen, P., Zhang, L., (2006) *Interaction and properties of highly exfoliated soy protein/montmorillonite nanocomposites, Biomacromolecules*, 7(6), 1700.

121 Guilherme, M.R., Mattoso, L.H.C., Gontard, N., Guilbert, S., Gastaldi, E., (2010) *Synthesis of nanocomposite films from wheat gluten matrix and MMT intercalated with different quaternary ammonium salts by way of hydroalcoholic solvent casting, Composites Part A*, 41(3), 375–382.

122 Kumar, P. (2009). Development of bio-nanocomposite films with enhanced mechanical and barrier properties using extrusion processing. North Carolina State University. PhD dissertation.

123 Kumar, P., Sandeep, K.P., Alavi, S., Truong, V.D., Gorga, R.E., (2010) *Effect of type and content of modified montmorillonite on the structure and properties of bio-nanocomposite films based on soy protein isolate and montmorillonite, J. Food Sci.* 75 (5), N46.

124 Kumar, P., Sandeep, K.P., Alavi, S., Truong, V.D., Gorga, R.E., (2010) *Preparation and characterization of bio-nanocomposite films based on soy protein isolate and montmorillonite using melt extrusion, J. Food Eng.* 100 (3), 480.

125 Lee, J.E., Kim, K.M., (2010) *Characteristics of soy protein isolate-montmorillonite composite films, J. Appl. Polym. Sci.*, 118 (4), 2257–2263.

126 Martucci, J.F., Ruseckaite, R.A., (2010) *Biodegradable three-layer film derived from bovine gelatin, J. Food Eng.* 99 (3), 377.

127 Sadare, O.O., Daramola, M.O., Afolabi, A.S., (2015) IAENG, Proceedings of the World Congress on Engineering, Vol II WCE 2015, July 1–3, London.

128 Nayak, P.L., Sasmal, A., Nayak, P., Sahoo, S., Mishra, J.K., Kang, S.C., Lee, J.W., Chang, Y.W., (2008) *nanocomposites from polycaprolactone (pcl)/soy protein isolate (spi) blend with organoclay, Polym. Plast. Technol. Eng.* 47, 6, 600.

129 Alig, I., Pötschke, P., Lellinger, D., Skipa T., Pegel, S., Kasaliwal, G.R., Villmow, T., (2012) *Establishment, morphology and properties of carbon nanotube networks in polymer melts, Polymer*, 53, 28.

130 Capek, I. (2006). Nanocomposite structures and dispersions: science and nanotechnology- fundamental principles and colloidal particles. Nanocomposite structures and dispersions, Vol. 23, p. 312. eBook ISBN: 9780080479590, Hardcover ISBN: 9780444527165. Elsevier Science, New York.

131 Herron, N., Thorn, D.L., (1998) *Nanoparticles: Uses and relationships to molecular cluster compounds, Adv. Mater.* 10, 1173.

132 Favier, V., Canova, G.R., Shrivastava, S.C., Cavaille, J.V., (1997) *Mechanical percolation in cellulose whisker nanocomposites, Polym. Eng. Sci.* 37(10), 1732.

133 Chazeau, L., Cavaille, J.Y., Canova, G., Dendievel, R., Boutherin, B., (1999) *Viscoelastic properties of plasticized PVC reinforced with cellulose whiskers, J. Appl. Polym. Sci.* 71(11), 1797.

134 Ogawa, M., Kuroda, K., (1997) *Preparation of inorganic–organic nanocomposites through intercalation of organoammonium ions into layered silicates, Bull. Chem. Soc. Jpn.*, 70 (11), 2593.

135 Theng, B.K.G., (1974) The Chemistry of Clay-Organic Reactions. New York, Wiley.

136 Schmidt, D., Shah, D., Giannelis, E.P., (2002) *New advances in polymer/layered silicate nanocomposites, Curr. Opin. Solid State Mater. Sci.* 6 (3), 205.

137 Alexandre, M., Dubois, P., (2000) *Polymer-layered silicate nanocomposites: preparation, properties and uses of a new class of materials, Mater. Sci. Eng.* 28 (1–2), 1.

138 Chen, L., Yang, W.J., Yang, C.Z., (1997) *Preparation of nanoscale iron and Fe3O4 powders in a polymer matrix, J. Mater. Sci.* 32 (13), 3571.

139 Muller, K., Bugnicourt, E., Latorre, M., Jorda, M., Echegoyen, S.Y., Lagaron, J.M., Miesbauer, O., Bianchin, A., Hankin, S., Bölz, U., Pérez, G., Jesdinszki, M., Lindner, M., Scheuerer, Z., Castelló, S., Schmid, M., (2017) *Review on the Processing and Properties of Polymer Nanocomposites and Nanocoatings and Their Applications in the Packaging, Automotive and Solar Energy Fields, Nanomaterials* 7, 74.

140 Tang, X.Z., Alavi, S., Herald, T.J., (2008) *Barrier and mechanical properties of starch-clay nanocomposite films, Cereal Chem.* 85 (3), 433.

141 Bagal, D.S., Karve, M.S., (2006) *Entrapment of plant invertase within novel composite of agarose- guar gum biopolymer membrane, Anal. Chim. Acta*, 555, 316.

142 Bagal-Kestwal, D.R., Kestwal, R.M., Chiang, B.H., (2015) *Invertase-nanogold clusters decorated plant membranes for fluorescence-based sucrose sensor, J. Nanotechnol.* 13, 1.

143 Barton, J., Niemczyk, A., Czaja, K., Korach, Ł., Sachermajewska, B., (2014) *Kompozyty, biokompozyty i nanokompozyty polimerowe. Otrzymywanie, skład, właściwości i kierunki zastosowań, Chemik* 68, 4, 280.

144 Nabi Saheb, D., Jog, J.P., (1999) *Natural fiber polymer composites: A review, Adv. Polym. Technol.* 18 (4), 351.

145 Gatenholm P., Bertilsson H., Mathiasson A., (1993) *The effect of chemical composition of interphase on dispersion of cellulose fibers in polymers. I. PVC-coated cellulose in polystyrene, J. Appl. Polym. Sci.* 49 (2), 197.

146 Dale, E.W., O'Dell, J., (1999) *Wood–polymer composites made with acrylic monomers, isocyanate, and maleic anhydride, J. Appl. Polym. Sci.* 73 (12), 2493.

147 Mani, P., Satyanarayan, K.G., (1990) *Effects of the surface treatments of lignocellulosic fibers on their debonding stress, J. Adhes. Sci. Technol.* 4 (1), 17.

148 Joseph, K., Thomas, S., Pavithran, C., (1996) *Effect of chemical treatment on the tensile properties of short sisal fibre-reinforced polyethylene composites, Polimery*, 37 (23), 5139.

149 Razi, P.S., Portier, R., Raman, A., (1999) *Studies on Polymer-wood interface bonding: effect of coupling agents and surface modification, J. Compos. Mater.* 33 (12), 1064.

150 Zare, Y., Shabani, I., Muraviev, D.N., Ruiz, P., Muñoz, M., Macanás, J., (2008) *Novel strategies for preparation and characterization of functional polymer-metal nanocomposites for electrochemical applications, Pure Appl. Chem.* 80, 2425.

151 Liu, Z., Jiao, Y., Wang, Y., Zhou, C., Zhang, Z., (2008) *Polysaccharides-based nanoparticles as drug delivery systems, Adv. Drug Deliv. Rev.* 60, 1650.

152 Vaia, R.A., Giannelis, E.P., (1997) Lattice model of polymer melt intercalation in organically-modified layered silicates, *Macromolecules*, 30, 7990.

153 Kumar, C.S.S.R., (2010) Nanocomposites, *Weinheim: Wiley-VCH*, 8, 466.

154 Cury Camargo P.H., Satyanarayana, K.G., Wypych, F., (2009) *Nanocomposites: synthesis, structure, properties and new application opportunities, Mat. Res.*, 12, 1.

155 Bagal-Kestwal, D.R., Kestwal, R., Chiang, B.H., (2013) Biosensors based on nano-materials and their applications, in *Applications of Nanomaterials*, (Eds. R.S. Chaughule, S.C. Watawe) American Scientific Publishers, Valencia, USA, pp. 1–52. ISBN: 1-58883-181-7.

156 Kestwal, R., Bagal-Kestwal, D.R., Chiang, B.H., (2012) *Analysis and enhancement of nutritional and antioxidant properties of vigna aconitifolia sprouts*, Plant Food Hum. *Nutr.*, 67, 136.

157 Bagal, D., Kestwal, R., Hsieh, B.C., Chen, R.L.C., Cheng, T.J., Chiang, B.H., (2010) *Electrochemical β(1 → 3)-d-glucan biosensors fabricated by immobilization of enzymes with gold nanoparticles on platinum electrode, Biosens. Bioelectron.* 26, 118.

158 Bagal-Kestwal, D.R., Kestwal, R., Chiang, B.H., (2011) *1,3-β-Glucanase from Vigna aconitifolia and its possible use in enzyme bioreactor fabrication, J. Biol. Macromol.*, 49, 894.

159 Noishiki, Y., Nishiyama, Y., Wada, M., Kuga, S., Magoshi, J., (2002) *Mechanical properties of silk fibroin–microcrystalline cellulose composite films, J. Appl. Polym. Sci.* 86, 3425.

160 Markovic, G., Visakh, P.M., (2017) Application of Chitin based rubber nanocom-posites, In *Rubber Based Bionanocomposites: Preparation*, Eds. Visakh, Springer Publishers, Cham. pp. 51.

161 Grabian, J., Gawdzińska, K., Przetakiewicz, W., Pijanowski, M., (2011) *Description of the particle distribution in the space of composite suspension casting by statistical methods,* Arch. *Foundry Eng.* 11(1), 31.

162 Moad, G., (2011) *Chemical modification of starch by reactive extrusion, Progr. Polym. Sci.* 36, 218.

163 García, N.L., D'Accorso, F.N.B., Goyanes, S., (2015) Biodegradable starch nanocomposites In *Eco-friendly Polymer Nanocomposites* (Eds. V.K. Thakur, M.K. Thakur), Advanced Structured Materials, pp. 75, Springer, India. DOI 10.1007/978-81-322-2470-9_2

164 Galicia-García, T., Martínez-Bustos, F., Jiménez-Arévalo, O.A., Arencón, D., Gámez-Pérez, J., Martínez, A.B., (2012) *Films of native and modified starch rein-forced with fiber: Influence of some extrusion variables using response surface methodology, J. Appl. Polym. Sci.* 126, 327.

165 Müller, C., Laurindo, J., Yamashita, F., (2012) *Composites of thermoplastic starch and nanoclays produced by extrusion and thermopressing, Carbohydr. Polym.* 89, 504.

166 Mitrus, M., Moscicki, L., (2014) *Extrusion-cooking of starch protective loose-fill foams, Chem. Eng. Res. Des.* 9, 778.

167 De Melo, C., Garcia, P.S., Grossmann, M.V.E., Yamashita, F., Da Antônia, L.H., Mali, S., (2011) *Properties of extruded xanthan-starch-clay nanocomposite films, Braz. Arch. Biol. Technol.* 54 (6), 1223.

168 Moigne, N., Sauceau, M., Benyakhlef, M., Jemai, R., Bénézet, J.C., Lopez-Cuesta, J.M., Rodier, E., Fages, J., (2014) *ECCM16 – 16th European Conference on Compos-ite Materials*, Seville, Spain, pp. 1.

169 Herrera, N., Roch, H., Salaberria, A.M., Pino-Orellana, M.A., Labidi, J., Fernandes, S.C.M., Radic, D., Leiva, A., Oksman, K., (2016) *Functionalized blown films of plasticized polylactic acid/chitin nanocomposite: Preparation and characterization,* Mater. Des. 92, 846.

170 Jiang, X., Drzal, L.T., (2012) *Reduction in percolation threshold of injection molded high-density polyethylene/exfoliated graphene nanoplatelets composites by solid state ball milling and solid state shear pulverization,* J. Appl. Polym. Sci., 124, 525.

171 Guo, W., Chen, G., (2014) *Fabrication of graphene/epoxy resin composites with much enhanced thermal conductivity via ball milling technique,* J. Appl. Polym. Sci., 131, 40565.

172 Sorrentino, A., Gorrasi, G., Tortora, M., Vittoria, V., Costantino, U., Marmottini, F., Padella, F., (2005) *Incorporation of Mg–Al hydrotalcite into a biodegradable Poly(ε-caprolactone) by high energy ball milling,* Polymer, 46, 1601.

173 Wu, H., Rook, B., Drzal, L.T., (2013) *Dispersion optimization of exfoliated graphene nanoplatelet in polyetherimide nanocomposites: Extrusion, precoating, and solid state ball milling,* Polym. Compos. 34, 426.

174 Wakabayashi, K., Brunner, P.J., Masuda, J., Hewlett, S.A., Torkelson, J.M., (2010) *Polypropylene-graphite nanocomposites made by solid-state shear pulverization: Effects of significantly exfoliated, unmodified graphite content on physical, mechanical and electrical properties,* Polymer, 51, 5525.

175 Oksman, K., Mathew, A.P., Bismarck, A., Rojas, O., Sain, M., (2014) Processing technologies, properties and applications, In *Handbook of Green Materials,* (Eds. K. Oksman, A.P. Mathew, A. Bismarck, O. Rojas, M. Sain) World Scientific, New Jersey (2014) pp. 2.

176 Torkelson, J., (2014) Patent number- 8734696, https://www.scholars.northwestern .edu/en/publications/polymer-graphite-nanocomposites-via-solid-state-shear-pulverizati-3.

177 Iwamoto, S., Yamamoto, S., Lee, S.H., Endo, T., (2014) *Solid-state shear pulverization as effective treatment for dispersing lignocellulose nanofibers in polypropylene composites,* Cellulose, 21, 1573.

178 Walker, A., Tao, Y., Torkelson J., (2007) *Polyethylene/starch blends with enhanced oxygen barrier and mechanical properties: Effect of granule morphology damage by solid-state shear pulverization,* Polymer, 48, 1066.

179 Iyer, K.A., Torkelson, J.M., (2013) *Novel, synergistic composites of polypropylene and rice husk ash: Sustainable resource hybrids prepared by solid-state shear pulverization,* Polym. Compos. 34, 1211.

180 Iyer, K.A., Torkelson, J.M., (2014) *Green composites of polypropylene and eggshell: Effective biofiller size reduction and dispersion by single-step processing with solid-state shear pulverization,* Compos. Sci. Technol. 102, 152.

181 Larrondo, R. Manley, J. St, (1981) *Electrostatic fiber spinning from polymer melts. I. Experimental observations on fiber formation and properties,* J. Polymer Sci. Polymer Physics Ed., 19, 909.

182 Sahay, R., Kumar, S.P., Sridhar R., Sundaramurthy, J., Venugopal J., Mhaisalkar, S.G., Ramakrishna S., (2012) *Electrospun composite nanofibers and their multifaceted applications,* J. Mater. Chem. 22, 12953.

183 Reneker, D.H., Yarin, A.L., Fong, H., Koombhangse, S., (2010) *Bending instability of electrically charged liquid jets of polymer solutions in electrospinning*, J. Appl. Phys., 87, 4531.

184 Reneker, D.H., Yarin, A.L., Zussman, E., Xu, H., (2007) Electrospinning of nanofibers from polymer solutions and melts, *Advances in Applied Mechanics*, Elsevier, Oxford, 2007, vol. 41, pp. 43.

185 Frenot, A., Chronakis, I. S., (2003) *Polymer nanofibers assembled by electrospinning*, *Curr. Opin. Colloid Interface Sci.*, 8, 64.

186 Huang, Z.M., Zhang, Y.Z., Kotaki, M., Ramakrishna S., (2003) *A review on polymer nanofibers by electro-spinning applications in nanocomposites, Compos. Sci. Technol.*, 63, 2223.

187 Dzenis, Y., (2004) *Spinning Continuous Fibers for Nanotechnology, Science*, 304, 1917.

188 Gora, A., Sahay, R., Thavasi, V., Ramakrishna, S., *Melt-electrospun fibers for advances in biomedical engineering, clean energy, filtration, and separation*, (2011) *Polym. Rev.* 51, 265.

189 Larrondo, R., Manley, J. St, (1981) *Electrostatic fiber spinning from polymer melts. I. Experimental observations on fiber formation and properties, J. Polymer Sci. Polymer Physics Ed.*, 19, 921.

190 Zhang, Y.Z., Venugopal, J., Huang, Z.M., Lim, C.T., Ramakrishna, S., (2005) *Characterization of the surface biocompatibility of the electrospun PCL-collagen nanofibers using fibroblasts, Biomacromolecules*, 6, 2583.

191 Xu, C.Y., Inai, R., Kotaki, M., Ramakrishna S., (2004) *Aligned biodegradable nanofibrous structure: a potential scaffold for blood vessel engineering, Biomaterials*, 25, 877.

192 Yang, F., Murugan, R., Wang, S., Ramakrishna, S., (2005) *Electrospinning of nano/micro scale poly(l-lactic acid) aligned fibers and their potential in neural tissue engineering, Biomaterials*, 26, 2603.

193 Lee, C.H., Shin, H.J., Cho, I.H., Kang, Y.M., Kim, I.A., Park, K.D., Shin, J.W., (2005) *Nanofiber alignment and direction of mechanical strain affect the ECM production of human ACL fibroblast, Biomaterials*, 26, 1261.

194 Khajavi, R., Abbasipour, M., (2012) *Electrospinning as a versatile method for fabricating coreshell, hollow and porous nanofibers, Scientia Iranica*, 19 (6), 2029.

195 Ye, S.H., Zhang, D., Liu, H.Q., Zhou, J.P., (2011) *ZnO nanocrystallites/cellulose hybrid nanofibers fabricated by electrospinning and solvothermal techniques and their photocatalytic activity, J. Appl. Polym. Sci.*, 121, 1757.

196 He, T.S., Ma, H.H., Zhou, Z.F., Xu, W.B., Ren, F.M., Shi, Z.F., Wang, J., (2009) *Preparation of ZnS–Fluoropolymer nanocomposites and its photocatalytic degradation of methylene blue, Polym. Degrad. Stab.* 94, 2251.

197 Liu X., Smith L.A., Hu J., Ma P.X., (2009) *Biomimetic nanofibrous gelatin/apatite composite scaffolds for bone tissue engineering, Biomaterials*, 30, 2252.

198 Li, C., Vepari, C., Jin, H.J., Kim, H., Kaplan, D., (2006) *Electrospun silk-BMP-2 scaffolds for bone tissue engineering, Biomaterials*, 27, 3115.

199 Ray, S.S., Okamoto, M., (2003) *Polymer/layered silicate nanocomposites: A review form preparation to processing, Prog. Polym. Sci.*, 28, 1539.

200 Pushparaj, V.L., Shaijumon, M.M., Kumar, A., Murugesan, S., Ci, L., Vajtai, R., Linhardt, R.J., Nalamasu, O., Ajayan, P.M., (2007) *Flexible energy storage devices based on nanocomposite paper, Proc. Nat. Acad. Sci. USA.* 104, 13574.

201 Kim, D.H., Park, S.Y., Kim, J., Park, M.J., (2010) *Preparation and properties of the single-walled carbon nanotube/cellulose nanocomposites using N-methylmorpholineN-oxide monohydrate. Appl. Polym. Sci.* 117, 3588.

202 Fernandes, S.C.M., Oliveira, L., Freire, C.S.R., Silvestre, A.J.D., Neto, C.P., Gandini, A., Desbrieresm, J., (2009) *Novel transparent nanocomposite films based on chitosan and bacterial cellulose, Green Chem.* 11, 2023.

203 Shimazaki, Y., Miyazaki, Y., Takezawa, Y., Nogi, M., Abe, K., Ifuku, S., Yano, H., (2007) *Excellent thermal conductivity of transparent cellulose nanofiber/epoxy resin nanocomposites, Biomacromolecules* 8, 2976.

204 Gorrasi, G., Bugatti, V., Vittoria, V., (2012) *Pectins filled with LDH-antimicrobial molecules: preparation, characterization and physical properties, Carbohydr. Polym.* 89, 132.

205 Stevens, K.N.J., Croes, S., Boersma, R.S., Stobberingh, E.E., van der Marel, C., van der Veen FH, Knetsch, M.L., Koole, L.H., (2011) *Hydrophilic surface coatings with embedded biocidal silver nanoparticles and sodium heparin for central venous catheters, Biomaterials* 32 (5), 1264.

206 Azeredo, H.M.C., Mattoso, L.H.C., Avena-Bustillos, R.J., Ceotto Filho, G., Munford, M.L., Wood, D., McHugh, T.H., (2010) *Nanocellulose reinforced chitosan composite films as affected by nanofiller loading and plasticizer content, J. Food Sci.* 75 (1), N1.

207 Ma, J., Zhu, W., Tian, Y., Wang, Z., (2016) *Preparation of Zinc oxide-starch nanocomposite and its application on coating, Nanoscale Res. Lett.* 11, 200.

208 Dehnad, D., Mirzaei, H., Emam-Djomeh, Z., Jafari, S.M., Dadashi, S., (2014) *Thermal and antimicrobial properties of chitosan-nanocellulose films for extending shelf life of ground meat, Carbohydr. Polym.* 109, 148.

209 Azizi, S., Ahmad, M.B., Ibrahim, N.A., Hussein, M.Z., Namvar, F., (2014) *Cellulose nanocrystals/ZnO as a bifunctional reinforcing nanocomposite for poly(vinyl alcohol)/chitosan blend films: Fabrication, characterization and properties, Int. J. Mol. Sci.* 15 (6), 11040.

210 Yan, E., Wang, C., Wang S., Sun L., Wang Y., Fan L., Zhang D., (2011) *Stimuli-responsive supramolecular polymeric materials, Mater. Sci. Eng.* 176, 458.

211 Salehi R., Arami M., Mahmoodi N.M., Bahrami H., Khorramfar S., (2010) Novel biocompatible composite (chitosan-zinc oxide nanoparticle): preparation, characterization and dye adsorption properties, *Colloids Surf.* 80, 86.

212 Zhu, H., Jiang R., Xiao L., Chang Y., Guan Y., Li X., Zeng G., (2009) *Photocatalytic decolorization and degradation of Congo Red on innovative crosslinked chitosan/nano-CdS composite catalyst under visible light irradiation, J. Hazard. Mater.* 169, 933.

213 Shakir, M., Jolly, R., M.S. Khan, S.M., Rauf, A., Kazmi, S., (2016) *Nano-hydroxyapatite/β-CD/chitosan nanocomposite for potential applications in bone tissue engineering, Int. J. Bio. Macromol.* 93, 276.

214 Sulak, M.T., Erhan, E., Keskinler, B., (2012) *Electrochemical phenol biosensor configurations based on nanobiocomposites, Sens. Mater.* 24, 141.

215 Bhiwankar, N.N., Weiss, R.A., (2006) *Melt intercalation/exfoliation of polystyreneesodium-montmorillonite nanocomposites using sulfonated polystyrene ionomer compatibilizers, Polymer,* 47, 6684e6691.

216 Vaia, R.A., Jandt, K.D., Kramer, E.J., Giannelis, E.P., (1996) *Kinetics of polymer melt intercalation, Chem. Mater.* 8, 2628.

217 Hunter, D.L., Kamena, K.W., Paul, D.R., (2007) *Processing and properties of polymers modified by clays, MRS Bull.,* 32, 323.

218 Choi, J.T., Kim, D.H., Ryu, K.S., Lee, H., Jeong, H.M., Shin, C.M., Kim, J.H., Kim, B.K., (2011) *Functionalized graphene sheet/polyurethane nanocomposites: Effect of particle size on physical properties, Macromol. Res.* 19, 809.

219 Chan, C.M., Wu, J., Li, J.X., Cheung, Y.K., (2002) *Polypropylene/calcium carbonate nanocomposites, Polymer,* 43, 2981.

220 Sumita, M., Tsukumo, Y., Miyasaka, K., Ishikawa, K., (1983) *Tensile yield stress of polypropylene composites filled with ultrafine particles, J. Mater. Sci.* 18, 1758.

221 Tien, Y.I., Wei, K.H., (2001) *High-tensile-property layered silicates/polyurethane nanocomposites by using reactive silicates as pseudo chain extenders, Macromolecules* 34, 9045.

222 Usuki, A., Kojima, Y., Kawasumi, M., Okada, A., Fukushima, Y., Kurauchi, T., Kamigaito, O., (1993) *Synthesis of nylon 6-clay hybrid, J. Mater. Res.,* 8, 1179.

223 LeBaron, P.C., Wang, Z., Pinnavaia, T.J., (1999) *Polymer-layered silicate nanocomposites: an overview, Appl. Clay Sci.,* 15, 11.

224 Greco, A., Maffezzoli, A., Calò, E., Massaro, C., Terzi, R., (2011) *An investigation into sintering of PA6 nanocomposite powders for rotational molding, J. Therm. Anal. Calorim.* 109, 1493.

225 Demitri, C., Moscatello, A., Giuri, A., Raucci, M.G., Corcione, C.E., (2015) *Preparation and characterization of EG-chitosan nanocomposites via direct exfoliation: a green methodology, Polymers,* 7, 2584.

226 Tabashi, S., Oromiehie, A., Bazgir, S., ECCM16 – 16th European Conference on Composite Materials, 22–26 June 2014, Seville, Spain.

227 Ghanbarzadeh, B., Almasi, H., Oleyaei, S.A., (2014) *A novel modified starch/carboxymethyl cellulose/montmorillonite bionanocomposite film: structural and physical properties, Int. J. Food Eng.* 10 (1), 121.

228 Swain, S.K., Jena, I., (2010) *Polymer/carbon nanotube nanocomposites: a novel material, Asian J. Chem.,* 22 (1), 1.

229 Seo, D.W., Yoon, W.J., Park, S.J., Jo, M.C., Kim, J.S., (2006) *The preparation of multi- walled CNT-PMMA nanocomposite, Carbon Lett,* 7 (4), 266.

230 Degner, B.M., Chung, C., Schlegel, V., Hutkins, R., McClements, D.J., (2014) *Factors influencing the freeze-thaw stability of emulsion-based foods, Compr. Rev. Food Sci. Food Saf.* 13 (2), 98.

231 Si, X., Chen, Q., Bi, J., Wu, X., Yi, J., Zhou, L., Li, Z., (2016) *Comparison of different drying methods on the physical properties, bioactive compounds and antioxidant activity of raspberry powders, J. Sci. Food Agric.* 96 (6), 2055.

232 Thomas, S., Visakh, P.M., Mathew, A.P., (2012) Advances in natural polymers: composites and nanocomposites, springer science and business media, Technology and Engineering, pp. 426.

233 Nair, K.G., Dufresne, A., (2003) *Crab shell chitin whisker reinforced natural rubber nanocomposites. 1. processing and swelling behavior, Biomacromolecules,* 4, 657.

234 Petersson, L., Kvien, I., Oksman, K., (2007) *Structure and thermal properties of poly(lactic acid)/cellulose whiskers nanocomposite materials, Compos. Sci. Technol.* 67, 2535.

235 Yuanyuan, Z., Song, L., (2012*) Preparation of chitosan/poly (lactic-co glycolic acid)(PLGA) nanocoposite for tissue engineering scaffold,* Optoelectron. Adv. Mater. 6 (3), 516.

236 Cui, Z., Zhao, H., Peng, Y., Han, J., Turng, L.S., Shen, C. (2014) *Fabrication and characterization of highly porous chitosan/poly(dl lactic-co-glycolic acid) nanocomposite scaffolds using electrospinning and freeze drying, J. Biobased Mater. Bioenergy,* 8, 281.

237 Ajalloueian, F., Tavanai, H., Hilborn, J., Donzel-Gargand, O., Leifer, K., Wickham, A., Arpanaei, A., (2014) *Emulsion electrospinning as an approach to fabricate plga/chitosan nanofibers for biomedical applications, Biomed. Res. Int.* 2014, 475280.

238 Sehaqui, H., Allais, M., Zhou, Q., Berglund, L., (2011) *Wood cellulose biocomposites with fibrous structures at micro- and nanoscale, Comp. Sci. Technol.* 71, 382.

239 Dufresne, A., (2010) *Processing of polymer nanocomposites reinforced with polysaccharide nanocrystals, Molecules,* 15, 4111.

240 Ben Elmabrouk, A., Thielemans, W., Dufresne, A., Boufi, S., (2009) *Preparation of poly(styrene-co-hexylacrylate)/cellulose whiskers nanocomposites via miniemulsion polymerization, J. Appl. Polym. Sci.*114, 2946.

241 Anglès, M.N., Dufresne, A., (2000) *Plasticized starch/tunicin whiskers nanocomposites. 1. structural analysis, Macromolecules,* 33, 8344.

242 Anglès, M.N., Dufresne, A., (2001) *Plasticized starch/tunicin whiskers nanocomposite materials. 2. mechanical behavior, Macromolecules,* 34, 2921.

243 Mathew, A.P., Dufresne, A., (2002) *Morphological investigation of nanocomposites from sorbitol plasticized starch and tunicin whiskers, Biomacromolecules,* 3, 609.

244 Orts, W.J., Shey, J., Imam, S.H., Glenn, G.M., Guttman, M.E., Revol, J.F., (2005) *Application of cellulose microfibrils in polymer nanocomposites, J. Polym. Env.,* 13, 301.

245 Angellier, H., Molina-Boisseau, S., Dole, P., Dufresne, A., (2006) *Thermoplastic starch—waxy maize starch nanocrystals nanocomposites, Biomacromolecules,* 7, 531.

246 Kvien, I., Sugiyama, J., Votrubec, M., Oksman, K., (2007) Characterization of starch based nanocomposites, *J. Mater. Sci.* 42, 8163.

247 Viguié, J., Molina-Boisseau, S., Dufresne, A., (2007) *Processing and characterization of waxy maize starch films plasticized by sorbitol and reinforced with starch nanocrystals, Macromol. Biosci.* 7, 1206.

248 Mathew, A.P., Thielemans, W., Dufresne, A., (2008) *Mechanical properties of nanocomposites from sorbitol plasticized starch and tunicin whiskers, J. Appl. Polym. Sci.* 109, 4065.

249 Svagan, A.J., Hedenqvist, M.S., Berglund, L., (2009) *Reduced water vapour sorption in cellulose nanocomposites with starch matrix, Sci. Technol.* 69, 500.

250 Azizi Samir, M.A.S., Alloin, F., Sanchez, J.Y., Dufresne, A., (2004) *Cellulose nanocrystals reinforced poly(oxyethylene), Polymer* 45, 4033.

251 Azizi Samir, M.A.S., Alloin, F., Gorecki, W., Sanchez, J.Y., Dufresne, A., (2004) *Nanocomposite polymer electrolytes based on poly(oxyethylene) and cellulose nanocrystals, J. Phys. Chem. B* 108, 10845.

252 Azizi Samir, M.A.S., Montero Mateos, A., Alloin, F., Sanchez, J.Y., Dufresne, A., (2004) *Plasticized nanocomposite polymer electrolytes based on poly(oxyethylene) and cellulose whiskers, Electrochim. Acta*, 49, 4667.

253 Azizi Samir, M.A.S., Chazeau, L., Alloin, F., Cavaillé, J.Y., Dufresne, A., Sanchez, J.Y., (2005) *POE-based nanocomposite polymer electrolytes reinforced with cellulose whiskers, Electrochim. Acta,* 50, 3897.

254 Azizi Samir, M.A.S., Alloin, F., Dufresne, A., (2006) *High performance nanocomposite polymer electrolytes, Compos. Interfaces*, 13, 545.

255 Zimmermann, T., Pöhler, E., Geiger, T., (2004) *Cellulose fibrils for polymer reinforcement, Adv. Eng. Mater.* 6, 754.

256 Roohani, M., Habibi, Y., Belgacem, N.M., Ebrahim, G., Karimi, A.N., Dufresne, A., (2008) *Cellulose whiskers reinforced polyvinyl alcohol copolymers nanocomposites, Eur. Polym. J.* 44, 2489.

257 Paralikar, S.A., Simonsen, J., Lombardi, J., (2008) *Poly(vinyl alcohol)/cellulose nanocrystal barrier membranes, J. Membr. Sci.* 320, 248.

258 Zimmermann, T., Pöhler, E., Schwaller, P., (2005) *Mechanical and morphological properties of cellulose fibril reinforced nanocomposites, Adv. Eng. Mater.* 7, 1156.

259 Lu, J., Wang, T., Drzal, L.T., (2008) *Preparation and properties of microfibrillated cellulose polyvinyl alcohol composite materials, Compos. Part A*, 39, 738.

260 Choi, Y.J., Simonsen, J., (2006) *Cellulose nanocrystal-filled carboxymethyl cellulose nanocomposites, J. Nanosci. Nanotechnol.* 6, 633.

261 Wang, Y., Cao, X., Zhang, L., (2006) *Effects of cellulose whiskers on properties of soy protein thermoplastics, Macromol. Biosci.* 6, 524.

262 Zheng, H., Ai, F., Chang, P.R., Huang, J., Dufresne, A., (2009) *Structure and properties of starch nanocrystal-reinforced soy protein plastics, Polym. Compos.* 30, 474.

263 Kemp, M.M., Linhardt, R.J., (2010) *Heparin-based nanoparticles, Nanomed. Nanobiotechnol.* 2, 77.

264 Simkovic, I., (2013) *Unexplored possibilities of all-polysaccharide composites, Carbohydr. Polym.* 95, 697

265 Lin, N., Huang, J., Dufresne, A., (2012) *Preparation, properties and applications of polysaccharide nanocrystals in advanced functional nanomaterials: A review, Nanoscale* 4, 3274.

266 Hubbe, M.A., Rojas, O.J., Lucia, L.A., Sain, M., (2008) *Cellulosic nanocomposites: A review, Bioresources* 3, 929.

267 Weiss, J., Takhistov, P., Mcclements, D.J., (2006) *Functional materials in food nanotechnology, J. Food Sci.* 71, R107.

268 Wang, X., Ramstrom, O., Yan, M., (2010) *Glyconanomaterials: Synthesis, characterization, and ligand presentation, Adv. Mater.* 22, 1946.

269 El-Boubbou, K., Huang, X., (2011) *Glyco-Nanomaterials: Translating insights from the "sugar-code" to biomedical applications, Curr. Med. Chem.* 18, 2060

270 Bhattacharya, A., Misra B.N., (2004) *Grafting: a versatile means to modify polymers: Techniques, factors and applications, Prog. Polym. Sci.,* 29 (8), 767.

271 Siqueira, G., Bras, J., Dufresne, A., (2010) *Cellulosic bionanocomposites: A review of preparation, properties and applications, Polymers*, 2 (4), 728.

272 Habibi, Y., Goffin, A.L., Schiltz, N., Duquesne, E., Dubois, P., Dufresne, A., (2008) *Bionanocomposites based on poly(ε-caprolactone)-grafted cellulose nanocrystals by ring-opening polymerization, J. Mater. Chem.* 18, 5002.

273 Lin, N., Chen, G., Huang, J., Dufresne, A., Chang, P.R., (2009) *Effects of polymer-grafted natural nanocrystals on the structure and mechanical properties of poly(lactic acid): A case of cellulose whisker-graft-polycaprolactone, J. Appl. Polym. Sci.* 113, 3417.

274 Gousse, C., Chanzy, H., Excoffier, G., Soubeyrand, L., Fleury, E., (2002) *Stable suspensions of partially silylated cellulose whiskers dispersed in organic solvents, Polymer*, 43, 2645.

275 Gousse, C., Chanzy, H., Cerrada, M.L., Fleury, E., (2004) *Surface silylation of cellulose microfibrils: preparation and rheological properties, Polymer*, 45, 1569.

276 Lu, J., Askeland, P., Drzal, L.T., (2008) *Surface modification of microfibrillated cellulose for epoxy composite applications, Polymer*, 49, 1285.

277 Montanari, S., Roumani, M., Heux, L., Vignon, M.R., (2005) *Topochemistry of carboxylated cellulose nanocrystals resulting from tempo-mediated oxidation, Macromolecules*, 38, 1665.

278 Montes, S., Azcune, I., Cabañero, G., Grande, H.J., Odriozola, I., Labidi, J., (2016) *Functionalization of cellulose nanocrystals in choline lactate ionic liquid, Materials*, 9, 499.

279 Pankratov, D.V., González-Arribas, E., Parunova, Y.M., Gorbacheva, M.A., Zeyfman, Y.S., Kuznetsov, S.V., Lipkin, A.V., Shleev S.V., (2015) *New nanobiocomposite materials for bioelectronics devices, Acta Nat.*, 7(1), 98.

280 Rosekoshy, R., Pothan, L.A., Thomas, S., (2013) Biopolymer nanocomposites: processing, properties, and applications, Starch-Based Bionanocomposite: Processing Techniques, pp. 203–226.

281 Ganji, Y., (2016) Polyethylene-based biocomposites and bionanocomposites, in *Thermoplastic Bionanocomposites Series* (Eds. P.M. Visakh, L. Sigrid) Scrivener Publishing, Wiley, pp. 69. ISBN-13: 978–1119–03845-0.

282 Czaja, W., Krystynowicz, A., Bielecki, S., Brown, R.M., (2006) *Microbial cellulose—the natural power to heal wounds, Biomaterials*, 27, (2), 145.

283 Gawdzińska, K., Maliński, M., (2005) *Metody charakteryzowania niejednorodności rozmieszczenia elementów strukturalnych w materiałach wielofazowych, Metallurgy*, 44, 1, 45.

284 Mujeeb Rahman, P., Abdul Mujeeb V.M., Muraleedharan K., Thomas S.K., (2018) *Chitosan/nano ZnO composite films: Enhanced mechanical, antimicrobial and dielectric properties, Arab. J. Chem.*, 11 (1), 120.

285 Araki, J., Wada, M., Kuga, S., (2001) *Steric stabilization of a cellulose microcrystal suspension by poly(ethylene glycol) grafting, Langmuir*, 17, 21.

286 Darder, M., Colilla, M., Ruiz-Hitzky, E., (2003) *Biopolymer—clay nanocomposites based on chitosan intercalated in montmorillonite, Chem. Mater.*15, 3774.

287 Darder, M., Colilla, M., Ruiz-Hitzky, E., (2005) *Chitosan—clay nanocomposites: application as electrochemical sensors, Appl. Clay Sci.* 28, 199.

288 Unalan I.U., Boyacı, D., Ghaani M., Trabattoni, S., Stefano F., (2016) *Graphene oxide bionanocomposite coatings with high oxygen barrier properties, Nanomaterials*, 6, 244.

289 Kumar, D., Saini, N., Pandit, V., Ali, S., (2012) *An Insight to Pullulan: A biopolymer in pharmaceutical approaches, Int. J. Basic Appl. Sci.* 1, 202.

290 Wang, J., Wan, Y., Huang, Y., (2012) *Immobilisation of heparin on bacterial cellulose-chitosan nano-fibres surfaces via the cross-linking technique, IET Nanobiotechnol.* 6, 52.

291 Volpato, F.Z., Almodovar, J., Erickson, K., Popat, K.C., Migliaresi, C., Kipper, M.J., (2012) *Preservation of FGF-2 bioactivity using heparin-based nanoparticles, and their delivery from electrospun chitosan fibers, Acta Biomater.* 8, 1551.

292 Star, A., Steuerman, D.W., Heath, J.R., Stoddart, J.F., (2002) *Starched carbon nanotubes Angew. Chem. Int. Ed.* 41, 2508.

293 Cao, X., Chen, Y., Chang, P.R., Huneault, M.A., (2007) *Preparation and properties of plasticized starch/multiwalled carbon nanotubes composites, J. Appl. Polym. Sci.* 106, 1431.

294 Liu, Z., Zhao, L., Chen, M., Yu, J., (2011) *Effect of carboxylate multi-walled carbon nanotubes on the performance of thermoplastic starch nanocomposites, Carbohydr. Polym.* 83, 447.

295 Yan, L., Chang, P.R., Zheng, P., (2011) *Preparation and characterization of starch-grafted multiwall carbon nanotube composites, Carbohydr. Polym.* 84, 1378.

296 Wu, D., Chang, P.R., Ma, X., *Preparation and properties of layered double hydroxide–carboxymethylcellulose sodium/glycerol plasticized starch nanocomposites, Carbohydr. Polym.* 86, 877.

297 Bagal-Kestwal, D.R., Kestwal, R., Hsieh, W.T., Chiang, B.H., (2014) *Chitosan–guar gum-silver nanoparticles hybrid matrix with immobilized enzymes for fabrication of beta-glucan and glucose sensing photometric flow injection system, J. Pharmaceut. Biomed.* 88, 571.

298 Bagal-Kestwal, D.R., Chiang, B.H., (2018) *Electrically nanowired-enzymes for probe modification and sensor fabrication, Biosens. Bioelectron.* 121, 223.

3

Biopolymeric Material-based Blends: Preparation, Characterization, and Applications

Muhammad Abdur Rehman[1, 3, 4] *and Zia ur Rehman*[2]

[1] *Department of Geology, Faculty of Science, Jalan University, Wilayah Persekutuan, 50603, Kula Lumpur, Malaysia*
[2] *University of Agriculture Faisalabad, Sub Campus, Toba Tek Singh, Pakistan*
[3] *Department of Chemistry, Faculty of Science, Jalan University, Wilayah Persekutuan, 50603, Kula Lumpur, Malaysia*
[4] *Room #1, Quality Control Laboratory, Commander Agro, Street #5A/5B, Industrial Estate, 60900, Multan, Pakistan*

3.1   Introduction

Biopolymers are produced in living organisms in the form of polymeric biomolecules through combining bio monomers by the formation of chemical bonds resulting in complex biological macromolecular structures [1]. There are three main classes of biopolymers according to the monomeric units combined to form the structure of the biopolymer: (i) polynucleotides, e.g. RNA and DNA, as they are long polymers composed of 13 or more nucleotide monomers, (ii) polypeptides, which are short polymers of amino acids, and (iii) polysaccharides, which comprise polymeric carbohydrate structures [2, 3]. The sources of biopolymers and common examples of biopolymers are rubber, suberin, melanin, and lignin [4].

A polymer blend or mixture is formed by combining two or more polymers to generate a new material with unique or desirable properties [5]. The blending technique evolved with the passage of time as new techniques for the modification of the available or existing polymers as well as biopolymers were developed [6]. The first biopolymer blends appeared with the valuable work of Thomas Hancock, who carried out mechanical mixing of rubber with trees from the genus *Palaquium*, e.g. *Palaquium gutta*, *Isonandra gutta*, and *Dichopsis gutta* based biopolymer blends [7]. The polymeric blends are divided into three broad categories: (i) immiscible polymer blends, (ii) compatible polymer blends, and (iii) miscible polymer blends [8–10]. The immiscible or heterogeneous polymers blends are formed by mixing two polymers and show two glass transition temperatures during thermal analysis [11]. The compatible polymer blends are formed by sufficient strong interaction among the component polymers [12]. Miscible polymeric blends exist as single-phase structure and display one glass transition temperature in the thermal analysis [13].

Preparation of biopolymeric blends is encouraged because of their biocompatible and biodegradable properties [14, 15]. The increased research in the preparation of versatile biopolymer blends contributes to their broad application, e.g. biomedical applications [16]. These novel blends demonstrate improved mechanical strength and biocompatibility as compared to the original unblended polymeric components

Bio Monomers for Green Polymeric Composite Materials, First Edition.
Edited by P.M. Visakh, Oguz Bayraktar and Gopalakrishnan Menon.
© 2020 John Wiley & Sons Ltd. Published 2020 by John Wiley & Sons Ltd.

[17]. These synthetic biomaterials are also known as biosynthetic and bioartificial materials [18]. It is a common perception that nature-derived polymeric materials are biocompatible as compared to the synthetic polymers [19, 20]. Biopolymer blends in this regard have emerged as promising materials with suitable thermal, biocompatible, and mechanical properties for use in the intended applications [21, 22]. The main biopolymers used in the preparation of blends for various applications include collagen, chitin, chitosan, keratin, silk, and elastin, which are all natural polymers derived from animals [23, 24]. Natural polymers are also derived from plants, e.g. cellulose and pectin [25]. Furthermore, various synthetic polymeric materials can be blended with naturally occurring polymers to construct novel blends [26]. These biopolymer blends are also enriched with minerals to allow them to nucleate and build up in the polymeric matrix to facilitate suitable size control, shape or morphology and distribution of crystals, like hard tissues in the living organisms [27, 28].

3.2 State of the Art in Biopolymeric Blends

The solubility of biopolymers, e.g. elastin and collagen, in water or organic solvents is a basic problem, except for collagen, which is obtained from the tissues of animals [29]. Collagen is reported to be soluble in acetic acid solutions [30]. Similarly, biopolymer chitosan shows solubility in diluted acetic acid [31]. However, the mean amount of the soluble form is relatively low and mainly depends on the molecular mass of the biopolymers [32]. This property of biopolymers is useful in blend formation with other soluble polymers. Polymers with little solvent affinity, e.g. elastin, silk, and keratin, have a problem during blend formation [33, 34]. This solubility problem may be resolved by the hydrolysis of biopolymers for their application in the blend formation process. A number of biopolymeric blends have been reported [35, 36]. Elastin has been used in biomaterial but in practice it is usually reported for blend making and only a few studies report blends of elastin [37, 38].

3.3 Preparative Methods for Blend Formation

A number of preparative methods are used to make biopolymeric blends from various renewable bioresources [39, 40]. The blends from biopolymers like starch, cellulose or rubber are used as raw materials to manufacture various products, e.g. decorative pieces and household utensils [41–43]. Nevertheless, preparative blends are underutilized. It is interesting to note that nature provides a huge amount of biopolymers, e.g. resins, plastics, gels, film, foam, coating materials, adhesive, and fibers, to meet the current and future needs of society [44–46], and their biodegradable properties mean that they are environmentally friendly [47]. In addition, these polymers have dominant hydrophobic properties, degrade after use, have mechanical properties, and withstand aqueous environments well, e.g. they resist corrosion and do not change or deform in water or moist environments [12, 48–50].

In principle, the various forms of biopolymers are used to make blends with other synthetic polymers to achieve the desired improvements in their properties (Table 3.1) [51, 52]. It is a common observation that blend formation provides a reasonable tool to

Table 3.1 The various forms of biopolymers.

#	Biopolymer	Resource	Available forms
1	Polysaccharides	Plants	Cellulose, starch, alginate, carrageenan, gums, konjac, pectin
		Bacteria	Chitosan, gellan, dextran, chitin, xanthan gum, polygalactosamine, curdlan, levan
		Fungi	Elsinan, scleroglucan, pulluan
		Animals	Hyluronic acid
2	Protein	Plant	Wheat gluten, corn gluten, soy, zein
		Bacteria	Polylysine
		Animal	Casein, albumin, collagen, gelatin, silk, resilin, polyamino acids, poly(g-glutamic acid), elastin, polyarginyl–polyaspartic acid
3	Lipids or surfactants		Emulsion, surfactants, waxes, acetoglycerides
4	Others		Natural rubber Shellac, lignin

make new materials with modified properties at relatively economical prices [53–55]. The main objective of making biopolymeric blends is to capitalize on the maximum performance of the blends by optimizing the properties [56]. In 1970–1980, numerous biopolymeric-based blends of starch were prepared with various polyolefins [12]. These early biopolymeric blends were non-biodegradable in nature [57], so they were not environmentally friendly.

In the preparation of biopolymeric blends, water is used as the solvent or dispersion medium because the majority of natural polymers are soluble in water [58, 59]. In these blends, the main components are proteins or polysaccharides, which form the basis for the structure or properties of these blends as analyzed by the glass transition temperature or thermogravimetric studies. These analyses provide strong evidence to support the role of water in the biopolymeric-based blend materials.

3.4 Blend Preparation by the Melting Process

The melt process for making biopolymeric blends is a way of overcoming the various shortcomings associated with the basic physicochemical properties of biopolymers [60]. For instance, starch is not suitable in its pure form due to its water-soluble properties, difficult processing and brittle nature of blend formations. These difficulties are overcome by using plasticizers as alternative materials to replace non-renewable petrochemical-derived plastics [61]. The formation of blends by combining two or more natural polymers with unique physicochemical properties is a practical way of overcoming the preparation and stability difficulties inherent in these blends. In this regard, natural rubber is melt blended with starch for various applications [12]. Similarly, biopolymeric blends have been prepared by melt blending of starch with 1,4-trans-polyisoprene, commonly known as gutta percha, to form biomedical or food packaging materials [62].

In this process of blend formation, the components are mixed at an adequate temperature of dispersion by thermal pressing. This is a useful technique for preparing a series of blends with various plasticizers or compatibilizers while preserving the biocompatible characteristics of the biopolymeric blends [62]. This technique is also used to prepare thermoplastics by the blending of starch, natural rubber, and natural latex or cornstarch. The mixing of the components in a batch mixer was carried out at 150 °C along with the series of blends formed by varying the rubber composition from 2.5% to 20% [63]. In this preparative method, homogenous mixing of the rubber and thermoplastics was achieved due to the use of water as the solvent medium with particle size 2–8 μm. However, measurements of tensile strength and modulus indicated a decrease in these properties of the biopolymeric blends, indicating their less brittle nature compared to the starting thermoplastics. It was interesting to note that the addition of more plasticizer facilitated the incorporation of a higher percentage of the rubber precursors into the blends. Phase separation at higher concentrations of rubber was observed, which limited the use of larger amounts of rubber in the blended materials. One advantage of this method is that allows blends to be prepared without any need for purification of the starting materials, i.e. latex or starch. The blend formation provided extra stability and improved compatibility between the blended phases [63].

New techniques have been developed to make biopolymeric blends, e.g. reaction extrusion technology to make starch–cellulose–acetate blends [12]. In this technology, a number of materials were used during the blend formation process. For instance, reactive processing technology was used to make cellulose-2,5-acetate precursors, which was grafted into cyclic lactones, polysaccharides, hydroxyl functional plasticizers or OH-functionalized fillers. Other biopolymers were added, i.e. lignin, cellulose, chitin or starch, to enhance the mechanical properties of the biopolymeric blends. The reactive extrusion technology provides a better solution to resolve the content compatibility issue resulting from the blend formation process [64]. The advantageous compatibility of this solution allows this method to make new types of functional biopolymeric blends.

3.5   Aqueous Blending Technology

Aqueous blending technology provides better options than the above technologies to make biopolymeric blends due to the natural properties of biopolymers [65]. It is a common observation that natural polymers decompose on heating and are unsuitable for making blends by the melting process [66]. The conventional softening process does not work either as the biopolymers degrade and cannot withstand the high processing temperatures. Many biopolymers, e.g. proteins and other natural polymers, are therefore unfit for melt processing technology. For this reason, a new technology known as aqueous blending technology was introduced. This technology is suitable for blend formation for use in various biomedical applications [67] because biopolymers are nontoxic and have biocompatible properties compatible with living tissues. Moreover, biopolymers embrace a rich renewable resource of biomaterials that are highly versatile and have chemistry that can be applied in a range of biomedical applications. The main natural resources include polysaccharides (e.g. alginate, chitin, chitosan, and starch), proteins (e.g. collagen from animals, soy proteins, and fibrin gel), and natural fibrous materials (e.g. lignocellulose materials) [12]. The natural biodegradation of starch-based biopolymers offers tremendous potential for use in biomedical applications. The renewable resources used in these materials and their natural abundance makes them competitive with synthetic material, which has limited resources and high prices.

Aqueous soluble blends of starches with cellulose acetate have been researched in a number of studies [68–70]. The biomedical applications of blends produced by aqueous technology include engineered tissue scaffolding, drug-delivery systems, and bone replacements [71]. For instance thermoplastic-based hydrogels made from biopolymer blends of starch and cellulose acetate were used as cementitious materials for bones or targeted biodelivery of drugs in the body [72]. The biodegradable hydrogels were prepared by biopolymeric blends formed by the combination of starch derived from corn starch and cellulose acetate. The blending reaction proceeds through free radical mechanization after reaction between methacrylate or acrylic acid monomers. The free radical reaction was initiated through a redox system consisting of 4-dimethylaminobenzyl alcohols and benzyl peroxide at ambient temperature [73]. The prepared blends have the dual benefits of the biodegradability of starch and the biostable character due to the incorporation of acrylic monomers and hydroxyapatite ceramics. A similar blend was reported by Espigares et al. as blended cornstarch–cellulose acetate and acrylic monomers to paper bone cement with partially biodegradable properties [74]. In addition, other mineral content, e.g. hydroxyapatite, was added to impart extra strength to make this material suitable for bone bonding in biomedical applications. The addition of hydroxyapatite in these blends resulted in a biocompatible character in the blended materials along with osteoconductive or oleophilic properties.

Aqueous blending technology was used to prepare blends of soluble starch and methyl cellulose (MC) by Arvanitoyannis and Biliaderis [75]. In these blends, sugars or glycerol were applied as the plasticizer materials. The as-prepared blends were subjected to casting, extrusion or hot pressing to get the biopolymeric blends. The water, sugars or glycerol contents result in a decrease in the T_g observed in the biopolymeric blends. However, these blends indicated a decrease in the mechanical strength. The tensile strength and flexural moduli in these blends were seen to decrease at a drastic rate with the corresponding increase in the total plasticizer content. It was observed that in this preparation method the addition of glycerol as plasticizer had a negative impact on the mechanical properties compared to another plasticizer, e.g. sorbitol, as it improves the mechanical properties, e.g. high level of elongation, in the biopolymeric blends.

As reported in various studies, the percentage content of each biopolymeric constituent plays an important role in controlling the properties of the final blends formed. Leopeniotis et al. prepared a number of blends by changing one constituent from starch acetate to cellulose acetate [76]. The blends were prepared with an acetone–water mixture to solve the solubility issues. The degree of substitution of starch acetate with cellulose acetate varied from 2.1 to 3.0 to make 25–30 wt% blends. The number of experiments and the controlling variables of the synthesis were planned with the help of statistically designed of experiments (DoEs) to reduce the time and the cost of the research. The DoE was based on five factors: acetyl content, acetone/water ratio, starch acetate/cellulose acetate ratio, temperature, and total solids in solution form. The DoE response was the stability or the strength of the blends formed from each set of experiments based on the central composite design (CCD). From the results and analysis of the statistical parameters, the CCD results indicated that the starch acetate content was the most influencing factor in this study. It is interesting to note that the DoE method of planning blend preparation to offer blend formation of the desirable properties with a limited number of experimental runs reduces the time and effort required for such research. It is also important to highlight that the DoE technique of blend preparation requires further exploration to efficiently prepare biopolymeric blends for future applications [76].

In another study, Psomiadou et al. used microcrystalline cellulose (MCC) or MC to develop blends from natural resources [77]. The MCC or MC precursors were blended with polyols using aqueous blending technology, extrusion or hot pressing. The effect of polyol on the thermal stability, mechanical strength, and water or gas permeability was monitored to establish the usability of these blends. The study suggests that the polyol or water content has a positive impact on the mechanical properties or stability of the biopolymeric blends. For instance, polyol content was directly related to the increase in the elongation or tensile strength of the biopolymeric blends. It was also noted that an increase in cellulose in the blend was related to a corresponding increase in tensile strength and a decrease in water vapor transmission properties. Moreover, the blends showed an increase in the degree of crystallinity with the passage of time resulting in a decrease in the permeability of gas or water in the biopolymeric blend. In a similar study, Peressini et al. studied the rheology of starch–methylcellulose blends [78]. The flow curves of the blends were investigated to account for shear-thinning characteristics. The steady shear viscosity or dynamic viscosity of the blends was monitored by mechanical spectra or Cox–Merz superposition. The results indicated a topological entanglement interaction of biopolymeric chains in the blends. Moreover, the blends showed improved properties, e.g. viscoelastic properties, at high strains compared to the precursors. It was observed that the incorporation of MC contents in the blends was the main factor responsible for the apparent viscosity or viscoelastic properties in the biopolymeric blends.

The problem of moisture sensitivity in starch-based biopolymeric blends was studied by Demirgoz et al. to control blend properties by chemical modification [79]. Modifications of starch-based blends were carried out by the chain crosslinking technique. The surface functional groups of starch, i.e. hydroxyl groups, were reacted with tri-sodium tri-meta phosphates. The chemical modification of the surface functional groups was monitored by Fourier transform infrared spectroscopy (FTIR). The other characterization method includes measurement of degree of hydration, degradation profiles, and contact angle measurements, and mechanical strength tests. The modification method was able to reduce moisture sensitivity by 15% in the blended materials. Moreover, the mechanical properties were also improved as the material showed resistance to degradation.

The concept of crosslinking or cross-bridging was also used to make polysaccharide cross-bridged proteins prepared by gene fusion techniques by Levy et al. on *Clostridium cellulovorans* and *Aspergillus niger* species [80]. The cross-bridging technique was able to construct cellulose/starch blends. The use of genetic modification is an attractive innovation as it requires no chemical reactions to make biopolymeric blends for use in various potential biomedical applications.

Aqueous blending technology has also been reported in the literature for the formation of chitosan biopolymeric blends. Kweon et al. prepared biopolymeric blends using a combination of proteins and chitosan [81]. The silk fibroin (SF) was blended with chitosan in the presence of acetic acids. The use of acetic acid promoted the formation of a sheet-like structure in the blended materials. The characterization results of the biopolymeric blends showed improved thermal stability and degradation stability. In another study, Lazaridou and Biliaderis studied the effect of blended materials on the thermal or mechanical properties [82]. The prepared materials include blended starch-chitosan or pullulan-chitosan, as analyzed by dynamic mechanical thermal analysis (DMTA). The blends were prepared in the presence of plasticizer, i.e. sorbitol. The glass transition temperature (T_g) decreased substantially because of the 10–30% of sorbitol used in the blend preparation. The results also indicated a single drop in the elastic modulus (E) as a single tan δ peak was recorded in the results. However, the increased amount of polyols in the blends was linked with a corresponding decrease in the tensile strength and large drop in Young's modulus. The results of the decrease in the modulus data were interpreted by modeling the analytical data using Fermi's equations [83].

Arvanitoyaniis et al. used aqueous blending technology to develop biopolymeric blends of gelatin and chitosan [84]. The preparation method requires pH $\sim$ 4.0 and 60 °C temperature in the presence of water or polyols. The as-prepared blends were characterized for changes in their physical properties, e.g. thermal stability, mechanical strength or permeation to gas or water. The blends showed an improved elongation but a notable decrease in elasticity modulus and tensile strength. Moreover, this method of blend formation required a relatively low temperature and promoted the development of crystallinity in the gelatin. The crystalline phases in the blends were less permeable to gases, e.g. oxygen or carbon dioxide. However, the use of water or polyols as the plasticizer allowed more gas permeability in the chitosan–gelatin biopolymeric blends [59].

Aqueous blending technology was also used to prepare antibacterial starch–chitosan biopolymeric blends [85]. The blends were irradiated by placing them in the path of an electron beam. The blends were characterized by scanning electron microscopy (SEM) and X-ray diffraction (XRD) techniques. The characterization by XRD indicated the interaction or microphases between the chitosan and starch moieties. The irradiation of the biopolymeric blends with an electron beam brought no change in the morphology or crystal structure. However, antibacterial properties have been noted in starch–chitosan biopolymeric blends.

3.6 Hydrophilic or Hydrophobic Biopolymeric Blends

Biopolymers are hydrophilic or water-loving in nature as they are rich in polar functional groups, e.g. –OH groups, that allow interaction with aqueous media [86]. Synthetic polymers, e.g. aliphatic polyesters, contain non-polar groups with hydrophobic properties. The preparation of biopolymeric blends from hydrophilic and hydrophobic precursors offers an opportunity to make new classes of materials with unique properties. The environmentally compatible properties and biodegradability of these biopolymeric blends have attracted the attention of researchers and a number of blends have been prepared as discussed below.

3.6.1 Biopolymeric Blends of Starch and Polylactic Acid

Biopolymeric blends prepared by the combination of starch and polylactic acid (PLA) have been extensively researched due to their natural renewable resources and non-toxic nature, and used to prepare packaging materials, avoid "green" taxes, and meet environmental regulations, e.g. in Japan. The green, environmentally friendly plastics or packaging were prepared by the use of PLA or starch as both are biodegradable in nature [69]. However, the high cost of PLA is an obstacle in the industrial scale production of such plastics. The emergence of blending technology, i.e. blending starch with PLA, resolves the issue of high cost in these blends due to the economical prices of starch. The competitive mechanical properties were introduced due to the use of PLA in the blends. The water-loving properties of starch make it absorb water and become swollen or gelatinized in aqueous media. Ke and Sun studied the mechanical strength of starch and PLA blends in the presence of different water contents [87]. The study results indicated that the initial water content of the starch has no visible impact on the mechanical properties of the biopolymeric blends but it has a definite effect on the ability of the blends to absorb or adsorb water. Moreover, the moisture content showed no effect on the crystal structure or thermal degradation properties. It was also observed that the method of blend formation has a significant impact on the blends formed, e.g. blends formed by compression molding displayed a higher level of crystallinity compared to blends formed by injection molding. However, blends formed by injection molding showed higher tensile strength, lower water absorption/adsorption, and longer elongation values [88]. The impact of annealing temperature was also seen in the blended materials, e.g. the level of crystallinity was improved in the blends prepared at 155 °C. The impact of temperature on thermal properties was studied using differential scanning calorimetry (DSC). The DSC thermal data was processed using the rate or kinetics model, i.e. Avrami kinetic modeling. The degree of crystallinity of the annealed samples and time for crystallization were deduced from the Avrami model [89]. The impact of nucleation agents, e.g. talc, was also studied when blended in a ratio of 1% (v/v) in the blends. The results also showed the improvement in the rate of PLA crystallization with a corresponding increase in the percentage of starch from 1% to 40%. The melting points and the degree of crystallinity were also affected by blend formation.

Cornstarch with various amylose contents was prepared and used to assess the impact of composition and temperature on biopolymeric blends [90]. The biopolymeric blends were prepared by blending various ratios of starch with PLA at 185 °C in a small twin-screw extruder. The blends were prepared in different batches, ground, dried, and added with 7.5% plasticizer before feeding into the injection mold, operating at 175 °C. The as-prepared biopolymeric blends were characterized for changes in morphology, water absorptivity, and mechanical properties. The characterization results indicated that the starch took the role of a filler in these blends. However, beyond 60% starch content the continuous PLA matrix phase became discontinuous with the excess of starch, along with a decreasing trend in the tensile strength of the blends. It is interesting to note that the water absorptivity in these blends increases with a corresponding increase in the content of starch in the blends [90]. Moreover, cornstarch with a high level of amylose starch showed relatively less water absorptivity than blends with normal cornstarch.

The impacts of the chemical and physical properties of starch on biopolymeric blend properties were studied by Park and Im [91]. The starch was gelatinized by the addition of a mixture of water and glycerol in a twin-screw mixer. The purpose of gelatinization was to break up the inter- or intra-molecular hydrogen bonds or reduce the level of crystallinity of the starch. The advantage of gelatinized starch compared to normal starch was seen in the emergence of voids in the normal starch blends, but not in the gelatinized starch blends. This phenomenon was explained on the basis that in the pure starch–PLA blends the starch particles separate out from the matrix, creating voids in the blends.

The twin-screw extruder technique was also used to prepare starch blends with thermoplastics resins [92]. The blends also contained poly(hydroxyester ether) (PHEE) to control the densities and radial expansion properties. In another study, PLA and PHEE were blended with high amylose starch, i.e. 70% amylose, wheat or potato starch. These blends showed reduced density but improved expansion properties.

The microstructural properties of starch blends with PLA, polyethylene, and vernonia oil were prepared by the melt processing technique and acid hydrolysis [93]. Surface fractures were observed during SEM examination of the blends. The pore size of the blends was recorded by atomic force microscopy (AFM) in the tapping mode [94]. The blends were porous in nature, which facilitated passage of water depending on the thickness of the blends.

3.6.1.1   Maleic Anhydride-grafted PLA Chains

The grafting of PLA chains with maleic anhydrides produced new interfacial compatibilizers known as maleic anhydride-grafted polylactic acids chains (MAG-PLA) [95]. MAG-PLA compatibilizers enhance the interfacial adhesions with hydrophilic starch granules.

3.6.1.2   Polycaprolactone-grafted Polysaccharide Copolymers

Polycaprolactone (PCL) compatibilizers were used to account for the properties of the biopolymeric blends. The effect of PCL on the morphological and mechanical properties of the blends was studied by comparing PCL-treated blends with control blends without any additives [96]. The PCL-treated nanocomposites displayed improved properties in the solid as well as the molten state. Thermogravimetric characterization techniques were used for stability studies of PCL-treated biopolymeric blends in the atmosphere. In one study, PCL additives were used to enhance the elongation of poly(L-lactide) (PLLA), but at the same time the strengths of the biopolymeric blends showed a decline in the use of PCL [97]. It was suggested that PLLA-based biopolymeric blends may be treated with surfactants, e.g. poly DL-lactic acid (PDLLA) via the solution blending method to improve mechanical strength, which is desirable in dental and orthopedic applications.

3.6.2   Hydrolytic Degradability of Biopolymeric Blends

A number of studies have focused on controlling the hydrolytic degradation of PLAs, along with the various components of the blending systems [98, 99]. For instance, Shinoda et al. applied poly(aspartic acid-co-lactide) (PAL) to promote the degradation of PLA-based biopolymeric blends [99]. The use of PAL promoted miscibility and the mechanical properties of the blends. The PAL-treated biopolymeric blends were transparent, an obvious advantage, but this did not affect the mechanical properties. The use of PAL as the antistatic agent results in improved hydrophilicity of the constituents of the blend. The improved hydrolytic degradation was observed in water, compost, and soil even with the addition of a small volume of PAL additives. PAL also improved the shelf life of the biopolymeric blends for extended use and various applications. The thermal and mechanical properties of the PAL-treated bio-polymeric blends were also appreciably improved.

3.6.3 Thermodynamics of Miscibility with Additives

The miscibility of hydrophobic and hydrophilic additives, e.g. PLA and starch, is a challenge during biopolymeric blend formations [100]. The thermodynamics of hydrophobic PLA and hydrophobic starch lead to poor adhesion of these two distinct chemical forms, causing poor performance. The chemistry of the dissimilar components can be played with by using various compatibilizers or additives to improve interfacial interactions in the biopolymeric blends [12].

3.6.3.1 Methylene Diphenyl Diisocyanate

The concept of using additives to overcome the challenges of miscibility of dissimilar components was resolved by the use of methylene diphenyl diisocyanate (MDI) [101, 102]. The role of MDI is to form of a block copolymer during blend formation in the intensive mixture for PLA and starch blend formation. Changes in the interfacial tension were observed by SEM micrographs. It has also been reported that the use of MDI compatibilizer improves the mechanical properties of biopolymeric blends. However, water absorption properties were not influenced by the control and MDI-treated blends. In a similar study, MDI was used to prepare starch–PLA blends at 175 °C by hot pressure molding. The MDI-treated blends showed an improvement in tensile strength up to 68 MPa compared with the control (62.7 MPa). However, the control showed better elongation, i.e. 6.5% compared to the MDI-treated blends with 5.1%. The impact of starch content in the blends was monitored by dynamic mechanical analysis (DMA). The DMA results indicated that the storage modulus increased with an increase in starch content up to 45% and then levelled off. Water absorptivity also improved with starch content in the blends. The study of the role of MDI suggested blends were suitable for potential disposable applications.

The starch moisture content, i.e. 10–20%, has an adverse effect on interfacial bonding between hydrophobic PLA and hydrophilic starch [103]. It was noted that an increase in water content, i.e. 20% moisture level, resulted in swollen starch granules in the PLA matrix, leading to weak mechanical strength due to water absorption by the blends. The physical gaining of the biopolymeric blends in the presence or absence of MDI was also evaluated. The blends prepared in the presence of MDI showed slower reduction in the moisture content with the passage of time compared to control blends. The decrease in the physical gaining of the MDI-treated biopolymeric blends was assigned to the enthalpy relaxation rather than the control blends. Changes in mechanical strength were more rapid in the control blends compared to the MDI-treated relatively stable blends. Analysis of the microstructures of the control and MDI-treated blends indicated a decrease in the level of interaction between PLA and starch over time around the interface.

3.6.3.2 Dioctyl Maleate

Dioctyl maleate (DOM) has also been reported as being a useful compatibilizer or additive to overcome the challenges of miscibility between the hydrophilic and hydrophobic components of biopolymeric blends [104]. As expected of a compatibilizer, DOM was reported to improve the mechanical properties, e.g. tensile strength, enhanced elongation, and thermal stability, reduce degradation, and improve water absorption.

A low concentration of DOM, i.e. below 5%, was effective in making marked improvements in the tensile strength of the biopolymeric blends [104]. The elongation properties of the biopolymer blends were also improved compared to the control blends. The DOM-treated blends also showed simultaneous compatibilization and plasticization, resulting in minimum thermal loss [105]. The water absorption of DOM-treated blends was also improved compared to control blends without any treatment.

3.6.3.3 Polyvinyl Alcohols

Polyvinyl alcohols (PVOHs) have also been used as compatibilizers in PLA–starch-based biopolymeric blends [106]. It was reported that the use of PVOHs enables preparation of blends with better compatibility and improved mechanical strengths. It was also reported that the chain length of PVOHs also plays an important role in refining the water absorption properties of the biopolymeric blends. As discussed earlier, gelatinized starch has better compatibility with the PLA and the resulting blends displayed higher tensile strengths. However, the use of gelatinized starches allowed the absorption of more water by the biopolymeric blends [107].

3.6.3.4 Poly(hydroxyester ether)

The use of PHEE compatibilizers provides a solution to the problem of compatibility when starch is blended with various types of thermoplastic resins to make foams in a twin-screw extruder. The use of PHEE additives was reported to lower the blend properties, e.g. density and radial expansion, compared with the control blend without any compatibilizer or additives [108].

3.6.3.5 Poly(β-hydroxybutyrate)-co-3-hydroxyvalerate

The scientific literature first reported poly(β-hydroxybutyrate)-co-3-hydroxyvalerate (PHB/VA) in 1901 [12]. These special polyesters were thought to be part of a survival strategy by microorganisms, acting as internal carbon or energy storage. The shortage of energy resources and need for plastics from renewable resources has attracted research on natural substances, including PHB. The properties of PHB/VA were unsuitable, e.g. brittleness, so these materials were functionalized, e.g. β-hydroxyvalerate, to prepare a new type of additive, poly(3-hydroxybutyric acid-3-hydroxyvaleric acid) [109].

3.6.3.6 Poly(3-hydroxybutyric acid-3-hydroxyvaleric acid)

The compatibilizer or additive poly(3-hydroxybutyric acid-3-hydroxyvaleric acid) (PHBV) was formed by the copolymerization of β-hydroxybutyrate and β-hydroxyvalerate by a research team at ICI in 1990 [110]. PHBV-treated blends are reported to show improved radial expansion and lower densities than control blends without PHBV additives. PHBV was further blended with starch and glycidyl methacrylate (starch-g-PGMA). PHBV-treated blends showed better tensile and flexural strengths than controls. However, it was observed that PHBV-treated blends show no change in the degree of elongation or the modulus. The blends displayed improved adhesion between the components of the blend, as evidenced by SEM.

3.6.4 Poly(hydroxyalkanoate)

Poly(hydroxyalkanoate) (PHA) is an extensively studied biopolymer that is produced by microorganisms [111, 112]. The bacterial resources of biopolymers have attracted much attention due to the high compatibility and renewable resource of thermoplastics [113]. It is interesting to note that more than 90 types of PHA formed by different monomers have been reported in the literature [114]. PHA has the potential to promote environmentally friendly blends for sustainable societies across the globe. PHA-derived plastics have similar properties to conventional polymers, e.g. polyethylene, polypropylene, and elastomers [115]. The blending technology extends the application of these materials, with great scope for future developments.

3.6.4.1 Poly(3-hydroxybutyrate)

PHB-based blends represent the most interesting biopolymeric blends because of their biodegradable and biocompatible properties. PHB blends are attractive materials for advanced tissue engineering, tissue scaffolding, and cardiovascular applications [116, 117].

3.7 Opportunities and Challenges

Currently, biopolymer blends are used in drug-delivery systems and the packaging industry [118, 119]. These two areas are not related, but the impact that biopolymer blends could have in these two industries and others is huge. One might question why there is need for such biopolymer blends when substitutes with far superior chemical and physical properties exist? The answer to this is that global petrochemical feedstock is depleting [113]. There is not a timeline for the exact depletion course but the shift of oil-based economies to other economic activities shows that the reality is looming. Dependence on one type of source which is both depleting drastically and impacting the environment is not sustainable. These days most consumers are well informed and care about the products they consume, often condemning products that impact the environment. These biopolymers answer many of the worries of industries and consumers: their eco-friendliness makes customers happy by making them feel that they are not destroying the environment and industries can reduce their heavy dependence on petrochemical polymers, which are often subject to international politics and are depleting at an alarming rate. The main hindrance to the large-scale adoption of biopolymer blends is their ability to match the industrial specifications. The use of biopolymer blends has already begun in the packaging industry and other research and development institutions for commercialization, but a lot more progress is still required. Even in case of controlled drug-delivery systems, biopolymer blends have to prove their consistent mechanical and biocompatible properties [120]. Research and development are continuing on but a faster approach is required before petrochemical resources are depleted to the level where the demands of industries around the world cannot be met. Once these obstacles have been removed, the future looks very bright. Already there has been a shift in the production packaging materials from using petrochemical polymer blends to using biopolymer blends mainly made from cellulose, hemicellulose lignin, silk or starch. These biopolymer blends will help in fast treatment by targeted drug delivery and quick degradation using the enzymatic response for successful drug delivery systems.

3.8 Summary

This chapter presents a comprehensive study of biopolymeric-based blends, state of the art, general preparative methods, aqueous blending technology, hydrophilic or hydrophobic nature, degradation problems, thermodynamics of miscibility, and opportunities and challenges. Biopolymeric blends have attracted academic, research, and industrial scientists' attention as they have properties that are desirable for various applications. The challenges arising from the new structural preparative methods and the resultant properties of the blends of natural polymeric materials are discussed, and examples are drawn from the scientific literature. The various forms of natural polymers, i.e. polysaccharides, proteins, lipids, natural rubber, chitosan, starch, and silk-based blends, are discussed with respect to preparative techniques, characterization methods, and various applications.

References

1 Kogan, G., Šoltés, L., Stern, R., and Gemeiner, P. (2007). Hyaluronic acid: a natural biopolymer with a broad range of biomedical and industrial applications. *Biotechnol. Lett.* 29: 17–25.

2 Hasırcı, V., Lewandrowski, K., Gresser, J.D. et al. (2001). Versatility of biodegradable biopolymers: degradability and an in vivo application. *J. Biotechnol.* 86: 135–150.

3 Van Krevelen, D.W. and Te Nijenhuis, K. (2009). *Properties of Polymers: Their Correlation with Chemical Structure; Their Numerical Estimation and Prediction from Additive Group Contributions.* Elsevier.

4 Pragya (2015). A theoretic study of degradable polymers. *Eur. J. Acad. Essays* 2: 7–10.

5 Ruzette, A.-V. and Leibler, L. (2005). Block copolymers in tomorrow's plastics. *Nat. Mater.* 4: 19–31.

6 He, W., Yong, T., Teo, W.E. et al. (2005). Fabrication and endothelialization of collagen-blended biodegradable polymer nanofibers: potential vascular graft for blood vessel tissue engineering. *Tissue Eng.* 11: 1574–1588.

7 Hancock, T. (1857). *Personal Narrative of the Origin and Progress of the Caoutchouc or India-Rubber Manufacture in England.* London: Longman, Brown, Green, Longmans, & Roberts.

8 Colby, R.H. (1989). Breakdown of time–temperature superposition in miscible polymer blends. *Polymer* 30: 1275–1278.

9 Rim, P.B. and Runt, J.P. (1984). Melting point depression in crystalline/compatible polymer blends. *Macromolecules* 17: 1520–1526.

10 Utracki, L.A. (1991). On the viscosity–concentration dependence of immiscible polymer blends. *J. Rheol.* 35: 1615–1637.

11 Paul, D.R. (2012). *Polymer blends.* Elsevier.

12 Yu, L., Dean, K., and Li, L. (2006). Polymer blends and composites from renewable resources. *Prog. Polym. Sci.* 31: 576–602.

13 Guirguis, O.W. and Moselhey, M.T.H. (2012). Thermal and structural studies of poly (vinyl alcohol) and hydroxypropyl cellulose blends. *Nat. Sci.* 4: 57.

14 Sionkowska, A. (2011). Current research on the blends of natural and synthetic polymers as new biomaterials. *Prog. Polym. Sci.* 36: 1254–1276.

15 Wu, R.-L., Wang, X.-L., Li, F. et al. (2009). Green composite films prepared from cellulose, starch and lignin in room-temperature ionic liquid. *Bioresour. Technol.* 100: 2569–2574.

16 Costa-Júnior, E.S., Barbosa-Stancioli, E.F., Mansur, A.A.P. et al. (2009). Preparation and characterization of chitosan/poly (vinyl alcohol) chemically crosslinked blends for biomedical applications. *Carbohydr. Polym.* 76: 472–481.

17 Sell, S.A., McClure, M.J., Garg, K. et al. (2009). Electrospinning of collagen/biopolymers for regenerative medicine and cardiovascular tissue engineering. *Adv. Drug Deliv. Rev.* 61: 1007–1019.

18 Ratner, B.D. (1993). New ideas in biomaterials science—a path to engineered biomaterials. *J. Biomed. Mater. Res. A* 27: 837–850.

19 Griffith, L.G. (2002). Emerging design principles in biomaterials and scaffolds for tissue engineering. *Ann. N. Y. Acad. Sci.* 961: 83–95.

20 Hubbell, J.A. (1995). Biomaterials in tissue engineering. *Nat. Biotechnol.* 13: 565–576.

21 Jawad, H., Lyon, A.R., Harding, S.E. et al. (2008). Myocardial tissue engineering. *Br. Med. Bull.* 87.

22 Triplett, R.G. and Budinskaya, O. (2017). New frontiers in biomaterials. *Oral Maxillofac. Surg. Clin.* 29: 105–115.

23 Chicatun, F., Griffanti, G., McKee, M.D., and Nazhat, S.N. (2017). Collagen/chitosan composite scaffolds for bone and cartilage tissue engineering. In: *Biomedical Composites*, 2e (ed. L. Ambrosio), 163–198. Elsevier.

24 Liu, M., Min, L., Zhu, C. et al. (2017). Preparation, characterization and antioxidant activity of silk peptides grafted carboxymethyl chitosan. *Int. J. Biol. Macromol.* 104 (Part A): 732–738.

25 Khalil, H.P.S., Tye, A.Y.Y., Saurabh, C.K. et al. (2017). Biodegradable polymer films from seaweed polysaccharides: a review on cellulose as a reinforcement material. *Express Polym. Lett.* 11: 244.

26 Torres-Giner, S., Wilkanowicz, S., Meléndez-Rodríguez, B., and Lagaron, J.M. (2017). Nanoencapsulation of aloe vera in synthetic and naturally occurring polymers by electro-hydrodynamic processing of interest in food technology and bioactive packaging. *J. Agric. Food Chem.* 65: 4439–4448.

27 Caridade, S.G. and Mano, J.F. (2017). Engineering membranes for bone regeneration. *Tissue Eng. A* 23: 1502–1533.

28 Harding, J.L. and Krebs, M.D. (2017). Bioinspired deposition-conversion synthesis of tunable calcium phosphate coatings on polymeric hydrogels. *ACS Biomater. Sci. Eng.* 3: 2024–2032.

29 Yang, X., Qiao, C., Li, Y., and Li, T. (2016). Dissolution and resourcfulization of biopolymers in ionic liquids. *React. Funct. Polym.* 100: 181–190.

30 Okazaki, E. and Osako, K. (2014). Isolation and characterization of acid-soluble collagen from the scales of marine fishes from Japan and Vietnam. *Food Chem.* 149: 264–270.

31 Binulal, N.S., Natarajan, A., Menon, D. et al. (2014). PCL–gelatin composite nanofibers electrospun using diluted acetic acid–ethyl acetate solvent system for stem cell-based bone tissue engineering. *J. Biomater. Sci. Polym. Ed.* 25: 325–340.

32 Kenawy, E., Imam Abdel-Hay, F., Mohy Eldin, M.S. et al. (2015). Novel aminated chitosan–aromatic aldehydes schiff bases: synthesis, characterization and bio-evaluation. *Int. J. Adv. Res.* 3: 563–572.

33 Chen, L., Zhou, M.-L., Qian, Z.-G. et al. (2017). Fabrication of protein films from genetically engineered silk-elastin-like proteins by controlled cross-linking. *ACS Biomater. Sci. Eng.* 3: 335–341.

34 Shavandi, A., Silva, T.H., Bekhit, A.A., and Bekhit, A.E.-D. (2017). Dissolution, extraction and biomedical application of keratin: methods and factors affecting the extraction and physicochemical properties of keratin. *Biomater. Sci.* 5: 1699–1735.

35 Abdul Khalil, H.P.S., Saurabh, C.K., Adnan, A.S. et al. (2016). A review on chitosan-cellulose blends and nanocellulose reinforced chitosan biocomposites: properties and their applications. *Carbohydr. Polym.* 150: 216–226.

36 Reddy, N., Reddy, R., and Jiang, Q. (2015). Crosslinking biopolymers for biomedical applications. *Trends Biotechnol.* 33: 362–369.

37 Ferraro, V., Anton, M., and Santé-Lhoutellier, V. (2016). The "sisters" α-helices of collagen, elastin and keratin recovered from animal by-products: functionality, bioactivity and trends of application. *Trends Food Sci. Technol.* 51: 65–75.

38 Ozsvar, J., Mithieux, S.M., Wang, R., and Weiss, A.S. (2015). Elastin-based biomaterials and mesenchymal stem cells. *Biomater. Sci.* 3: 800–809.

39 Fortunati, E. (2016). Multifunctional films, blends, and nanocomposites based on chitosan: use in antimicrobial packaging. In: *Antimicrobial Food Packaging* (ed. J. Barros-Velázquez), 467–477. Elsevier.

40 Peres, L.B., Peres, L.B., Faria, T.J. et al. (2017). PLLA/PMMA blend in polymer nanoparticles: influence of processing methods. *Colloid Polym. Sci.* 295: 1621–1633.

41 Mohammad, A. (2016). *3D Printing with Biopolymers on Textile Knitted Structures*. GRIN Verlag.

42 Sam, S.T., Nuradibah, M.A., Chin, K.M., and Hani, N. (2016). Current application and challenges on packaging industry based on natural polymer blending. In: *Natural Polymers*. Springer.

43 Wang, X., Tang, R., Zhang, Y. et al. (2016). Preparation of a novel chitosan based biopolymer dye and application in wood dyeing. *Polymers* 8: 338.

44 Klemm, D., Kramer, F., Moritz, S. et al. (2011). Nanocelluloses: a new family of nature-based materials. *Angew. Chem. Int. Ed.* 50: 5438–5466.

45 Nayak, P.L. (1999). Biodegradable polymers: opportunities and challenges. *J. Macromol. Sci.*, Part C 39 (3): 481–505.

46 Plackett, D. (2011). *Biopolymers: New Materials for Sustainable Films and Coatings*. Wiley.

47 Hernández, N., Williams, R.C., and Cochran, E.W. (2014). The battle for the "green" polymer. Different approaches for biopolymer synthesis: bioadvantaged vs. bioreplacement. *Org. Biomol. Chem.* 12: 2834–2849.

48 Averous, L., Fauconnier, N., Moro, L., and Fringant, C. (2000). Blends of thermoplastic starch and polyesteramide: processing and properties. *J. Appl. Polym. Sci.* 76: 1117–1128.

49 Rhim, J.-W. and Wang, L.-F. (2013). Mechanical and water barrier properties of agar/κ-carrageenan/konjac glucomannan ternary blend biohydrogel films. *Carbohydr. Polym.* 96: 71–81.

50 Phan The, D., Debeaufort, F., Voilley, A., and Luu, D. (2009). Biopolymer interactions affect the functional properties of edible films based on agar, cassava starch and arabinoxylan blends. *J. Food Eng.* 90: 548–558.

51 Siracusa, V., Rocculi, P., Romani, S., and Rosa, M.D. (2008). Biodegradable polymers for food packaging: a review. *Trends Food Sci. Technol.* 19: 634–643.

52 Zhang, Y., Ouyang, H., Lim, C.T. et al. (2005). Electrospinning of gelatin fibers and gelatin/PCL composite fibrous scaffolds. *J. Biomed. Mater. Res. B Appl. Biomater.* 72: 156–165.

53 Di Martino, A., Sittinger, M., and Risbud, M.V. (2005). Chitosan: a versatile biopolymer for orthopaedic tissue-engineering. *Biomaterials* 26: 5983–5990.

54 Kaplan, D.L. (1998). Introduction to biopolymers from renewable resources. In: *Biopolymers from Renewable Resources* (ed. D. Kaplan), 1–29. Springer.

55 Mijovic, J., Shen, M., Sy, J.W., and Mondragon, I. (2000). Dynamics and morphology in nanostructured thermoset network/block copolymer blends during network formation. *Macromolecules* 33: 5235–5244.

56 Mekonnen, T.H., Misra, M., and Mohanty, A.K. (2015). Processing, performance, and applications of plant and animal protein-based blends and their biocomposites. In: *Biocomposites* (ed. M.P. Misra, K.M. Jitendra and K. Amar), 201–235. Elsevier.

57 Parulekar, Y. and Mohanty, A.K. (2006). Biodegradable toughened polymers from renewable resources: blends of polyhydroxybutyrate with epoxidized natural rubber and maleated polybutadiene. *Green Chem.* 8: 206–213.

58 Rogina, A. (2014). Electrospinning process: versatile preparation method for biodegradable and natural polymers and biocomposite systems applied in tissue engineering and drug delivery. *Appl. Surf. Sci.* 296: 221–230.

59 Vieira, M.G.A., da Silva, M.A., dos Santos, L.O., and Beppu, M.M. (2011). Natural-based plasticizers and biopolymer films: a review. *Eur. Polym. J.* 47: 254–263.

60 Soroudi, A. and Jakubowicz, I. (2013). Recycling of bioplastics, their blends and biocomposites: a review. *Eur. Polym. J.* 49: 2839–2858.

61 Pietrini, M., Roes, L., Patel, M.K., and Chiellini, E. (2007). Comparative life cycle studies on poly(3-hydroxybutyrate)-based composites as potential replacement for conventional petrochemical plastics. *Biomacromolecules* 8: 2210–2218.

62 Arvanitoyannis, I., Kolokuris, I., Nakayama, A., and Aiba, S.-I. (1997). Preparation and study of novel biodegradable blends based on gelatinized starch and 1,4-trans-polyisoprene (gutta percha) for food packaging or biomedical applications. *Carbohydr. Polym.* 34: 291–302.

63 Jha, A. and Bhowmick, A.K. (1997). Thermoplastic elastomeric blends of nylon-6/acrylate rubber: influence of interaction on mechanical and dynamic mechanical thermal properties. *Rubber Chem. Technol.* 70: 798–814.

64 Hong, S.M., Hwang, S.S., Seo, Y. et al. (1997). Reactive extrusion of in-situ composite based on PET and LCP blends. *Polym. Eng. Sci.* 37: 646–652.

65 Siva Kumar, N., Venkata Subbaiah, M., Subba Reddy, A., and Krishnaiah, A. (2009). Biosorption of phenolic compounds from aqueous solutions onto chitosan–abrus precatorius blended beads. *J. Chem. Technol. Biotechnol.* 84: 972–981.

66 Zuo, M., Song, Y., and Zheng, Q. (2009). Preparation and properties of wheat gluten/methylcellulose binary blend film casting from aqueous ammonia: a comparison with compression molded composites. *J. Food Eng.* 91: 415–422.

67 Meinel, A.J., Germershaus, O., Luhmann, T. et al. (2012). Electrospun matrices for localized drug delivery: current technologies and selected biomedical applications. *Eur. J. Pharm. Biopharm.* 81: 1–13.

68 Guan, J. and Hanna, M.A. (2004). Functional properties of extruded foam composites of starch acetate and corn cob fiber. *Ind. Crops Prod.* 19: 255–269.

69 Nampoothiri, K.M., Nair, N.R., and John, R.P. (2010). An overview of the recent developments in polylactide (PLA) research. *Bioresour. Technol.* 101: 8493–8501.

70 Xu, Y.X., Dzenis, Y., and Hanna, M.A. (2005). Water solubility, thermal characteristics and biodegradability of extruded starch acetate foams. *Ind. Crops Prod.* 21: 361–368.

71 Malafaya, P.B., Silva, G.A., and Reis, R.L. (2007). Natural–origin polymers as carriers and scaffolds for biomolecules and cell delivery in tissue engineering applications. *Adv. Drug Deliv. Rev.* 59: 207–233.

72 Pereira, C.S., Cunha, A.M., Reis, R.L. et al. (1998). New starch-based thermoplastic hydrogels for use as bone cements or drug-delivery carriers. *J. Mater. Sci. Mater. Med.* 9: 825–833.

73 Elvira, C., Mano, J.F., San Roman, J., and Reis, R.L. (2002). Starch-based biodegradable hydrogels with potential biomedical applications as drug delivery systems. *Biomaterials* 23: 1955–1966.

74 Espigares, I., Elvira, C., Mano, J.F. et al. (2002). New partially degradable and bioactive acrylic bone cements based on starch blends and ceramic fillers. *Biomaterials* 23: 1883–1895.

75 Arvanitoyannis, I. and Biliaderis, C.G. (1999). Physical properties of polyol-plasticized edible blends made of methyl cellulose and soluble starch. *Carbohydr. Polym.* 38: 47–58.

76 Hanger, L.Y., Lotz, A., and Lepeniotis, S. (1996). Descriptive profiles of selected high intensity sweeteners (HIS), HIS blends, and sucrose. *J. Food Sci.* 61: 456–459.

77 Psomiadou, E., Arvanitoyannis, I., and Yamamoto, N. (1996). Edible films made from natural resources; microcrystalline cellulose (MCC), methylcellulose (MC) and corn starch and polyols—part 2. *Carbohydr. Polym.* 31: 193–204.

78 Peressini, D., Bravin, B., Lapasin, R. et al. (2003). Starch–methylcellulose based edible films: rheological properties of film-forming dispersions. *J. Food Eng.* 59: 25–32.

79 Demirgöz, D., Elvira, C., Mano, J.F. et al. (2000). Chemical modification of starch based biodegradable polymeric blends: effects on water uptake, degradation behaviour and mechanical properties. *Polym. Degrad. Stab.* 70: 161–170.

80 Levy, I. and Shoseyov, O. (2002). Cellulose-binding domains: biotechnological applications. *Biotechnol. Adv.* 20: 191–213.

81 Park, J.H., Saravanakumar, G., Kim, K., and Kwon, I.C. (2010). Targeted delivery of low molecular drugs using chitosan and its derivatives. *Adv. Drug Deliv. Rev.* 62: 28–41.

82 Biliaderis, C.G., Lazaridou, A., and Arvanitoyannis, I. (1999). Glass transition and physical properties of polyol-plasticised pullulan–starch blends at low moisture. *Carbohydr. Polym.* 40: 29–47.

83 Lazaridou, A. and Biliaderis, C.G. (2002). Thermophysical properties of chitosan, chitosan–starch and chitosan–pullulan films near the glass transition. *Carbohydr. Polym.* 48: 179–190.

84 Arvanitoyannis, I.S., Nakayama, A., and Aiba, S.-I. (1998). Chitosan and gelatin based edible films: state diagrams, mechanical and permeation properties. *Carbohydr. Polym.* 37: 371–382.

85 Kuorwel K.K., Cran, M.J., Sonneveld, K. et al. (2011). Antimicrobial activity of biodegradable polysaccharide and protein-based films containing active agents. *J. Food Sci.* 76: R90–R102.

86 Mohanty, A.K., Misra, M., and Drzal, L.T. (2005). *Natural Fibers, Biopolymers, and Biocomposites.* CRC Press.

87 Ke, T. and Sun, X. (2000). Physical properties of poly (lactic acid) and starch composites with various blending ratios. *Cereal Chem.* 77: 761–768.

88 Kiss, G. (1987). In situ composites: blends of isotropic polymers and thermotropic liquid crystalline polymers. *Polym. Eng. Sci.* 27: 410–423.

89 Di Lorenzo, M.L. and Silvestre, C. (1999). Non-isothermal crystallization of polymers. *Prog. Polym. Sci.* 24: 917–950.

90 Ren, J., Fu, H., Ren, T., and Yuan, W. (2009). Preparation, characterization and properties of binary and ternary blends with thermoplastic starch, poly(lactic acid) and poly(butylene adipate-co-terephthalate). *Carbohydr. Polym.* 77: 576–582.

91 Park, J.W. and Im, S.S. (2002). Phase behavior and morphology in blends of poly(L-lactic acid) and poly(butylene succinate). *J. Appl. Polym. Sci.* 86: 647–655.

92 Sarazin, P., Li, G., Orts, W.J., and Favis, B.D. (2008). Binary and ternary blends of polylactide, polycaprolactone and thermoplastic starch. *Polymer* 49: 599–609.

93 Princen, L.H. and Rothfus, J.A. (1984). Development of new crops for industrial raw materials. *J. Am. Oil Chem. Soc.* 61: 281–289.

94 Bar, G., Thomann, Y., Brandsch, R. et al. (1997). Factors affecting the height and phase images in tapping mode atomic force microscopy. Study of phase-separated polymer blends of poly(ethene-co-styrene) and poly(2, 6-dimethyl-1, 4-phenylene oxide). *Langmuir* 13: 3807–3812.

95 Zhou, C., Shi, Q., Guo, W. et al. (2013). Electrospun bio-nanocomposite scaffolds for bone tissue engineering by cellulose nanocrystals reinforcing maleic anhydride grafted PLA. *ACS Appl. Mater. Interfaces* 5: 3847–3854.

96 Dubois, P., Krishnan, M., and Narayan, R. (1999). Aliphatic polyester-grafted starch-like polysaccharides by ring-opening polymerization. *Polymer* 40: 3091–3100.

97 Li, Y. and Shimizu, H. (2009). Improvement in toughness of poly(l-lactide)(PLLA) through reactive blending with acrylonitrile–butadiene–styrene copolymer (ABS): morphology and properties. *Eur. Polym. J.* 45: 738–746.

98 Little, U., Buchanan, F., Harkin-Jones, E. et al. (2009). Accelerated degradation behaviour of poly(ε-caprolactone) via melt blending with poly(aspartic
acid-co-lactide) (PAL). *Polym. Degrad. Stab.* 94: 213–220.

99 Shinoda, H., Asou, Y., Kashima, T. et al. (2003). Amphiphilic biodegradable
copolymer, poly(aspartic acid-co-lactide): acceleration of degradation rate and
improvement of thermal stability for poly(lactic acid), poly(butylene succinate) and
poly(ε-caprolactone). *Polym. Degrad. Stab.* 80: 241–250.

100 Ren, J. (2010). *Biodegradable Poly(Lactic Acid): Synthesis, Modification, Processing*.
Springer.

101 Wang, H., Sun, X., and Seib, P. (2001). Strengthening blends of poly(lactic
acid) and starch with methylenediphenyl diisocyanate. *J. Appl. Polym. Sci.* 82:
1761–1767.

102 Wang, H., Sun, X., and Seib, P. (2002). Mechanical properties of poly(lactic acid)
and wheat starch blends with methylenediphenyl diisocyanate. *J. Appl. Polym. Sci.*
84: 1257–1262.

103 Avérous, L. (2004). Biodegradable multiphase systems based on plasticized starch:
a review. *J. Macromol. Sci. Polym. Rev.* 44: 231–274.

104 Avérous, L. (2008). Polylactic acid: synthesis, properties and applications. In:
Monomers, Polymers and Composites from Renewable Resources (ed. M. Belgacem
and A. Gandini), 1. Elsevier.

105 Zhang, J.-F. and Sun, X. (2004). Mechanical and thermal properties of poly(lactic
acid)/starch blends with dioctyl maleate. *J. Appl. Polym. Sci.* 94: 1697–1704.

106 Mungara, P., Chang, T., Zhu, J., and Jane, J. (2002). Processing and physical properties of plastics made from soy protein polyester blends. *J. Polym. Environ.* 10:
31–37.

107 Bryant, C.M. and Hamaker, B.R. (1997). Effect of lime on gelatinization of corn
flour and starch. *Cereal Chem.* 74: 171–175.

108 Garlotta, D., Doane, W., Shogren, R. et al. (2003). Mechanical and thermal properties of starch-filled poly(D,L-lactic acid)/poly(hydroxy ester ether) biodegradable
blends. *J. Appl. Polym. Sci.* 88: 1775–1786.

109 Baek, J.-Y., Xing, Z.-C., Kwak, G. et al. (2012). Fabrication and characterization of
collagen-immobilized porous PHBV/HA nanocomposite scaffolds for bone tissue
engineering. *J. Nanomater.* 2012: 1.

110 Orts, W.J., Nobes, G.A.R., Kawada, J. et al. (2008). Poly(hydroxyalkanoates):
biorefinery polymers with a whole range of applications. The work of Robert H.
Marchessault. *Can. J. Chem.* 86: 628–640.

111 Akaraonye, E., Keshavarz, T., and Roy, I. (2010). Production of polyhydroxyalkanoates: the future green materials of choice. *J. Chem. Technol. Biotechnol.* 85:
732–743.

112 Rehm, B.H.A. (2010). Bacterial polymers: biosynthesis, modifications and applications. *Nat. Rev. Microbiol.* 8: 578–592.

113 Mohanty, A.K., Misra, M., and Drzal, L.T. (2002). Sustainable bio-composites from
renewable resources: opportunities and challenges in the green materials world. *J.
Polym. Environ.* 10: 19–26.

114 Zinn, M., Witholt, B., and Egli, T. (2001). Occurrence, synthesis and medical application of bacterial polyhydroxyalkanoate. *Adv. Drug Deliv. Rev.* 53: 5–21.

115 Al-Malaika, S. and Amir, E.J. (1989). Thermoplastic elastomers: Part III—Ageing and mechanical properties of natural rubber-reclaimed rubber/polypropylene systems and their role as solid phase dispersants in polypropylene/polyethylene blends. *Polym. Degrad. Stab.* 26: 31–41.

116 Barham, P.J., Keller, A., Otun, E.L., and Holmes, P.A. (1984). Crystallization and morphology of a bacterial thermoplastic: poly-3-hydroxybutyrate. *J. Mater. Sci.* 19: 2781–2794.

117 Gogolewski, S., Jovanovic, M., Perren, S.M. et al. (1993). Tissue response and in vivo degradation of selected polyhydroxyacids: Polylactides (PLA), poly(3-hydroxybutyrate) (PHB), and poly(3-hydroxybutyrate-co-3-hydroxyvalerate) (PHB/VA). *J. Biomed. Mater. Res. A* 27: 1135–1148.

118 Wang, X., Yucel, T., Lu, Q. et al. (2010). Silk nanospheres and microspheres from silk/PVA blend films for drug delivery. *Biomaterials* 31: 1025–1035.

119 Malesu, V.K., Sahoo, D., and Nayak, P.L. (2011). Chitosan–sodium alginate nanocomposites blended with cloisite 30b as a novel drug delivery system for anticancer drug curcumin. *Int. J. Appl. Bio. Pharm. Tech.* 2: 402–411.

120 Ravi Kumar, M.N.V. and Kumar, N. (2001). Polymeric controlled drug-delivery systems: perspective issues and opportunities. *Drug Dev. Ind. Pharm.* 27: 1–30.

4

Applications of Biopolymeric Gels in Medical Biotechnology

Zulal Yalinca[1] and Şükrü Tüzmen[2]

[1]Department of Chemistry, Eastern Mediterranean University, North Cyprus via Mersin 10, Famagusta, Turkey
[2]Molecular Biology and Genetics Program, Department of Biological Sciences, Eastern Mediterranean University, North Cyprus via Mersin 10, Famagusta, Turkey

4.1 Introduction

4.1.1 Historical Background

The term "hydrogel" first appeared in the literature in 1894 [1]. Owing to Wichterle and Lim's work pertaining to hydrogels in 1960, hydrogels have attracted a lot of attention from biomaterial scientists ever since [1, 2]. Based on natural and synthetic polymers, hydrogels can be used in cell encapsulation research, especially the establishment of a novel field in tissue engineering [3–9]. Advanced formulations in the hydrogel field gave rise to self-regulated drug-delivery complexes referred to as *smart hydrogels*. There are many applications for smart hydrogels because of their ability to respond to chemical, physical, and biological stimuli sensitively. However, even these types of hydrogel systems have had drawbacks in clinical administration. Thus, there needs to be further investigation into designing systems of smarter hydrogels [4, 8, 10]. For more than two decades, scientists have been working to overcome the poorresponse rate of hydrogels in biomedical applications. Hydrogels demonstrate a significant class of biomaterials in medical biotechnology. These categories may include tissue engineering and regeneration, diagnostics, cellular immobilization, cellular biomolecular separation, and utilization of barrier materials for biological adhesion for regulation. Due to their biocompatibility hydrogels may cause minimum inflammatory outcome and tissue damage. There have been many attempts to enhance the response rate by the introduction of certain techniques, including size reduction of gel particles, surface modifications of hydrogels, and modification of gel porosity. Applications are extensive and include biosensors, contact lenses, and artificial implants [10–13].

4.1.2 Classification of Hydrogels

Hydrogels are invaluable in their three-dimensional (3D) network capacity to capture and release active compounds and biomolecules. They resemble soft tissues due to their ability to hold a high concentration of water or biological fluids [8, 14, 15]. Hydrogel

Bio Monomers for Green Polymeric Composite Materials, First Edition.
Edited by P.M. Visakh, Oguz Bayraktar and Gopalakrishnan Menon.
© 2020 John Wiley & Sons Ltd. Published 2020 by John Wiley & Sons Ltd.

technology has brought considerable advances in the biopharmaceutical field. Hydrogels have a sponge-like capacity to absorb water due to their hydrophilic functional groups. The water absorbed into hydrogels permits diffusion of certain molecules while the polymer component of hydrogels acts as a matrix for holding water molecules together. An additional ability of hydrogels involves a single polymer that interacts with itself to form large-scale materials. The final form of the gel is considered as neither liquid nor solid. This type of the property enables the gels to have flexible behavior, which is not found in either form of the physical state. Polymer-derived hydrogels, which are amongst many biomaterials of drugs, are exploited in therapeutic areas of pharmacogenomics. Hydrogels have the ability to imitate the physical, chemical, biological, and electrical features of many tissues. These features earn them the ability to serve as potential candidates for biomaterials. The ability to take up 20–40-fold more water molecules related to dried weight is possible because of the 3D structure of the hydrogel network. Their physical characteristics enable hydrogels to be molded into different shapes and sizes. New methodologies in designing hydrogels have given impetus to this field of biomaterial research. These novel approaches involve super porous hydrogels [16], comb-like grafted hydrogels [17–19], self-assembling hydrogels [20, 21], and recombinant triblock copolymers [22–24]. To construct hydrogel systems with well-defined chemical characteristics, information regarding polymer chemistry and synthesis, features of the materials to be utilized, parameters of mode of interaction, material release capability, and delivery systems need to be taken into consideration [14, 25, 26]. Not only bio-based but also synthetic material can be used in the preparation of hydrogels for biological and pharmaceutical applications [1, 10, 14, 27–30]. In order to acquire hydrogels with improved features, natural and synthetic polymers can be blended to facilitate formation of hybrid hydrogels, which in turn will help accommodate high drug concentration and controlled release in the targeted area. Hydrogels can swell not only in aqueous environments, but also in lipid media [11]. Overall, they are defined by their capacity to absorb fluids, increasing their volume. It is expected that hydrogels that are utilized in drug delivery and biopharmaceuticals should have the ability to be biocompatible and biodegradable (i.e. poly-α-esters, polycaprolactone, polyhydroxyalkanoates, polycarbonates, polyphosphazenes, polyurethanes, synthetic polyethers, polyamino acids, natural polyaminoacids, and polysaccharides such as chitosan, alginate, dextran, agarose, mannan, and inulin) (Figure 4.1) [2–4, 14, 25, 31–35]. A number of significant features need to be taken

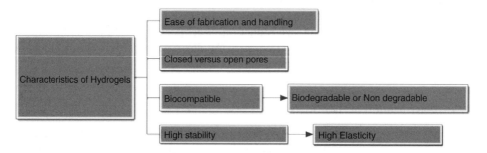

Figure 4.1 Significant characteristics of hydrogels.

into consideration prior to configuration of hydrogels. Certain criteria should be met, including:

a) the inflammatory response should not be triggered
b) controlled degradation time
c) desired functional response
d) desirable application properties [36].

In turn, the above parameters facilitate novel crosslinking methodologies for designing appropriate materials. Additionally, it should be noted that the features and potential uses of hydrogels depend on the method of choice for preparation and the type of monomers utilized in the preparation of polymeric networks [14, 37]. The diversity of hydrogel structures illustrates the variety of monomers that can make up the hydrogel [38–40]. Preparation techniques can facilitate the configuration of hydrogels [41]:

a) homopolymer-derived hydrogels
b) copolymer-derived hydrogels
c) multipolymer interpenetrating polymeric hydrogels.

Hydrogels can be classified according to several categories [10]:

a) Crystallinity
   i) Amorphous
   ii) Semicrystalline
   iii) Crystalline
b) Type of crosslinking
   i) Physical
   ii) Chemical
c) Physical configuration
   i) Matrix
   ii) Film
   iii) Microspheres
d) Network polarity
   i) Neutral
   ii) Anionic
   iii) Cationic
   iv) Amphoteric
   v) Zwitterionic
e) Degradability
   i) Biodegradable
   ii) Non-degradable
f) Functional reactions
   i) Biological
   ii) Physical
   iii) Chemical
g) Source of origin
   i) Natural hydrogels
   ii) Synthetic hydrogels
   iii) Hybrid hydrogels

External environmental conditions can be taken as a base to prepare hydrogels to respond according to the need, which may include expanding or shrinking [3, 11, 40].

4.1.3 Preparation Methods of Hydrogels

Crosslinking is the most versatile method to facilitate biopolymeric deficiencies [42]. Mechanical properties and the stability of biomaterials can be ameliorated by crosslinking agents. However, limitations exist due to reduced degradability and lack of functional groups and potential cytotoxicity introduced by crosslinking agents [41, 43]. Based on the biopolymers to be utilized and the enhancement of desired features, different kinds of crosslinking agents and techniques can be used [11, 43]. The following are several examples of common crosslinking agents that are utilized in hydrogel making: glutaraldehyde, glyoxal, ethylene glycol diglycidyl ether, methylene bis acrylamide, tripolyphosphate, *N*-hydroxysuccinimide, epichlorohydrin, citric acid, sodium metaphosphate, and carbodiimides [11, 41–46]. The chemical and physical nature of hydrogels depends on the concentration of crosslinking agents, the degree of crosslinking, the kind of crosslinker/monomer, and the method of preparation used [4, 11, 42]. All preparations of hydrogels go through crosslinking methodologies (Figure 4.2) [10, 41, 47]. This creates stabilization and assists in multidimensional expanding, which results in the network structure. Creation of crosslinking within hydrogels alters their physical appearance to prevent dissolution. Mechanical strength, heat resistance, dissolution, and disintegration can be controlled by the degree and type of crosslinking in designing hydrogels [42]. Nevertheless, there exist some disadvantages in processing since their insolubility and infusibility related to the degree of crosslinking may cause inflexibility.

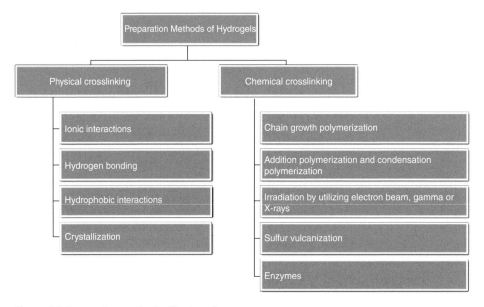

Figure 4.2 Preparation methods of hydrogels.

4.1.3.1 Physical Crosslinked Hydrogels

This type of crosslinking of hydrogels can be formed by applying diverse environmental conditions such as pH, temperature, and ionic strength [11, 14]. Additionally, distinct physicochemical interactions, including ionic interactions, hydrogen bonding, and crystallization, can result in physical crosslinking [14, 41, 42] (Figure 4.2).

4.1.3.2 Chemical Crosslinked Hydrogels

Chemical crosslinking can be synthesized by covalent bonding formation [3, 41, 42] (Figure 4.2) via the following:

1) Chain growth polymerization
   a) Free radical polymerization
   b) Controlled free radical polymerization
   c) Anionic polymerization
   d) Cationic polymerization
2) Addition polymerization and condensation polymerization
3) Irradiation by utilizing electron beam, gamma, or X-rays
4) Sulfur vulcanization
5) Enzymatic crosslinking

Chemical crosslinking is a permanent outcome due to the configurational changes, and it is irreversible. There are two methods to prepare chemically crosslinked hydrogels:

a) polymerization between monomer and polyfunctional crosslinker
b) direct crosslinking of hydrophilic polymers.

However, 3D polymerization possesses certain disadvantages, e.g. it gives rise to residual monomers and these monomers are required to go through stringent purification to prevent unreacted monomers from serving as toxic materials. A common hydrogel preparation technique may involve 3D polymerization, where the water-soluble monomers are polymerized with a multifunctional crosslinker or by directly crosslinking hydrophilic polymers [41]. A comparison of physical and chemical crosslinking is given in Table 4.1 [41, 42, 47].

4.1.3.3 General Properties of Hydrogels

Interpretation of the preparation methodologies of hydrogels is significant in predicting physical and chemical properties [15, 48–50]. Ideal hydrogel materials based on functional properties are listed in Table 4.2 [3, 11]. Hydrogels are excellent candidates for biotechnological applications, including drug-delivery systems, since they possess soft, hydrophilic, high swelling ability and a viscoelastic nature, are biodegradable, and have biocompatible characteristics [10].

4.2 Types of Biopolymeric Gels

Factors affecting the properties of hydrogels include characteristics such as the extent of crosslinking, porosity, pore size, and pore heterogeneity (Figure 4.3) [8, 10, 51–56].

Table 4.1 Comparison of physical and chemical crosslinking.

Physical crosslinking	Chemical crosslinking
Reversible	Permanent and irreversible
Achieved by association, aggregation, complexation, crystallization, and hydrogen bonding	Achieved by covalent bonding
No need for chemical modification or additional entities	Prevents dilution of hydrogel matrix and diffusion of polymers from the site of injections
Can prevent the undesired effects of chemical crosslinking by electron beam method and irradiation utilization Radiation crosslinking results in a cleaner process and unwanted side products	Undesired reactions can result with bioactive substance in the hydrogel matrix

Table 4.2 Ideal hydrogel materials based on functional properties.

Maximum absorption capacity

Desired rate of absorption by controlled particle size and porosity

Maximum resistance to solvent

Environmentally friendly

Long shelf life

High stability

Maximum durability

Photostable

Photodurable

pH neutrality

Elasticity

Decreased viscosity

Maximum resistance to heat

Rewettability

Biologically compatible

Cost effective

The volume of water absorbed will be determined by the material absorption and diffusion through the hydrogel. During the formation of hydrogel matrices there various pore sizes may form within the hydrogels. Thus far, numerous hydrogel matrix networks have been configured to facilitate the requirements of the pharmaceutical and medical industries [5]. Hydrogels are categorized according to four major subclasses: non-porous, micro-porous, macro-porous, and super-porous (Table 4.3) [52, 57].

There exists a need for lifetime extension, relative non-toxicity, and improving the properties of hard/soft materials. Micro/macro-scale cracks could be created by mechanical deformation. This outcome can result in loss of mechanical properties

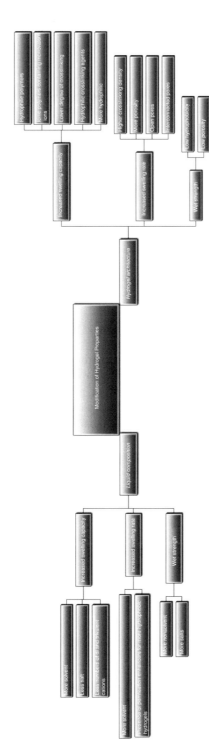

Figure 4.3 Factors affecting modification of hydrogel characteristics.

Table 4.3 Absorption framework of porosity-related characteristics of hydrogels.

Subclass	Morphology	Swelling rate
Non-porous	Network porosity absent	Highly reduced
Micro-porous	Wide porosity range between 100 and 1000 Å with a closed-cell structure	Reduced
Macro-porous	Wide porosity range between 0.1 and 1.0 μm with a closed-cell structure	Rapid
Super-porous	Excessive porosity with a open cell structure	Very rapid

Table 4.4 Pros and cons of bio-based and synthetic polymers.

Bio-based polymers	Synthetic polymers
Lower mechanical strength	No natural bioactive properties
Risk of pathogenic response	Well-defined chemistry
Natural biocompatibility/biodegradability	Lack of natural biocompatibility/biodegradability
Biologically identifiable parts	Lack of biologically identifiable parts

and decay [58, 59]. These hurdles can be prevented by the "self-healing" properties of hydrogels [59–61]. The chemical composition and mechanical properties of hydrogels facilitate the release of chemical and biological molecules. This property can be beneficial for time release of therapeutics. Polymeric networks of hydrogels can be enhanced by their water content. Considering implanted hydrogels, their degradability property serves a significant function in reducing the immune response [62–64]. As illustrated in Tables 4.2 and 4.3, physical and mechanical properties when synthesizing hydrogels involve sturdiness, architecture of mesh size and porosity, and the 3D network. According to the applications of hydrogels their physical and mechanical characteristics can be adapted as necessary. Construction of hydrogels can be performed from either bio-based or synthetic polymers. The pros and cons of bio-based and synthesis polymers are summarized in Table 4.4 [11, 65].

4.3 Applications of Biopolymeric Gel

Amongst the various polymeric hydrogels are the earliest biopolymers utilized in clinical settings [66, 67]. Hydrogels can provide protection for small biological molecules and chemical compounds within stringent environmental conditions (Table 4.5) [11, 38].

The mode of response of hydrogels may depend on the chemical composition of the polymeric networks. Hydrogels can be modified to respond to environmentally triggered stimuli, including changes to pH, ionic strength, and temperature (Figure 4.4) [3, 11, 14, 39]. Various characteristics, including ionic strength, the nature of the counter ions, and the composition of the polymeric backbone, facilitate the swelling of ionic hydrogels [11, 14].

Table 4.5 World-wide hydrogels resources.

Manufacturer	Product	Application
Pharmaplast	• Low adherent absorbing dressings • First-aid bandages • Transparent film dressings • Medical adhesive tapes • Compression products • Casting and bandaging • Swabs • Post-operative dressings • Non-adherent dressings • Hydrogel products • Hydrocolloid products • Foam products • Soft silicone products • Deodorizing wound dressings • Antimicrobial wound care products • Hemostatic wound dressings • Incise drapes • Skin-closure strips • Foot care products • Heat–cool patch • Hand sanitizer gel • Miscellaneous	Wound care
Pharmaplast	• Crack care • Wipes • Black head products • Depilatory products	Cosmetics
Medline	• Sterile procedure trays • Drapes, gowns, and accessories • Personal protective apparel • Surgical gloves • Surgical devices • Facemasks • Sterilization products • Safety devices • Perioperative supplies • Examination gloves • Patient hygiene • Patient wear and textiles • Advanced skin and wound care • Minor procedure trays • Ice bags • Electrodes • Pressure ulcer prevention	Medical tools
Polymer Science Inc.	• P-DERM silicone gel adhesive • P-DERM acrylic adhesive • P-DERM hydrogels and hydrocolloids	Medical tools

(Continued)

Table 4.5 (Continued)

Manufacturer	Product	Application
Dermasciences	• Medihoney • Xtrasorb • Bioguard • Sponges • Compression bandages • Dermagren • Alginate dressings • Aquasite hydrogels	Burn and wound care
Ocular Therapeutix	• ReSure® sealant	Ocular
Covidien/Medtro	• Hydrogel wound dressing	Wound care
Elta/Swiss-Amer	• Elta Dry Hydrogen TD wound dressing • Elta Hydrovase wound gel • Elta wound hydrogel	Wound care
Southwest Techn	• Southwest Elasto-Gel plus sterile wound dressing with tape • Southwest Elasto-Gel horseshoe-shaped dressing	Wound care
MPM Medical In	• MPM CoolMagic hydrogel polymer sheet dressing • MPM Regenecare HA wound care hydrogel dressing • MPM Regenecare HA topical anesthetic hydrogel spray • MPM RadiaPlex Rx wound gel dressing with hyaluronic acid	Wound care

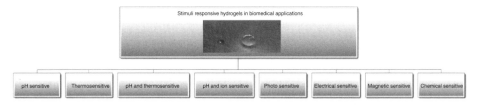

Figure 4.4 Stimuli responsive hydrogels in biomedical applications.

4.3.1 Applications of Hydrogels in Drug-delivery Systems

Hydrogels have facilitated the expansion of smart drug-delivery protocols. These cellular systems inhibit the degradation of encapsulated drugs from harmful surroundings including environments such as low pH and digestive enzymes. According to the targeted sites, hydrogels can also regulate drug release by altering their structure [14]. Nanotechnology offers appropriate routes for the time-controlled and site-specific delivery of biologically active molecules [47, 68]. The efficacy of hydrogels can be enhanced by the incorporation of conditionally triggered specifications. These may include various stimuli such as growth factors or immune specific agents. Panigrahi et al. [69] have reported drug delivery related to the central nervous system. This research has led to advantages including stability of the drug assisting in drug release and minimizing major side effects [70]. A vital goal of targeted drug delivery lies within the protocols of site-specific drug release. Novel treatments for site-specific drug release utilize specifications such as disease-specific enzymatic activities to prompt the release of drugs from hydrogels [15]. Encapsulated inactive precursors of drug molecules can be activated *in vivo* utilizing enzymatic hydrolysis. One such example includes cancer-specific enzymes originally released by tumor cells, but that can also be used to initiate the release of therapeutic molecules to avoid or decrease metastasis [15]. Hydrophilic and hydrophobic drugs can also be encapsulated in hydrogels and released depending on the site of action of the environment [4]. There have been attempts to form tailormade hydrogels to improve the efficiency of hydrogel-based drug-delivery systems. Currently, controlled stimuli responsive drug release is still a big challenge [4]. To date, the objective of chemotherapeutic applications has been to destroy cancer cells but leave healthy cells unaffected. The present aim is to develop a smart system for drug delivery in order to increase efficacy and decrease drug toxicity [71]. A useful area of drug delivery utilizing hydrogels has been reported in research pertaining to ocular drug-delivery carriers [31]. An ideal criterion of drug-delivery protocol combines the physiological conditions of patients with environmentally controlled drug release [72]. Preferably, a drug-delivery platform should be suitable for the bioenvironment of a patient, where the release rate will be facilitated by surrounding parameters [31, 72]. Hence, hydrogels could provide feasible drug release due to their chemical composition in response to environmental stimuli [72]. Hydrogel-based products are potential candidates for drug-delivery systems if they can provide the necessary conditions for drug release [31].

4.3.2 Applications of Hydrogels in siRNA and Peptide-based Therapeutics

siRNAs are small double-stranded RNA molecules with defined lengths consisting of approximately 21–27 nucleotides. They participate in post-transcriptional regulation of mRNAs [73–75]. When present in a biological system, siRNAs cause specific gene silencing of targeted transcripts by destroying the mRNA of interest. siRNAs exhibit a potential capacity to be utilized as biotherapeutics toward targeting any cell or tissue type [76], but there are many hurdles to be overcome, including serum nuclease degradation, low transfectability efficiency, and short half-life in blood [77]. Additionally, the naked siRNAs have various drawbacks, such as their large size and anionic chemical compositions, which inhibit their uptake by the cellular membranes. Considering different delivery vectors, nanogels illustrate a promising cell-type specific delivery of siRNAs [78]. In order to facilitate siRNA loading efficiency, positively charged polymeric-based materials are used to increase the affinity between siRNA and the composition of the hydrogels. Amongst the cationic polymers chitosan has been reported to facilitate the cellular uptake of siRNAs both *in vitro* and *in vivo*. Additionally, chitosan may protect rapid degradation of siRNAs *in vivo* [79]. The rate of siRNA/drug release could be facilitated by environmental pH differences between healthy and cancerous tissues. Utilization of certain cell-type specific delivery systems of siRNAs, such as antibodies, aptamers, and peptides, is worthy of further investigation [12, 78]. Numerous potential outcomes have been reported thus far pertaining to clinical applications of siRNAs. However, the validation period of the siRNA-based therapies is worth the time and effort required in order to enhance the contribution of RNAi-based products to future medical applications [78].

4.3.3 Applications of Hydrogels in Wound Healing, Tissue Engineering, and Regenerative Medicine

The objective of tissue engineering is to restore damaged tissues and organs in order to generate artificial agents for transplantation [15, 40]. Biopolymers have the advantage of designing 3D matrices and developing new applications for tissue engineering [15, 80]. Hydrogels are considered to be one of the best-qualified materials for this purpose. The successful applications of tissue engineering are facilitated by the generation of functional hydrogels [15, 81]. Porous scaffolds allow colonization of cells, imitating extracellular matrix (ECM) and furnishing enhanced cell adhesion, cell migration, and cell proliferation [40]. Ideal preparations of these materials should resemble the ECM. This condition can be achieved by the utilization of materials with high swelling capacity, sufficient mechanical stability, and bioactive capability [15, 70, 82, 83]. The function of hydrogels can be enhanced within a living entity by the macroscopic design of its matrix. Pores can be constructed depending on the chemistry of hydrogels [70, 84]. The intentional incorporation of pores into the hydrogel structure can facilitate cell-specific proliferation. Numerous hydrogel applications in tissue engineering, such as production of tissues (cardiovascular tissues), involve host tissue vascularization. This aids in delivering the hydrogels loaded with cells into the tissue, to facilitate delivery of nutrients and oxygen to the cells, encapsulated by the hydrogels [70]. Anderson and Shive [85] and Fournier et al. [86] have reported the benefit of an increased surface area of hydrogels. This property improves the biocompatibility of these materials due to better interaction between tissue and hydrogels to provide accessibility for certain cell types. In order to facilitate functional combinations of biological entities together with electronic devices, it is critical to establish seamless connections between these two fields [87]. A number of protocols have been presented to achieve more efficacious attachment between implanted electrodes and surrounding tissues [88–90]. To date, there have been many areas of hydrogel applications of polysaccharides, especially chitosan- and alginate-based hydrogels due to their biocompatibilities [36].

4.4 Conclusions and Future Perspectives

To date, numerous scientific findings have illustrated that basic and translational research pertaining to hydrogels has promising future applications. Future investigation in this research area shows potential in further optimization of biocompatible bio-based polymeric networks [15, 91–93]. Interpretation of molecular orchestration techniques has already been observed in cell biology for the development of functional biomaterials. This will establish the way forward for future novel hydrogel applications. However, a number of challenges still need to be overcome in order to enhance the clinical applications of hydrogels in biomedicine. To achieve the optimum conditions for a targeted delivery, continued improvements in all of these challenges will promote hydrogel-based drug-delivery applications [4, 94]. In spite of the advantages, there remain many obstacles to resolving problems in translational medicine. Advanced technologies in material science and electronics will facilitate our understanding of the way in which we can combine biology with electronics.

References

1 Buwalda, S.J., Boere, K.W.M., Dijkstra, P.J. et al. (2014). Hydrogels in a historical perspective: from simple networks to smart materials. *J. Control. Release* 190: 254–273. https://doi.org/10.1016/j.jconrel.2014.03.052.

2 Hoffman, A.S. (2002). Hydrogels for biomedical applications. *Adv. Drug Delivery Rev.* 54 (1): 3–12. https://doi.org/10.1016/s0169-409x(01)00239-3.

3 Ahmed, E.M. (2015). Hydrogel: preparation, characterization, and applications: a review. *J. Adv. Res.* 6 (2): 105–121. https://doi.org/10.1016/j.jare.2013.07.006.

4 Chirani, N., Yahia, L.H., Gritsch, L. et al. (2015). History and applications of hydrogels. *J. Biomed. Sci. 4: 1–23.*

5 Hoffman, A.S. (2012). Hydrogels for biomedical applications. *Adv. Drug Delivery Rev.* 64: 18–23. https://doi.org/10.1016/j.addr.2012.09.010.

6 Lim, F. and Sun, A.M. (1980). Microencapsulated islets as bioartificial endocrine pancreas. *Science* 210 (4472): 908–910. https://doi.org/10.1126/science.6776628.

7 Rosiak, J.M. and Yoshii, F. (1999). Hydrogels and their medical applications. *Nucl. Instrum. Methods Phys. Res., Sect. B* 151 (1–4): 56–64. https://doi.org/10.1016/s0168-583x(99)00118-4.

8 Vermonden, T. and Klumperman, B. (2015). The past, present and future of hydrogels. *Eur. Polym. J.* 72: 341–343. https://doi.org/10.1016/j.eurpolymj.2015.08.032.

9 Yannas, I.V., Lee, E., Orgill, D.P. et al. (1989). Synthesis and characterization of a model extracellular-matrix that induces partial regeneration of adult mammalian skin. *Proc. Natl. Acad. Sci. U.S.A.* 86 (3): 933–937. https://doi.org/10.1073/pnas.86.3 .933.

10 Rizwan, M., Yahya, R., Hassan, A. et al. (2017). pH sensitive hydrogels in drug delivery: brief history, properties, swelling, and release mechanism, material selection and applications. *Polymers* 9 (4): https://doi.org/10.3390/polym9040137.

11 Kabiri, K., Omidian, H., Zohuriaan-Mehr, M.J., and Doroudiani, S. (2011). Superabsorbent hydrogel composites and nanocomposites: a review. *Polym. Compos.* 32 (2): 277–289. https://doi.org/10.1002/pc.21046.

12 Miyata, T., Uragami, T., and Nakamae, K. (2002). Biomolecule-sensitive hydrogels. *Adv. Drug Delivery Rev.* 54 (1): 79–98. https://doi.org/10.1016/s0169-409x(01)00241-1.

13 Vert, M. (2007). Polymeric biomaterials: strategies of the past vs. strategies of the future. *Prog. Polym. Sci.* 32 (8–9): 755–761. https://doi.org/10.1016/j.progpolymsci .2007.05.006.

14 Qiu, Y. and Park, K. (2012). Environment-sensitive hydrogels for drug delivery. *Adv. Drug Delivery Rev.* 64: 49–60. https://doi.org/10.1016/j.addr.2012.09.024.

15 Ulijn, R.V., Bibi, N., Jayawarna, V. et al. (2007). Bioresponsive hydrogels. *Mater. Today* 10 (4): 40–48. https://doi.org/10.1016/s1369-7021(07)70049-4.

16 Ullah, F., Othman, M.B.H., Javed, F. et al. (2015). Classification, processing and application of hydrogels: a review. *Mater. Sci. Eng. C – Mater. Biol. Appl.* 57: 414–433. https://doi.org/10.1016/j.msec.2015.07.053.

17 Chen, S.Q., Li, J.M., Pan, T.T. et al. (2016). Comb-type grafted hydrogels of PNI-PAM and PDMAEMA with reversed network-graft architectures from controlled radical polymerizations. *Polymers* 8 (2): https://doi.org/10.3390/polym8020038.

18 Gonzalez-Gomez, R., Ortega, A., Lazo, L.M., and Burillo, G. (2014). Retention of heavy metal ions on comb-type hydrogels based on acrylic acid and 4-vinylpyridine, synthesized by gamma radiation. *Radiat. Phys. Chem.* 102: 117–123. https://doi.org/10.1016/j.radphyschem.2014.04.026.

19 Miao, L., Lu, M.G., Yang, C.L. et al. (2013). Preparation and microstructural analysis of poly(ethylene oxide) comb-type grafted poly(N-isopropyl acrylamide) hydrogels crosslinked by poly(epsilon-caprolactone). *J. Appl. Polym. Sci.* 128 (1): 275–282. https://doi.org/10.1002/app.38172.

20 Ryan, D.M. and Nilsson, B.L. (2012). Self-assembled amino acids and dipeptides as noncovalent hydrogels for tissue engineering. *Polym. Chem.* 3 (1): 18–33. https://doi.org/10.1039/c1py00335f.

21 Sun, T.L., Wu, Z.L., and Gong, J.P. (2012). Self-assembled structures of a semi-rigid polyanion in aqueous solutions and hydrogels. *Sci. China-Chem.* 55 (5): 735–742. https://doi.org/10.1007/s11426-012-4497-x.

22 Alexander, A., Ajazuddin, Khan, J., and Saraf, S. (2013). Poly(ethylene glycol)-poly(lactic-co-glycolic acid) based thermosensitive injectable hydrogels for biomedical applications. *J. Control. Release* 172 (3): 715–729. https://doi.org/10.1016/j.jconrel.2013.10.006.

23 Khodaverdi, E., Tekie, F.S.M., Hadizadeh, F. et al. (2014). Hydrogels composed of cyclodextrin inclusion complexes with PLGA-PEG-PLGA triblock copolymers as drug delivery systems. *Aaps Pharmscitech* 15 (1): 177–188. https://doi.org/10.1208/s12249-013-0051-1.

24 Tabassi, S.A.S., Tekie, F.S.M., Hadizadeh, F. et al. (2014). Sustained release drug delivery using supramolecular hydrogels of the triblock copolymer PCL-PEG-PCL and alpha-cyclodextrin. *J. Sol-Gel Sci. Technol.* 69 (1): 166–171. https://doi.org/10.1007/s10971-013-3200-9.

25 Gupta, P., Vermani, K., and Garg, S. (2002). Hydrogels: from controlled release to pH-responsive drug delivery. *Drug Discovery Today* 7 (10): 569–579. https://doi.org/10.1016/s1359-6446(02)02255-9.

26 Li, Y.L., Rodrigues, J., and Tomas, H. (2012). Injectable and biodegradable hydrogels: gelation, biodegradation and biomedical applications. *Chem. Soc. Rev.* 41 (6): 2193–2221. https://doi.org/10.1039/c1cs15203c.

27 Huynh, C.T., Nguyen, M.K., and Lee, D.S. (2011). Injectable block copolymer hydrogels: achievements and future challenges for biomedical applications. *Macromolecules* 44 (17): 6629–6636. https://doi.org/10.1021/ma201261m.

28 Nguyen, M.K. and Alsberg, E. (2014). Bioactive factor delivery strategies from engineered polymer hydrogels for therapeutic medicine. *Prog. Polym. Sci.* 39 (7): 1235–1265. https://doi.org/10.1016/j.progpolymsci.2013.12.001.

29 Ozdil, D. and Aydin, H.M. (2014). Polymers for medical and tissue engineering applications. *J. Chem. Technol. Biotechnol.* 89 (12): 1793–1810. https://doi.org/10.1002/jctb.4505.

30 Webber, M.J. and Anderson, D.G. (2015). Smart approaches to glucose-responsive drug delivery. *J. Drug Targeting* 23 (7–8): 651–655. https://doi.org/10.3109/1061186x .2015.1055749.

31 Calo, E. and Khutoryanskiy, V.V. (2015). Biomedical applications of hydrogels: a review of patents and commercial products. *Eur. Polym. J.* 65: 252–267. https://doi .org/10.1016/j.eurpolymj.2014.11.024.

32 Jiang, Y.J., Chen, J., Deng, C. et al. (2014). Click hydrogels, microgels and nanogels: emerging platforms for drug delivery and tissue engineering. *Biomaterials* 35 (18): 4969–4985. https://doi.org/10.1016/j.biomaterials.2014.03.001.

33 Koetting, M.C., Peters, J.T., Steichen, S.D., and Peppas, N.A. (2015). Stimulus-responsive hydrogels: theory, modern advances, and applications. *Mater. Sci. Eng. R-Rep.* 93: 1–49. https://doi.org/10.1016/j.mser.2015.04.001.

34 Manavitehrani, I., Fathi, A., Badr, H. et al. (2016). Biomedical applications of biodegradable polyesters. *Polymers* 8 (1): https://doi.org/10.3390/polym8010020.

35 Sannino, A., Demitri, C., and Madaghiele, M. (2009). Biodegradable cellulose-based hydrogels: design and applications. *Dent. Mater.* 2 (2): 353–373. https://doi.org/10 .3390/ma2020353.

36 Ulery, B.D., Nair, L.S., and Laurencin, C.T. (2011). Biomedical applications of biodegradable polymers. *J. Polym. Sci. B – Polym. Phys.* 49 (12): 832–864. https://doi .org/10.1002/polb.22259.

37 Qiu, Y. and Park, K. (2001). Environment-sensitive hydrogels for drug delivery. *Adv. Drug Delivery Rev.* 53 (3): 321–339. https://doi.org/10.1016/s0169-409x(01)00203-4.

38 Peppas, N.A. (1997). Hydrogels and drug delivery. *Curr. Opin. Colloid Interface Sci.* 2 (5): 531–537. https://doi.org/10.1016/s1359-0294(97)80103-3.

39 Peppas, N.A., Bures, P., Leobandung, W., and Ichikawa, H. (2000). Hydrogels in pharmaceutical formulations. *Eur. J. Pharm. Biopharm.* 50 (1): 27–46. https://doi .org/10.1016/s0939-6411(00)00090-4.

40 Peppas, N.A., Hilt, J.Z., Khademhosseini, A., and Langer, R. (2006). Hydrogels in biology and medicine: from molecular principles to bionanotechnology. *Adv. Mater.* 18 (11): 1345–1360. https://doi.org/10.1002/adma.200501612.

41 Hoare, T.R. and Kohane, D.S. (2008). Hydrogels in drug delivery: progress and challenges. *Polymer* 49 (8): 1993–2007. https://doi.org/10.1016/j.polymer.2008.01.027.

42 Akhtar, M.F., Hanif, M., and Ranjha, N.M. (2016). Methods of synthesis of hydrogels … A review. *Saudi Pharm. J.* 24 (5): 554–559. https://doi.org/10.1016/j.jsps.2015.03 .022.

43 Reddy, N., Reddy, R., and Jiang, Q.R. (2015). Crosslinking biopolymers for biomedical applications. *Trends Biotechnol.* 33 (6): 362–369. https://doi.org/10.1016/j.tibtech .2015.03.008.

44 Berger, J., Reist, M., Mayer, J.M. et al. (2004). Structure and interactions in covalently and ionically crosslinked chitosan hydrogels for biomedical applications. *Eur. J. Pharm. Biopharm.* 57 (1): 19–34. https://doi.org/10.1016/s0939-6411(03)00161-9.

45 Yalinca, Z., Yilmaz, E., and Bullici, F.T. (2012). Evaluation of chitosan tripolyphosphate gel beads as bioadsorbents for iron in aqueous solution and in human blood in vitro. *J. Appl. Polym. Sci.* 125 (2): 1493–1505. https://doi.org/10.1002/app.34911.

46 Yilmaz, E., Yalinca, Z., Yahya, K., and Sirotina, U. (2016). pH responsive graft copolymers of chitosan. *Int. J. Biol. Macromol.* 90: 68–74. https://doi.org/10.1016/j .ijbiomac.2015.10.003.

47 Hamidi, M., Azadi, A., and Rafiei, P. (2008). Hydrogel nanoparticles in drug delivery. *Adv. Drug Delivery Rev.* 60 (15): 1638–1649. https://doi.org/10.1016/j.addr.2008.08 .002.

48 Lim, H.L., Hwang, Y., Kar, M., and Varghese, S. (2014). Smart hydrogels as functional biomimetic systems. *Biomater. Sci.* 2 (5): 603–618. https://doi.org/10.1039/ c3bm60288e.

49 Park, S. and Park, K.M. (2016). Engineered polymeric hydrogels for 3D tissue models. *Polymers* 8 (1): https://doi.org/10.3390/polym8010023.

50 Singhal, R. and Gupta, K. (2016). A review: tailor-made hydrogel structures (classifications and synthesis parameters). *Polym. – Plast. Technol. Eng.* 55 (1): 54–70. https://doi.org/10.1080/03602559.2015.1050520.

51 Kyburz, K.A. and Anseth, K.S. (2013). Three-dimensional hMSC motility within peptide-functionalized PEG-based hydrogels of varying adhesivity and crosslinking density. *Acta Biomater.* 9 (5): 6381–6392. https://doi.org/10.1016/j.actbio.2013.01 .026.

52 Okay, O. (2000). Macroporous copolymer networks. *Prog. Polym. Sci.* 25 (6): 711–779. https://doi.org/10.1016/s0079-6700(00)00015-0.

53 Omidian, H., Park, K., and Rocca, J.G. (2007). Recent developments in superporous hydrogels. *J. Pharm. Pharmacol.* 59 (3): 317–327. https://doi.org/10.1211/jpp.59.3 .0001.

54 Omidian, H., Rocca, J.G., and Park, K. (2005). Advances in superporous hydrogels. *J. Control. Release* 102 (1): 3–12. https://doi.org/10.1016/j.jconrel.2004.09.028.

55 Seliktar, D. (2012). Designing cell-compatible hydrogels for biomedical applications. *Science* 336 (6085): 1124–1128. https://doi.org/10.1126/science.1214804.

56 Yu, L. and Ding, J.D. (2008). Injectable hydrogels as unique biomedical materials. *Chem. Soc. Rev.* 37 (8): 1473–1481. https://doi.org/10.1039/b713009k.

57 Ganji, F., Vasheghani-Farahani, S., and Vasheghani-Farahani, E. (2010). Theoretical description of hydrogel swelling: a review. *Iran. Polym. J.* 19 (5): 375–398.

58 Garcia, S.J. (2014). Effect of polymer architecture on the intrinsic self-healing character of polymers. *Eur. Polym. J.* 53: 118–125. https://doi.org/10.1016/j.eurpolymj .2014.01.026.

59 Strandman, S. and Zhu, X.X. (2016). Self-healing supramolecular hydrogels based on reversible physical interactions. *Gels* 2: 1–31.

60 Okay, O. (2015). Self-healing hydrogels formed via hydrophobic interactions. In: *Supramolecular Polymer Networks and Gels*, vol. 268 (ed. S. Seiffert), 101–142.

61 Wu, D.Y., Meure, S., and Solomon, D. (2008). Self-healing polymeric materials: a review of recent developments. *Prog. Polym. Sci.* 33 (5): 479–522. https://doi.org/10 .1016/j.progpolymsci.2008.02.001.

62 Fernandez-Garcia, L., Mari-Buye, N., Barios, J.A. et al. (2016). Safety and tolerability of silk fibroin hydrogels implanted into the mouse brain. *Acta Biomater.* 45: 262–275. https://doi.org/10.1016/j.actbio.2016.09.003.

63 Hou, L.D., Li, Z., Pan, Y. et al. (2016). A review on biodegradable materials for cardiovascular stent application. *Front. Mater. Sci.* 10 (3): 238–259. https://doi.org/10 .1007/s11706-016-0344-x.

64 Singh, A. and Peppas, N.A. (2014). Hydrogels and scaffolds for immunomodulation. *Adv. Mater.* 26 (38): 6530–6541. https://doi.org/10.1002/adma.201402105.

65 Kabiri, K., Omidian, H., Hashemi, S.A., and Zohuriaan-Mehr, M.J. (2003). Synthesis of fast-swelling superabsorbent hydrogels: effect of crosslinker type and concentration on porosity and absorption rate. *Eur. Polym. J.* 39 (7): 1341–1348. https://doi .org/10.1016/s0014-3057(02)00391-9.

66 Kopecek, J. (2009). Hydrogels: from soft contact lenses and implants to self-assembled nanomaterials. *J. Polym. Sci. A – Polym. Chem.* 47 (22): 5929–5946. https://doi.org/10.1002/pola.23607.

67 Kopecek, J. and Yang, J.Y. (2007). Review – Hydrogels as smart biomaterials. *Polym. Int.* 56 (9): 1078–1098. https://doi.org/10.1002/pi.2253.

68 Moghimi, S.M., Hunter, A.C., and Murray, J.C. (2001). Long-circulating and target-specific nanoparticles: theory to practice. *Pharmacol. Rev.* 53 (2): 283–318.

69 Panigrahi, M., Das, P.K., and Parikh, P.M. (2011). Brain tumor and Gliadel wafer treatment. *Indian J. Cancer* 48 (1): 11–17. https://doi.org/10.4103/0019-509x.76623.

70 Aurand, E.R., Lampe, K.J., and Bjugstad, K.B. (2012). Defining and designing polymers and hydrogels for neural tissue engineering. *Neurosci. Res.* 72 (3): 199–213. https://doi.org/10.1016/j.neures.2011.12.005.

71 Salehi, R., Rasouli, S., and Hamishehkar, H. (2015). Smart thermo/pH responsive magnetic nanogels for the simultaneous delivery of doxorubicin and methotrexate. *Int. J. Pharm.* 487 (1–2): 274–284. https://doi.org/10.1016/j.ijpharm.2015.04.051.

72 Lee, P.I. and Kim, C.J. (1991). Probing the mechanisms of drug release from hydrogels. *J. Control. Release* 16 (1–2): 229–236. https://doi.org/10.1016/0168-3659(91)90046-g.

73 Savas, S., Azorsa, D.O., Jarjanazi, H. et al. (2011). NCI60 cancer cell line panel data and RNAi analysis help identify EAF2 as a modulator of simvastatin and lovastatin response in HCT-116 cells. *PLoS One* 6 (4): https://doi.org/10.1371/journal.pone .0018306.

74 Tuzmen, S., Tuzmen, P., Arora, S. et al. (2011). RNAi-based functional pharmacogenomics. *Methods Mol. Biol.* 700: 271–290. https://doi.org/10.1007/978-1-61737-954-3_18.

75 Tüzmen, S., Azorsa D.,Weaver D., Caplen N., Kallioniemi O., and Mousses S. (2004). Validation of siRNA knockdowns by real-time quantitative PCR. In: International qPCR Symposium and Application Workshop.

76 Reischl, D. and Zimmer, A. (2009). Drug delivery of siRNA therapeutics: potentials and limits of nanosystems. *Nanomed. – Nanotechnol. Biol. Med.* 5 (1): 8–20. https://doi.org/10.1016/j.nano.2008.06.001.

77 Morrissey, D.V., Lockridge, J.A., Shaw, L. et al. (2005). Potent and persistent in vivo anti-HBV activity of chemically modified siRNAs. *Nat. Biotechnol.* 23 (8): 1002–1007. https://doi.org/10.1038/nbt1122.

78 Lares, M.R., Rossi, J.J., and Ouellet, D.L. (2010). RNAi and small interfering RNAs in human disease therapeutic applications. *Trends Biotechnol.* 28 (11): 570–579. https://doi.org/10.1016/j.tibtech.2010.07.009.

79 Mao, S.R., Sun, W., and Kissel, T. (2010). Chitosan-based formulations for delivery of DNA and siRNA. *Adv. Drug Delivery Rev.* 62 (1): 12–27. https://doi.org/10.1016/j .addr.2009.08.004.

80 Hersel, U., Dahmen, C., and Kessler, H. (2003). RGD modified polymers: biomaterials for stimulated cell adhesion and beyond. *Biomaterials* 24 (24): 4385–4415. https://doi.org/10.1016/s0142-9612(03)00343-0.

81 Perale, G., Rossi, F., Sundstrom, E. et al. (2011). Hydrogels in spinal cord injury repair strategies. *ACS Chem. Neurosci.* 2 (7): 336–345. https://doi.org/10.1021/cn200030w.

82 Lieleg, O. and Ribbeck, K. (2011). Biological hydrogels as selective diffusion barriers. *Trends Cell Biol.* 21 (9): 543–551. https://doi.org/10.1016/j.tcb.2011.06.002.

83 Rosso, F., Giordano, A., Barbarisi, M., and Barbarisi, A. (2004). From cell-ECM interactions to tissue engineering. *J. Cell. Physiol.* 199 (2): 174–180. https://doi.org/10.1002/jcp.10471.

84 Lin, S., Sangaj, N., Razafiarison, T. et al. (2011). Influence of physical properties of biomaterials on cellular behavior. *Pharm. Res.* 28 (6): 1422–1430. https://doi.org/10.1007/s11095-011-0378-9.

85 Anderson, J.M. and Shive, M.S. (1997). Biodegradation and biocompatibility of PLA and PLGA microspheres. *Adv. Drug Delivery Rev.* 28 (1): 5–24. https://doi.org/10.1016/s0169-409x(97)00048-3.

86 Fournier, E., Passirani, C., Montero-Menei, C.N., and Benoit, J.P. (2003). Biocompatibility of implantable synthetic polymeric drug carriers: focus on brain biocompatibility. *Biomaterials* 24 (19): 3311–3331. https://doi.org/10.1016/s0142-9612(03)00161-3.

87 Wallace, G.G. and Moulton, S.E. (2009). Organic bionics: molecules, materials and medical devices. *Chem. Aust.* 76: 3–8.

88 Aregueta-Robles, U.A., Woolley, A.J., Poole-Warren, L.A. et al. (2014). Organic electrode coatings for next-generation neural interfaces. *Front. Neuroeng.* 7: 15. https://doi.org/10.3389/fneng.2014.00015.

89 Hassarati, R.T., Marcal, H., John, L. et al. (2016). Biofunctionalization of conductive hydrogel coatings to support olfactory ensheathing cells at implantable electrode interfaces. *J. Biomed. Mater. Res. B – Appl. Biomater.* 104 (4): 712–722. https://doi.org/10.1002/jbm.b.33497.

90 Lan, S., Veiseh, M., and Zhang, M. (2005). Surface modification of silicon and gold-patterned silicon surfaces for improved biocompatibility and cell patterning selectivity. *Biosens. Bioelectron.* 20 (9): 1697–1708. https://doi.org/10.1016/j.bios.2004.06.025.

91 Ito, A., Mase, A., Takizawa, Y. et al. (2003). Transglutaminase-mediated gelatin matrices incorporating cell adhesion factors as a biomaterial for tissue engineering. *J. Biosci. Bioeng.* 95 (2): 196–199. https://doi.org/10.1016/s1389-1723(03)80129-9.

92 Kim, S., Chung, E.H., Gilbert, M., and Healy, K.E. (2005). Synthetic MMP-13 degradable ECMs based on poly(N-isopropylacrylamide-co-acrylic acid) semi-interpenetrating polymer networks. I. Degradation and cell migration. *J. Biomed. Mater. Res. A* 75A (1): 73–88. https://doi.org/10.1002/jbm.a.30375.

93 Yang, H., Morris, J.J., and Lopina, S.T. (2004). Polyethylene glycol-polyamidoamine dendritic micelle as solubility enhancer and the effect of the length of polyethylene glycol arms on the solubility of pyrene in water. *J. Colloid Interface Sci.* 273 (1): 148–154. https://doi.org/10.1016/j.jcis.2003.12.023.

94 Patenaude, M., Smeets, N.M.B., and Hoare, T. (2014). Designing injectable, covalently cross-linked hydrogels for biomedical applications. *Macromol. Rapid Commun.* 35 (6): 598–617. https://doi.org/10.1002/marc.201300818.

5

Introduction to Green Polymeric Membranes

Mohamad Azuwa Mohamed, Nor Asikin Awang, Wan Norharyati Wan Salleh and Ahmad Fauzi Ismail

Advanced Membrane Technology Research Centre (AMTEC), School of Chemical and Energy Engineering, Faculty of Engineering, Universiti Teknologi Malaysia, 81310 UTM Johor Bahru, Malaysia

5.1 Introduction

"Green technology" is a technology intended to mitigate or reverse the effects of human activity on the environment. Basically, this technology encompasses endless development in terms of methods and materials, from techniques for generating energy to non-toxic cleaning products, creating a sustainable society for future generations. Green technology can be divided into five main subject areas: renewable energy (e.g. utilizing solar energy as an alternative energy source), green building, green purchasing, green chemistry, and green nanotechnology. Overall, this technology creates a center of economic activity around technologies and products that benefit the environment, speeding their implementation and creating new careers that truly protect the planet.

In recent years, the comprehensive applications of green polymeric membranes have attracted attention from researchers and industry in the effort to reduce and control the white pollution caused by the usage of synthetic polymers. White pollution has a detrimental effect on soil structure, water, nutrient transport, and crop growth, thereby disrupting the agricultural environment and reducing crop production [1]. The application of green polymers is currently more accepted since it is environmentally friendly, causing no harm to humans or the environment. As they are derived from natural sources, these biodegradable non-toxic polymers are now the focus among researchers who wish to explore and apply green polymeric membranes in various applications. Membrane technology sourced from green polymers or biopolymers as raw materials is now widely used in various applications. As well as cost savings, green-based membranes will not harm the environment and living things. Green polymeric membranes have been extensively commercialized in the field of membrane science and technology by using various membrane separation processes such as microfiltration, ultrafiltration, nanofiltration, reverse osmosis (RO), gas separation, pervaporation, and renewable energy [2–11].

This chapter examines commonly used green polymeric membranes, including cellulose and chitosan (CS) in the field of water and wastewater treatment. In addition, an explanation of membrane fabrication methods, structure, and configuration is also given. Moreover, the desired properties of the fabricated membrane are discussed

Bio Monomers for Green Polymeric Composite Materials, First Edition.
Edited by P.M. Visakh, Oguz Bayraktar and Gopalakrishnan Menon.
© 2020 John Wiley & Sons Ltd. Published 2020 by John Wiley & Sons Ltd.

further. An overview of the potential administration of green polymeric membranes in the field of water and wastewater treatment is provided, such as heavy metal removal, water purification, dye removal, and biomedical applications.

5.2 Types of Green Polymeric Membranes

There is a wide range of naturally occurring polymers produced from renewable resources that are available for numerous material applications. These renewable resources include cellulose and chitosan, which are widely employed in most manufacturing today. Green polymeric membranes are also referred to as biodegradable polymers. In this section, cellulose and chitosan polymeric membranes are discussed.

5.2.1 Cellulose Polymeric Membranes

Cellulose is the major natural biomacromolecule that is widely applied in packaging, biosorption, and biomedical. The structure of cellulose consists of β-D-glucopyranose joined by a β-1,4-glycosidic linkage. Cellulose is a semicrystalline polymer, which consists of crystalline and amorphous regions. Normally, the sources of cellulose that are derived from cellulose come from wood fibers, annual plants, microbes, seed fibers (cotton), and grasses (bagasse, bamboo) [12]. Furthermore, cellulose-based polymers (cellulose acetates [CA] and regenerated cellulose [RC]) are mostly used for commercial ultrafiltration membranes [13, 14]. The acetylation of cellulose leads to the preparation of cellulose acetate. It has been suggested that the acetylation of cellulose is successfully accomplished using acetic acid and acetic anhydride in the presence of a catalyst (sulfuric acid, superacids like SO_4^{2-}/ZrO_2, dialkylcarbodiimide, N,N-carbonyldiimidazole, iminium chlorides, and iodine) [15].

Zhang et al. prepared well-constructed cellulose acetate membranes for the forward osmosis application. In addition, cellulose acetate nanofiltration hollow fiber membranes have been formulated and tested in the forward osmosis process [16]. Interestingly, Mohamed et al. utilized recycled newspaper (RNP) as the cellulose source for regenerated cellulose membrane (RCM) fabrication. The RCM was successfully manufactured from RNP via an NaOH/urea aqueous system [1]. The production of RCM sourced from RNP proved that it is low in cost and environmentally friendly. Xia et al. also developed cellulose-based films that were directly prepared from waste newspaper [17]. Alternatively, these researchers fabricated the films via ionic liquid for cellulose dissolution. Moreover, Biganska and Navard also prepared RCM by using an N-methylmorpholine N-oxide–water solvent system [18]. A summary of the properties and the pure water flux of RCM is given in Table 5.1.

The surface morphology of RCM demonstrated by Mohamed et al. is illustrated in Figure 5.1. A scanning electron microscopy (SEM) analysis illustrated that the surface of the transparent cellulose membrane was smooth [1]. The cross-sectional image showed that the fabricated membrane consists of a homogeneous dense symmetric membrane structure composed of a skin layer. The skin layer was observed due to the increase in polymer concentration. As viewed by Li et al., cellulose configured in the hollow fiber was formed and applied for oil–water separation [25]. The structure and morphology of the cellulose hollow fiber was shown in Figure 5.2. The cross-sectional structure (a)

Table 5.1 Summary of the properties and pure water flux of regenerated cellulose membrane (RCM).

Cellulose sources	Solvent aqueous system	Coagulation bath	Structure properties	Mean pore size (nm)	Porosity (%)	Permeability (ml h^{-1} m^{-2} mmHg^{-1})	Ref.
RNP	7 wt% NaOH/12 wt% urea	5% H_2SO_4 at 25 °C	Homogeneous symmetric dense structure	2.48 ± 0.41	41.03 ± 2.37	0.47 ± 0.02	[1]
Cotton linter	7.5 wt% NaOH/11 wt% urea	5% H_2SO_4	Homogeneous mesh network structure	30.6	87	41.50	[19]
Cellulose carbamate	7 wt% NaOH/1.6 wt% ZnO	3 wt% H_2SO_4	Homogeneous mesh network structure	115–624	87–91	10.93–23.66	[20]
Cotton linter	6 wt% NaOH/4 wt% urea	$CaCl_2$ aqueous solution + 3 wt% HCl	Homogeneous mesh network structure	120–140	84–86	66.5–82.9	[21]
Softwood pulp	N-methylmorpholine-N-oxide (NMMO)/water mixture	H_2O	Dense asymmetric structure	13.4–20.2	24–41	28.17–47.67	[22]
Cellulose pulp	N-methylmorpholine-N-oxide (NMMO)/water mixture	H_2O	Finger-like structure	—	—	—	[18]
Bacteria cellulose	4 wt% NaOH/3 wt% urea	5 wt% CaCl + 1 wt% HCl	Dense porous structure	1.26	—	—	[23]
Cotton pulp	Ionic liquid (1-butyl-3-methylimidazolium chloride)	H_2O	Homogeneous nonporous surface	—	—	—	[24]

Source: Adapted from Ref. [1].

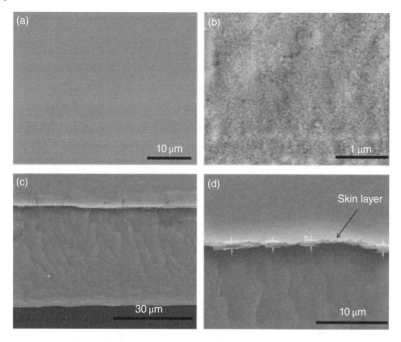

Figure 5.1 SEM images of the RCM: (a) and (b) surface; (c) and (d) cross-section [1].

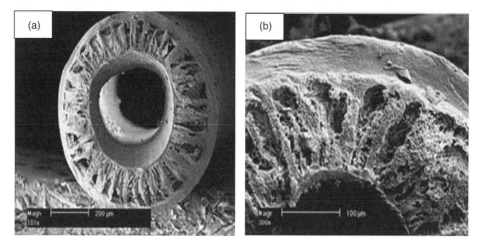

Figure 5.2 SEM images of a cellulose hollow fiber membrane [25].

was clearly observed in the SEM image and (b) possessed finger-like microvoids which appeared from the inner surface of the fiber.

5.2.2 Chitosan Polymeric Membranes

Chitosan is a renewable resource-based material procured from chitin. Chitin is the main building component of crustacean shells. It is a copolymer of *N*-acetyl-D-glucosamine and D-glucosamine. Chitosan sourced from chitin can be obtained via a deacetylation reaction [26]. The process to extract chitosan from chitin involves the removal of the acetate moiety from chitin. The extraction process is conducted using hydration, which involves amide hydrolysis. Under alkaline treatment, sodium hydroxide (NaOH) is used in the presence of chitin deacetylase. To improve chitosan polymeric membrane practicability and performance, the current research trend tends to incorporate nanomaterial and hybridization/composite by impregnating various types of polymers within the chitosan polymeric matrix. The latest advances in chitosan-based membranes are tabulated in Table 5.2.

Chitosan-based membranes can be prepared by an immersion–precipitation process which applies silica particles as porogen [34]. Chitosan powder is first dissolved in 2 wt% HNO_3 before being casted as a film. The resultant porous membrane of chitosan is successfully fabricated after being immersed in 5 wt% of NaOH solution at 80 °C for two hours. Chitosan-based ceramic membrane can also be prepared via the dip coating technique [35]. Bierhalz et al. formulated chitosan from white mushroom and shrimp shells. The fabricated membrane with a dense structure is illustrated in Figure 5.3. From the SEM analysis, the dense membrane displays a lamellar structure with an adequate cohesion layer. Meanwhile, the membrane, consisting of medium and high molecular mass chitosan, demonstrated a striated structure morphology [36]. As detailed by Bierhalz and colleagues, as the molecular mass of polysaccharides increases, the viscosity of the polymeric solution is enhanced, resulting in a high stability of foam

Table 5.2 Latest advancement in chitosan-based membranes in various applications.

Type of chitosan-based membrane	Preparation technique	Application	Ref.
Chitosan/poly(ethylene oxide) nanofibrous membrane	Electrospinning	Adsorption of divalent heavy metal ion	[27]
Blend polyvinyl alcohol-chitosan/polyvinylidene fluoride composite membrane	Blending and thin-layer coating	Pervaporation dehydration of binary and ternary mixtures of n-butyl acetate, n-butanol and water	[28]
Graphene oxide-modified chitosan/polyvinyl pyrrolidone nanocomposite membranes	Electrospinning	Wound-healing tissue-engineering applications	[29]
Cellulose–chitosan porous membrane	Blending	Forensic DNA analysis	[30]
Collagen/alginate–chitosan hydrogel membrane	Blending	Controlling cell–cell interactions	[31]
Grafted-chitosan membrane	Grafting	Periodontal regeneration surgery	[32]
Electrospunpolycaprolactone membrane coated with chitosan–silver nanoparticles	Electrospinning	Wound-healing tissue-engineering applications	[33]
Chitosan/organo-montmorillonite membrane	Blending	Direct methanol fuel cell	[11]

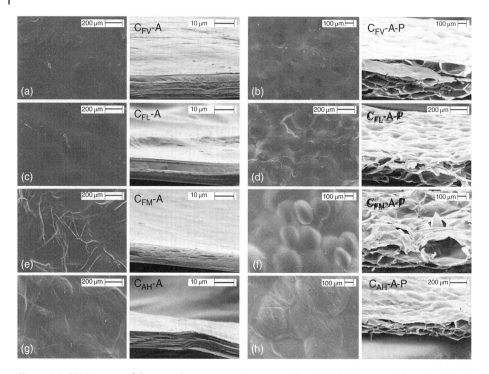

Figure 5.3 SEM images of dense and porous membranes produced with chitosan of fungal origin of very low (A and B), low (C and D) and medium (E and F) molecular mass, and with chitosan animal sources with high molecular mass (G and H) [36].

produced with the surfactant, leading to the production of pores. This explains why the membrane-based animal-derived chitosan, which possesses the highest molecular mass, exhibits the greatest porosity, whereas fungal chitosan-based membranes have high porosity due to polyelectrolyte complex (PEC) formation between polysaccharides and alginate.

5.3 Properties of Green Polymeric Membranes

The ability of cellulose and chitosan to form membranes is widely commercialized for various types of separation processes, including RO and nanofiltration. Furthermore, cellulose-based polymers (CA and RC) are mostly used for commercial ultrafiltration membranes [37]. This section discusses the special properties of green polymeric membranes in the field of membrane separation.

5.3.1 Film-forming Properties

Celluloses and chitosan can form cohesive films since they are flexible. They can be casted into different sizes of films. Moreover, there is a strong interaction between cellulose chain molecules due to intramolecular and intermolecular hydrogen bonding within the cellulose molecule. Intramolecular hydrogen bonding refers to the bonding between D-glucopyranose rings within the polymer chains, while intermolecular hydrogen bonding refers to bonding with another polymer chain. These two types of hydrogen bonds allow the formation of a very packed molecular arrangement that leads to the creation of three-dimensional crystals and web-like structures.

5.3.2 Mechanical Properties

The mechanical properties of membranes are very important in membrane separation, especially in membranes with a porous structure. For membranes with a porous structure, the mechanical stability is crucial to preserve the shape and size of the pores. Cellulose has remarkable mechanical strength. It has been reported that cellulose films displayed an average Young's modulus of 14 GPa [38], and the pure cellulose matrix film about 85.1 MPa in tensile strength and about 6223 MPa in tensile modulus [39]. Moreover, it has been reported that the tensile strength and elastic modulus of the nanocomposite film could be increased up to 124 MPa and 5 GPa, respectively, by varying the ratio of cellulose whiskers to regenerated cellulose matrix (cellulose II) [40]. However, in the case of chitosan-based membranes, application in water and gas separation processes was hindered due to its poor mechanical resistance. Thus, it is very important to utilize suitable support as a way to improve the mechanical resistance of chitosan flat sheet membranes [41].

5.3.3 Thermal Stability Properties

Thermal stability properties are vital in almost all applications. Photocatalytic membranes require the heat from light irradiation to start the photodegradation process. Cellulose can be considered to possess moderate thermal stability. It has been reported that rapid chemical decomposition of cellulose occurs between 315 and 400 °C [42], while decomposition of chitosan occurs between 175 and 400 °C [11], as shown in Figure 5.4. This indicates that the thermal stability of these green polymeric membranes

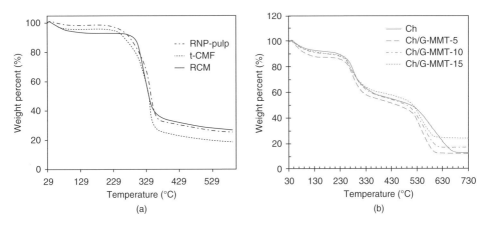

Figure 5.4 (a) Comparison of the thermal stability of RNP pulp, treated cellulose microfiber (t-CMF), and regenerated cellulose membrane [1]. (b) Comparison of the thermal stability pristine chitosan and its composite [11].

will preserve the integrity of the pore structures in the nanometer dimension. However, it is important to identify the temperature process before membrane selection is made.

5.3.4 Chemical Stability

The chemical stability of a membrane includes the resistance of the membrane at extreme pH values and other chemical conditions such as strong acids, strong bases, or oxidation agents. The chemical stability is crucially important since it enables the reuse of the membrane, which saves cost. Basically, natural cellulose is chemically stable since it is an insoluble, crystalline microfibril that is difficult to react with other compounds. Most reactions with cellulose are heterogeneous. A study on the effect of UV/O_3 on the surface properties of nanofibrillated cellulose (NFC) dry film was carried out in [43]. The effect was observed to be long lasting, and the treatment did not degrade the cellulose film, with no significant chemical changes observed in the Fourier transform infrared spectroscopy (FTIR) spectra (Figure 5.5). It was suggested that the amphiphilic character of the cellulose molecule has the ability to rearrange the outermost surface molecules, hence minimizing its surface free energy. This behavior may be one reason for the overall stability of cellulose materials [44]. This method, based on the oxidative properties of UV radiation and the ozone, can be considered as one of the most effective methods to remove contaminants or foulants from the nanocellulosic surface.

5.3.5 Hydrophilicity–Hydrophobicity Balance Properties

The hydrophilicity–hydrophobicity balance is correlated with the wettability of the membrane. Cellulose is naturally hydrophilic and will wet out with water. If we know the surface energy of the membrane, we can distinguish between hydrophobic and hydrophilic membranes. If the surface energy is more than 70 dynes cm^{-1}, a hydrophilic polymer below 70 dynes cm^{-1} will be hydrophobic. Modification of the surface chemistry of the cellulose also allows us to tailor the surface groups to specific applications. For example, the surface silylation of cellulose will increase the hydrophobicity. It has been reported that cellulose films with high water contact angle (117–146°) can be prepared from modified cellulose by solution casting [45]. Furthermore, a previous study showed that the contact angle of chitosan-based membrane (27–85°) can be modified by tuning the ratio of chitosan and poly(ethylene oxide) concentration during

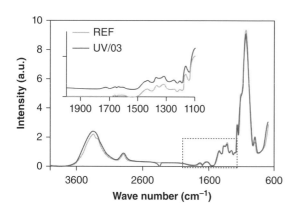

Figure 5.5 The ATR IR spectrum of the cellulose film illustrates that the sample is not chemically modified due to the ozonation [43].

the fabrication process [27]. These results indicate that the lower contact angle will lead to a higher heavy metal ion adsorption capability. Therefore, an optimized membrane has a more hydrophilic and smoother surface than a commercial forward osmosis membrane, which results in lower fouling. Thus, potential applicability was suggested in treating wastewater. In the photocatalytic membrane application hydrophilic properties are important since they allow the aqueous pollutant solutions to adsorb the surface membrane containing the catalyst. Moreover, they will increase the photodegradation of the pollutants on the catalyst surface [10].

5.4 Applications of Green Polymeric Membranes

There are various applications that have been established regarding the uses of green polymeric membranes. These applications utilize various types of biopolymers regenerated from different sources. Some of these biopolymers are employed in the field, which involves heavy metal removal, water purification, dye removal, etc., as tabulated in Table 5.3.

5.4.1 Heavy Metal Removal

The contamination of drinking water due to heavy metal is a serious issue among researchers. Lead, for example, is one of the harmful heavy metals able to damage a human's fetal brain, kidneys, circulatory system, and nervous system [52]. Due to the harmful effects of heavy metals, the US Environmental Protection Agency set the maximum contaminant level (MCL) of $15\,\mu g$-Pb l^{-1} for drinking water [52]. Normally, the technologies, such as coagulation and ion exchange, that are applied in heavy metal removal work well, but they have several disadvantages, such as low efficiency and high cost for energy and materials. Thus, the membrane technology that can overcome these drawbacks will have enhanced performance. The adsorption capacities for different types of modified biopolymers for heavy metal removal are shown in Table 5.4.

As demonstrated by a previous study, the removal of heavy metal ions (Cu^{2+}, Zn^{2+}, Cd^{2+}, and Pb^{2+}) is successfully accomplished via the development of cellulosic biopolymer [46] developed from okra fiber. Okra fiber is an agricultural waste biomass that contains cellulose, hemicelluloses, and lignin. Chemical modification was performed to improve the adsorption capacity for heavy metal removal. Cellulosic okra fiber was grafted with acrylonitrile and methyacrylic acids. The grafting process was done in a way that increased the graft yield. The performance analysis revealed that the grafted cellulosic okra fiber has a promising adsorption ability to remove heavy metal from wastewater. The adsorption isotherm was studied by using Langmuir and the Dubinin–Radushkevich model. The development of the biopolymer-based nanosystem from biodegradable poly-γ-glutamic acid (γ-PGA) was successfully fabricated and used for the removal of ferric ion from water [54]. The amount of γ-PGA, Fe^{3+} concentrations, their proportions, and the extent of crosslinking of γ-PGA are the studied parameters. The findings proved that γ-PGA biopolymer can bind with ferric ions, creating stable particulate complexes, thus successfully producing water with a low concentration of Fe^{3+}. The mechanism for how the membrane removes the heavy metal is illustrated in Figure 5.6. The membrane can work in two ways: crossflow

Table 5.3 Applications of green polymeric membranes.

Applications	Sources of green polymeric	Configuration	Performance	Ref.
Heavy metal removal	Cellulosic okra	Grafted okra fibers	Alternative adsorbent for removal of Cu^{2+}, Zn^{2+}, Cd^{2+}, Pb^{2+} with maximum monolayer capacity q_m of 1.209, 0.9623, 1.2609, and 1.295 $mmol\,g^{-1}$ respectively	[46]
	Chitosan	Composite nanofibers chitosan/ MWCNT/Fe_3O_4	Removal of Cr(VI) with maximum adsorption 360.1 $mg\,g^{-1}$	[26]
Water purification	Cellulose whiskers from banana fibers	Hollow fiber membranes	Able to separate pure water from mixtures of CCl_4, $CHCl_3$, and CH_2Cl_2	[26]
Dye removal	Chitosan (extracted from snow crab shell)	Powder	Removal of Basic Blue 41 (BB41) and Basic Red (BR18) with maximum adsorption capacities of 217.39 and 158.73 $mg\,g^{-1}$, respectively	[47]
	Tamarind hull	Powder	Removal of Basic Violet 16 and Basic Red 18 with maximum adsorption capacities of 45.454 and 66.667 $mg\,g^{-1}$, respectively	[48]
	Chitosan	Flat sheet membrane	Removal of Direct Blue 71 and Direct Red 31 with maximum adsorption capacities of 77.18 and 61.77 $mg\,g^{-1}$, respectively	[49]
Biomedical	Cellulose cuprammonium solution	Regenerated hollow fiber	DNA molecule with MW of 1×10^8 completely passed through the virus removal filter	[50]
	α-chitin/nanosilver composite	Composite scaffolds	Excellent performance for antibacterial activities toward *S. aureus*, *E. coli* and also possesses great blood clotting ability	[51]

mode or dead-end mode. Figure 5.4a shows dead-end mode filtration, in which the aqueous solution passes through the membrane in a perpendicular direction, while in Figure 5.6b the aqueous solution of heavy metal moves parallel to the direction of the membrane.

Besides cellulose, chitosan is another significant biopolymer that can be obtained from chitin by deacetylation [26]. Chitosan is biocompatible, non-toxic, and biodegradable

Table 5.4 Adsorption capacities of modified biopolymers for heavy metals [53].

Adsorbent	Adsorption capacity (mg g^{-1})					
	Pb^{2+}	Cd^{2+}	Zn^{2+}	Cu^{2+}	Cr^{6+}	As^{6+}
Crosslinked chitosan		150		164		230
Crosslinked starch gel	433			135		
Alumina/chitosan composite				200		

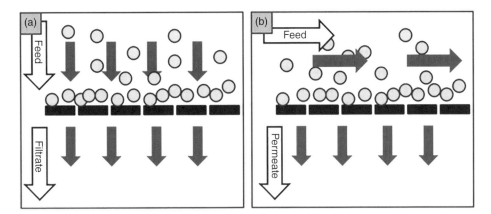

Figure 5.6 (a) Dead-end filtration and (b) crossflow filtration.

in nature [35]. It differs from cellulose since cellulose will dissolve in most organic solvents, whereas chitosan is soluble in water in which a small amount of acetic acid is present [26]. Thakur and Voicu reported the development of the composite nanofiber chitosan/multi walled carbon nanotubes/Fe$_3$O$_4$ for the removal of Cr (VI) [26]. By studying the different sorption parameters (temperature, initial concentration, and contact time), the maximum retention of Cr (VI) was noted at an equilibrium time of 30 minutes, pH 2, temperature 45 °C, and had a value of 360.1 mg g^{-1}. Another application of heavy metal removal with the use of chitosan was demonstrated by Li et al. Polyacrylamide–chitosan acts to remove heavy metal ions from aqueous solutions and is also used as a coating for magnetic nanoparticles. The separation performances were 43.35 mg g^{-1} for Cu^{2+}, 63.67 mg g^{-1} for Pb^{2+}, and 263.9 mg g^{-1} for Hg^{2+} [55].

5.4.2 Water Purification

Effluent consisting of pharmaceutical compounds such as antibiotics, vasodilators, β-blockers, organic pollutants (e.g. phenolic compounds), and anti-epileptics has been found in most wastewater, sewage, groundwater, and drinking water [56]. The trace level of pharmaceutical compounds in drinking water has become a critical issue since it contributes to negative human health effects, including endocrine disruption and genotoxicity. Researchers have been attempting to develop membranes to separate pharmaceuticals from water from many years, but these compounds are still present in the environment and drinking water at high levels in both developed and non-developed countries.

This situation shows that there is poor management and lack of efficiency in wastewater treatment plants (WWTPs) [57]. As WWTPs are not good at removing pharmaceutical compounds from water, researchers are trying to overcome this problem by making some modifications in terms of membrane selectivity for toxins and drugs. Thus, new technologies and materials are being researched for water purification. Various kinds of bio-based polymers are used nowadays since they are environmentally friendly and low in cost. Mohamed et al. have successfully prepared RCM with photocatalytic properties in an effort to produce green portable photocatalysts and photocatalytic membranes from the degradation of organic pollutants [10, 58].

Thakur and Voicu reviewed water purification involving the separation of pure water from various mixtures of tetrachloride (CCl_4), chloroform ($CHCl_3$), and dichloromethane (CH_2Cl_2) [26]. The separation process used a highly compact poly(ethylene-co-vinyl acetate)/cellulose membrane in which the cellulose membrane was fabricated by using cellulose whiskers from banana fibers. Following the same principle, Kim et al. demonstrated the application of polyether sulfone (PES) membrane coated with a thick layer of cellulose acetate (CA)/poly(ethylene glycol), which was applied in the removal of water vapor from H_2O/N_2 mixtures [59]. The findings proved that the water vapor's permeation through the membrane increased because of the polymer chain's mobility, which resulted from the increased total free volume. The composite membrane was applied to the separation of an ethanol–water mixture. The preparation of the membrane was done via impregnation of bacterial cellulose (BC) hydrogel with sodium alginate before crosslinking with calcium chloride solution. The total permeate flux of water through the membrane was measured at $33 \, g \, m^{-2} \, h^{-1}$. Table 5.5 compares the separation abilities of several polymeric membranes.

Along with cellulose, chitosan-based composite membranes for water purification have been investigated. Prasad et al. successfully synthesized solvent-resistant chitosan/poly(ether-block-amide) composite membranes that were used for pervaporation of n-methyl-2-pyrrolidone/water mixtures [26]. The fabricated composite membrane was crosslinked with tetraethyl ortisilicate (TEOS) to minimize swelling and, at the same time, improve the selectivity of the membrane. The water flux for non-modified chitosan was $0.024 \, kg \, m^{-2} \, h^{-1}$, compared with modified chitosan, which had a water flux of $0.019 \, kg \, m^{-2} \, h^{-1}$, while the selectivities for the membranes were 182 and 225, respectively. Chitosan was also applied in the removal of salts from water

Table 5.5 Comparison of separation performance with other polymeric membranes [59].

Polymeric membrane	Temperature (°C)	Relative humidity (%)	Water vapor permeance (GPU)	Selectivity
PAN nanofiber-1	50	50	13 498	27 [H_2O/O_2]
PAN nanofiber-2	50	50	3 143	387[H_2O/O_2]
PSf/NaAlg	25	0.13 wt% in feed gas	1 130	54[H_2O/O_2]
PEI/PEBAX1657	21	47	260	274[H_2O/O_2]
PES/CA + PEG2000(5 wt%)	30	20	444	176[H_2O/O_2]

PAN, polyacrylonitrile.

[26]. In this application, chitosan acts as a coating layer for poly(1,4-phenylene ether sulfone) (PPEES) membrane.

5.4.3 Dye Removal

Generally, dyes can be categorized based on their chemical structure (e.g. azo, anthraquinone, indigo, triphenylmethane, etc.), which depends on the method and domain of usage [60]. Effluent which consists of dyes that come from industrial processes will pollute the environment and harm aquatic life. This is because the dye itself contain carcinogens, teratogens, and mutagens [61]. As mentioned by Vakili et al., the presence of the dye, even in low concentrations, will lower the photosynthetic activity of aquatic life because it will prevent penetration of sunlight and oxygen [61]. Waste that contains colored substances normally comes from the textile industries and other dyeing fields, such as paper, printing, food, and plastics. The approximate volume of discharged wastewater, especially from textile processes, is between 40 and $65 \, kg^{-1}$ of the textile product [61]. To improve the performance of chitosan in various applications, additives are added to change the membrane surface area. The additives enhance the electrostatic and/or chemical bonding for the adsorption of dyes. As represented in Figure 5.7, electrostatic interaction and hydrogen bonding for adsorption of direct Blue 71 have been proposed [62].

The application of chitosan to dye removal has two main advantages: it is cheap compared to other adsorbents and it has powerful chelating potential against pollutants [60]. The efficiency of chitosan as a green adsorbent was proven when it shown to absorb almost all classes of dyes except basic dyes, which can be related to its natural cationic properties. Sadeghi-Kiakhani et al. concluded that chitosan–ethyl acrylate (Ch-g-Ea) could be developed as a biopolymer adsorbent and applied for the removal of Basic Blue

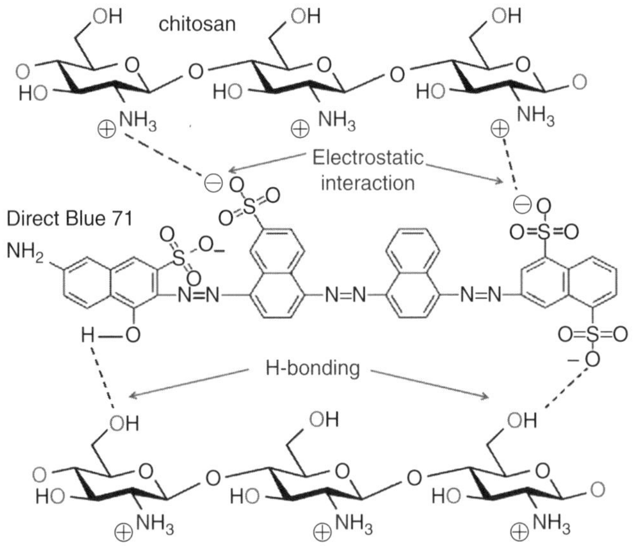

Figure 5.7 Electrostatic interaction and hydrogen bonding (H-bonding) of a direct dye with chitosan chains [62].

Table 5.6 Comparison of BB41 and BR18 removal between Ch-g-Ea and others in the literature [47].

Dye	Adsorbents	q_0 (mg g^{-1})
Basic Blue 9	Banana peel (raw)	20.8
Basic Blue 9	Yellow passion fruit	44.7
Basic Blue 9	Hazel nut shell	76.9
Basic Blue 9	Sawdust-walnut	59.17
Basic Red 18	Tamarind hull	66.66
Basic Red 9	Polyethyelene terephthalate (PET) carbond	33.4
BB41	Ch-g-Ea	217.39
BR18	Ch-g-Ea	158.73

41 (BB41) and Basic Red 18 (BR18) from colored solution [47]. Their findings indicated that this adsorbent is capable of removing dyes since the maximum adsorption capacities are 217.39 and 158.73 mg g^{-1} for BB41 and BR18, respectively. The dye removal ability was attributed to the large number of carboxyl groups attached to the backbone of the chitosan. The comparison of the removal of BB41 and BR18 is displayed in Table 5.6.

The modification of chitosan was accomplished to improve its capability and potential for the removal of dyes. As reported by He et al., the elimination of dyes (Blue 71 and Red 31) from aqueous solution was achieved using a chitosan/oxidized starch/silica (CS/OSR/silica) hybrid membrane [49]. This hybrid membrane was fabricated using OSR and 3-aminopropyltriethoxysilane (APTES) as crosslinking agents. The hybrid CS/OSR/silica membrane produced showed an improvement in terms of thermal stability, as well as the ability to lower the degree of swelling in the water. The adsorption capacity of CS/OSR/silica hybrid membrane was found to be optimal at pH 9.82 and 60 °C. The adsorption capacity of pure CS/OSR membrane was lower (75.53 mg g^{-1} for Blue 71, 60.19 mg g^{-1} for Red 31) than that of CS/OSR/silica hybrid membrane (77.18 mg g^{-1} for Blue 71, 61.77 mg g^{-1} for Red 31). The capability and success of the hybrid membrane to remove these dyes can be seen from the analysis of the morphology of the membrane via SEM micrographs. The morphology demonstrates that there is an obvious difference in terms of surface roughness between pure CS/OSR membrane and the CS/OSR/silica hybrid membrane. As displayed in Figure 5.8, the surface roughness of the CS/OSR/silica hybrid membrane is more obvious compared with the unmodified CS, CS/OSR membrane. When the surface roughness is pronounced, the surface will be able to capture dye particles better since surface roughness increases the surface area.

The applications of biopolymers in dye removal not only focus on chitosan, but also involved the use of cellulose. As we know, cellulose can be abundantly modified and is a renewable biopolymer in nature. The structure of cellulose with a hydroxyl group allows the cellulose to undergo chemical modification that can improve its efficiency for dye removal from water. As formulated by De Castro Silva et al., the biopolymer cellulose was incorporated with chloroethyl phosphate for the removal of green dye via an adsorption mechanism [63]. The results proved that cellulose–phosphate had higher

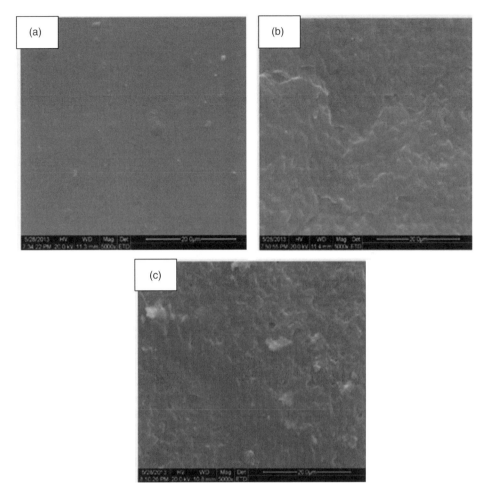

Figure 5.8 SEM micrographs of different membranes: (a) pure CS, (b) CS/OSR, and (c) CS/OSR/silica hybrid [49].

adsorption performance compared to pristine cellulose adsorbent. The removal levels of brilliant green dye by cellulose phosphate at 25, 35, and 45 °C were 46.7, 58.42, and 90.5 mg g^{-1}, while for cellulose modified with chloroethyl phosphate, at the same temperatures, removal levels were 113.6, 114.2, and 112.1 mg g^{-1}.

5.4.4 Biomedical Applications

Biopolymers have also been applied in the biomedical field. As studied by Hirasaki et al., the permeation mechanism of DNA molecules was achieved via the porous structure of regenerated cellulose hollow fiber (BMM™) [50]. The application of BMM was proposed to prevent virus infection after the injection of blood plasma. The filtration of virus using BMM resulted in infection by the virus being minimized. Analysis by SEM showed that DNA molecules become elongated like a filament inside the hollow fiber, which was oriented in the flow direction. It was found that DNA molecules with

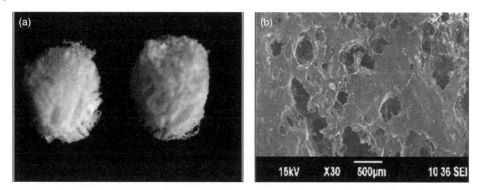

Figure 5.9 (a) Chitosan scaffold (left) and chitin/nanosliver composite scaffold (right). (b) SEM image of chitin/nanosilver composite scaffold [51].

a molecular weight of 1×10^8 easily passed through the BMM filter. Tsurumi et al. reported very similar research on the use of cuprammonium regenerated cellulose hollow fiber (BMM hollow fiber) in virus removal. However, Tsurumi et al. highlighted the membrane structure of the BMM hollow fiber and reported the basic concept for improving virus removal performance [64]. The effect of the surface roughness on the cellulose hollow fiber performance for dialysis was also studied. This research concluded that a membrane with a rough surface had a high platelet adhesion ratio and a lack of hemocompatibility, while a membrane with a smooth surface showed lower platelet adhesion ratios and revealed better hemocompatibility. A novel α-chitin/nanosilver composite scaffold (Figure 5.9) was developed and used for wound-healing applications. On analysis, it was found that these nanocomposites have excellent antibacterial performance toward *Staphylococcus aureus* and *Escherichia coli*. They also possess good blood-clotting abilities [51]. In addition, graphene oxide-modified chitosan/polyvinyl pyrrolidone nanocomposite membranes [29] and polycaprolactone membrane coated with chitosan-silver nanoparticles [33] prepared by electrospinning demonstrated remarkable potential in wound-healing tissue-engineering applications.

5.4.5 Renewable Energy

The application of green polymeric membranes in renewable energy, especially in direct methanol fuel cells (DMFCs) has also been explored. The DMFC is a type of fuel cell that uses methanol as fuel to generate electricity. In 2015, Jiang et al. successfully prepared a novel nanocomposite membrane intended for both proton-exchange membrane fuel cells (PEMFCs) and DMFCs [65]. This membrane was based on blending BC polymers and Nafion. Their results indicated that the increase in water uptake, Young's modulus, thermal, and area swelling stabilities was due to the characteristics of BC. In addition, the BC/Nafion blended membranes exhibited maximum power densities of 106 and $20.4 \, \text{mW cm}^{-2}$ for PEMFC and DMFC, respectively. In 2016, Purwanto et al. successfully fabricated polyelectrolyte membranes from chitosan incorporated with montmorillonite (MMT)-crosslinked 3-glicidoxy propyltrimethoxysilane (G) for DMFCs [11]. The chitosan/G-MMT with 5 wt% G-MMT loading exhibited the best methanol permeability of $3.03 \times 10^{-7} \, \text{cm}^2 \, \text{s}^{-1}$ and proton conductivity of $4.66 \, \text{mS cm}^{-1}$ (Figure 5.10).

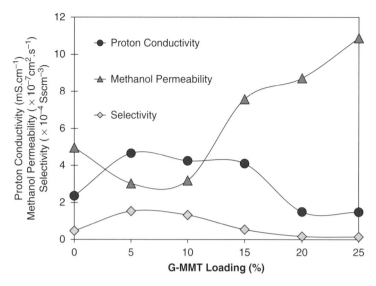

Figure 5.10 Proton conductivity, methanol permeability, and relative selectivity of pure chitosan and chitosan/G-MMT composite membranes with different G-MMT loadings (%).

5.5 Conclusion

The use of green polymeric materials is not a new concept. Their production has been feasible on a large scale for more than a decade. Green polymeric membranes can be used as an alternative approach to produce environmentally friendly membranes and to promote sustainability in the field of membrane technology. In this chapter, types of green polymeric membranes, their physicochemical properties, and their potential applications are described. Their applicability in various industries is induced by their significant physicochemical properties. However, despite these advancements, there are still some drawbacks which prevent the wider commercialization of green polymeric membranes in many applications. The main drawback is that they still cannot compete with conventional synthetic polymeric membranes in terms of performance, therefore more research is needed to improve their feasibility and commercialization in various applications that require superior performance. Current research involving green polymeric membranes is focused on fabrication methods, processing, and surface and structure modifications. In addition, the long-term stability and durability of green polymeric membranes for specific applications has also been a challenge for researchers all around the world. The introduction of nanostructure fillers (e.g. graphene oxide, metal oxides, carbon nanotubes, nano-clay, etc.) and blending with other polymers, or making new copolymers, has significantly improved their overall properties and performance. These improvements are generally found at low filler content, and this nano-reinforcement is a very attractive route to generate new functional green polymeric membranes for various applications. It should be noted that the development of green polymeric membranes with physicochemical properties for specific functionality is crucial for practicability in industry. Green polymeric membranes with various physicochemical properties have a promising contribution to make in various applications.

References

1 Mohamed, M.A., Salleh, W.N.W., Jaafar, J. et al. (2015). Feasibility of recycled newspaper as cellulose source for regenerated cellulose membrane fabrication. *J. Appl. Polym. Sci.* 132: 1–10. https://doi.org/10.1002/app.42684.

2 Ramesh Babu, P. and Gaikar, V.G. (2001). Membrane characteristics as determinant in fouling of UF membranes. *Sep. Purif. Technol.* 24: 23–34. https://doi.org/10.1016/S1383-5866(00)00207-0.

3 Dogan, H. and Hilmioglu, N.D. (2010). Zeolite-filled regenerated cellulose membranes for pervaporative dehydration of glycerol. *Vacuum* 84: 1123–1132. https://doi.org/10.1016/j.vacuum.2010.01.043.

4 Xiong, X., Duan, J., Zou, W. et al. (2010). A pH-sensitive regenerated cellulose membrane. *J. Membr. Sci.* 363: 96–102. https://doi.org/10.1016/j.memsci.2010.07.031.

5 Yang, Q., Fukuzumi, H., Saito, T. et al. (2011). Transparent cellulose films with high gas barrier properties fabricated from aqueous alkali/urea solutions. *Biomacromolecules* 12: 2766–2771. https://doi.org/10.1021/bm200766v.

6 Singh, N., Chen, Z., Tomer, N. et al. (2008). Modification of regenerated cellulose ultrafiltration membranes by surface-initiated atom transfer radical polymerization. *J. Membr. Sci.* 311: 225–234. https://doi.org/10.1016/j.memsci.2007.12.036.

7 Fukuzumi, H., Saito, T., Iwata, T. et al. (2009). Transparent and high gas barrier films of cellulose nanofibers prepared by TEMPO-mediated oxidation. *Biomacromolecules* 10: 162–165. https://doi.org/10.1021/bm801065u.

8 Ma, H., Burger, C., Hsiao, B.S., and Chu, B. (2012). Nanofibrous microfiltration membrane based on cellulose nanowhiskers. *Biomacromolecules* 13: 180–186. https://doi.org/10.1021/bm201421g.

9 Zhu, T., Lin, Y., Luo, Y. et al. (2012). Preparation and characterization of TiO_2-regenerated cellulose inorganic–polymer hybrid membranes for dehydration of caprolactam. *Carbohydr. Polym.* 87: 901–909. https://doi.org/10.1016/j.carbpol.2011.08.088.

10 Mohamed, M.A., Salleh, W.N.W., Jaafar, J. et al. (2016). Physicochemical characteristic of regenerated cellulose/N-doped TiO_2 nanocomposite membrane fabricated from recycled newspaper with photocatalytic activity under UV and visible light irradiation. *Chem. Eng. J.* 284: 202–215. https://doi.org/10.1016/j.cej.2015.08.128.

11 Purwanto, M., Atmaja, L., Mohamed, M.A. et al. (2016). Biopolymer-based electrolyte membranes from chitosan incorporated with montmorillonite-crosslinked GPTMS for direct methanol fuel cells. *RSC Adv.* 6: 2314–2322. https://doi.org/10.1039/C5RA22420A.

12 Hokkanen, S., Bhatnagar, A., and Sillanpää, M. (2016). A review on modification methods to cellulose-based adsorbents to improve adsorption capacity. *Water Res.* 91: 156–173. https://doi.org/10.1016/j.watres.2016.01.008.

13 Murphy, D. and de Pinho, M.N. (1995). An ATR-FTIR study of water in cellulose acetate membranes prepared by phase inversion. *J. Membr. Sci.* 106: 245–257. https://doi.org/10.1016/0376-7388(95)00089-U.

14 Zhang, S., Wang, K.Y., Chung, T.-S. et al. (2010). Well-constructed cellulose acetate membranes for forward osmosis: minimized internal concentration polarization with an ultra-thin selective layer. *J. Membr. Sci.* 360: 522–535. https://doi.org/10.1016/j.memsci.2010.05.056.

15 Das, A.M., Ali, A.A., and Hazarika, M.P. (2014). Synthesis and characterization of cellulose acetate from rice husk: eco-friendly condition. *Carbohydr. Polym.* 112: 342–349. https://doi.org/10.1016/j.carbpol.2014.06.006.

16 Su, J., Yang, Q., Teo, J.F., and Chung, T.-S. (2010). Cellulose acetate nanofiltration hollow fiber membranes for forward osmosis processes. *J. Membr. Sci.* 355: 36–44. https://doi.org/10.1016/j.memsci.2010.03.003.

17 Xia, G., Wan, J., Zhang, J. et al. (2016). Cellulose-based films prepared directly from waste newspapers via an ionic liquid. *Carbohydr. Polym.* 151: 223–229. https://doi.org/10.1016/j.carbpol.2016.05.080.

18 Biganska, O. and Navard, P. (2009). Morphology of cellulose objects regenerated from cellulose-N-methylmorpholine N-oxide-water solutions. *Cellulose* 16: 179–188. https://doi.org/10.1007/s10570-008-9256-y.

19 Mao, Y., Zhou, J., Cai, J., and Zhang, L. (2006). Effects of coagulants on porous structure of membranes prepared from cellulose in NaOH/urea aqueous solution. *J. Membr. Sci.* 279: 246–255. https://doi.org/10.1016/j.memsci.2005.07.048.

20 Fu, F., Guo, Y., Wang, Y. et al. (2014). Structure and properties of the regenerated cellulose membranes prepared from cellulose carbamate in NaOH/ZnO aqueous solution. *Cellulose* 21: 2819–2830. https://doi.org/10.1007/s10570-014-0297-0.

21 Zhou, J., Zhang, L., Cai, J., and Shu, H. (2002). Cellulose microporous membranes prepared from NaOH/urea aqueous solution. *J. Membr. Sci.* 210: 77–90. https://doi.org/10.1016/S0376-7388(02)00377-0.

22 Ichwan, M. and Son, T. (2011). Preparation and characterization of dense cellulose film for membrane application. *J. Appl. Polym. Sci.* 124: 1409–1418. https://doi.org/10.1002/app.

23 Phisalaphong, M., Suwanmajo, T., and Sangtherapitikul, P. (2008). Novel nanoporous membranes from regenerated bacterial cellulose. *J. Appl. Polym. Sci.* 107: 292–299. https://doi.org/10.1002/app.27118.

24 Liu, Z., Wang, H., Li, Z. et al. (2011). Characterization of the regenerated cellulose films in ionic liquids and rheological properties of the solutions. *Mater. Chem. Phys.* 128: 220–227. https://doi.org/10.1016/j.matchemphys.2011.02.062.

25 Li, H., Cao, Y., Qin, J., and Jie, X. (2006). Development and characterization of anti-fouling cellulose hollow fiber UF membranes for oil – water separation. *J. Membr. Sci.* 279: 328–335. https://doi.org/10.1016/j.memsci.2005.12.025.

26 Thakur, V.K. and Voicu, S.I. (2016). Recent advances in cellulose and chitosan based membranes for water purification: a concise review. *Carbohydr. Polym.* 146: 148–165. https://doi.org/10.1016/j.carbpol.2016.03.030.

27 Shariful, M.I., Bin Sharif, S., Lee, J.J.L. et al. (2017). Adsorption of divalent heavy metal ion by mesoporous-high surface area chitosan/poly (ethylene oxide) nanofibrous membrane. *Carbohydr. Polym.* 157: 57–64. https://doi.org/10.1016/j.carbpol.2016.09.063.

28 Zhang, S., Zou, Y., Wei, T. et al. (2017). Pervaporation dehydration of binary and ternary mixtures of n-butyl acetate, n-butanol and water using PVA-CS blended membranes. *Sep. Purif. Technol.* 173: 314–322. https://doi.org/10.1016/j.seppur.2016.09.047.

29 Mahmoudi, N. and Simchi, A. (2017). On the biological performance of graphene oxide-modified chitosan/polyvinyl pyrrolidone nanocomposite membranes: in vitro

and in vivo effects of graphene oxide. *Mater. Sci. Eng. C* 70: 121–131. https://doi
.org/10.1016/j.msec.2016.08.063.

30 Chumwangwapee, S., Chingsungnoen, A., and Siri, S. (2016). A plasma modified
cellulose-chitosan porous membrane allows efficient DNA binding and provides
antibacterial properties: a step towards developing a new DNA collecting card.
Forensic Sci. Int. Genet. 25: 19–25. https://doi.org/10.1016/j.fsigen.2016.07.020.

31 Song, Y., Zhang, D., Lv, Y. et al. (2016). Microfabrication of a tunable
collagen/alginate-chitosan hydrogel membrane for controlling cell–cell interactions. *Carbohydr. Polym.* 153: 652–662. https://doi.org/10.1016/j.carbpol.2016.07
.058.

32 Lee, B.-S., Lee, C.-C., Lin, H.-P. et al. (2016). A functional chitosan membrane with
grafted epigallocatechin-3-gallate and lovastatin enhances periodontal tissue regeneration in dogs. *Carbohydr. Polym.* 151: 790–802. https://doi.org/10.1016/j.carbpol
.2016.06.026.

33 Nhi, T.T., Khon, H.C., Hoai, N.T.T. et al. (2016). Fabrication of electrospun polycaprolactone coated with chitosan-silver nanoparticles membranes for wound
dressing applications. *J. Mater. Sci. Mater. Med.* 27: 156. https://doi.org/10.1007/
s10856-016-5768-4.

34 Zhang, X., Yu, H., Yang, H. et al. (2015). Graphene oxide caged in cellulose
microbeads for removal of malachite green dye from aqueous solution. *J. Colloid
Interface Sci.* 437: 277–282. https://doi.org/10.1016/j.jcis.2014.09.048.

35 Jana, S., Saikia, A., Purkait, M.K., and Mohanty, K. (2011). Chitosan based ceramic
ultrafiltration membrane: preparation, characterization and application to remove
Hg(II) and As(III) using polymer enhanced ultrafiltration. *Chem. Eng. J.* 170:
209–219. https://doi.org/10.1016/j.cej.2011.03.056.

36 Bierhalz, A.C.K., Westin, C.B., and Moraes, Â.M. (2016). Comparison of the properties of membranes produced with alginate and chitosan from mushroom and from
shrimp. *Int. J. Biol. Macromol.* 91: 496–504. https://doi.org/10.1016/j.ijbiomac.2016
.05.095.

37 Cartwright, P.S. (2010). *The Science and Technology of Industrial Water Treatment.*
CRC Press Taylor & Francis Group.

38 Henriksson, M. and Berglund, L.A. (2007). Structure and properties of cellulose
nanocomposite films containing melamine formaldehyde. *J. Appl. Polym. Sci.* 106:
2817–2824. https://doi.org/10.1002/app.26946.

39 Jayaramudu, J., Reddy, G.S.M., Varaprasad, K. et al. (2013). Preparation and properties of biodegradable films from Sterculia urens short fiber/cellulose green
composites. *Carbohydr. Polym.* 93: 622–627. https://doi.org/10.1016/j.carbpol.2013
.01.032.

40 Qi, H., Cai, J., Zhang, L., and Kuga, S. (2009). Properties of films composed of cellulose nanowhiskers and a cellulose matrix regenerated from alkali/urea solution.
Biomacromolecules 10: 1597–1602. https://doi.org/10.1021/bm9001975.

41 Daraei, P., Madaeni, S.S., Salehi, E. et al. (2013). Novel thin film composite membrane fabricated by mixed matrix nanoclay/chitosan on PVDF microfiltration
support: preparation, characterization and performance in dye removal. *J. Membr.
Sci.* 436: 97–108. https://doi.org/10.1016/j.memsci.2013.02.031.

42 Yang, H., Yan, R., Chen, H. et al. (2007). Characteristics of hemicellulose, cellulose
and lignin pyrolysis. *Fuel* 86: 1781–1788. https://doi.org/10.1016/j.fuel.2006.12.013.

43 Österberg, M., Peresin, M.S., Johansson, L.S., and Tammelin, T. (2013). Clean and reactive nanostructured cellulose surface. *Cellulose* 20: 983–990. https://doi.org/10.1007/s10570-013-9920-8.

44 Johansson, L.-S., Tammelin, T., Campbell, J.M. et al. (2011). Experimental evidence on medium driven cellulose surface adaptation demonstrated using nanofibrillated cellulose. *Soft Matter* 7: 10917. https://doi.org/10.1039/c1sm06073b.

45 Andresen, M., Johansson, L.S., Tanem, B.S., and Stenius, P. (2006). Properties and characterization of hydrophobized microfibrillated cellulose. *Cellulose* 13: 665–677. https://doi.org/10.1007/s10570-006-9072-1.

46 Singha, A.S. and Guleria, A. (2014). Chemical modification of cellulosic biopolymer and its use in removal of heavy metal ions from wastewater. *Int. J. Biol. Macromol.* 67: 409–417. https://doi.org/10.1016/j.ijbiomac.2014.03.046.

47 Sadeghi-Kiakhani, M., Arami, M., and Gharanjig, K. (2013). Preparation of chitosan-ethyl acrylate as a biopolymer adsorbent for basic dyes removal from colored solutions. *J. Environ. Chem. Eng.* 1: 406–415. https://doi.org/10.1016/j.jece.2013.06.001.

48 Khorramfar, S., Mahmoodi, N.M., Arami, M., and Gharanjig, K. (2010). Equilibrium and kinetic studies of the cationic dye removal capability of a novel biosorbent Tamarindus indica from textile wastewater. *Color. Technol.* 126: 261–268. https://doi.org/10.1111/j.1478-4408.2010.00256.x.

49 He, X., Du, M., Li, H., and Zhou, T. (2016). Removal of direct dyes from aqueous solution by oxidized starch cross-linked chitosan/silica hybrid membrane. *Int. J. Biol. Macromol.* 82: 174–181. https://doi.org/10.1016/j.ijbiomac.2015.11.005.

50 Hirasaki, T., Sato, T., Tsuboi, T. et al. (1995). Permeation mechanism of DNA molecules in solution through cuprammonium regenerated cellulose hollow fiber (BMMtm). *J. Membr. Sci.* 106: 123–129.

51 Jayakumar, R., Menon, D., Manzoor, K. et al. (2010). Biomedical applications of chitin and chitosan based nanomaterials—a short review. *Carbohydr. Polym.* 82: 227–232. https://doi.org/10.1016/j.carbpol.2010.04.074.

52 Fu, F. and Wang, Q. (2011). Removal of heavy metal ions from wastewaters: a review. *J. Environ. Manag.* 92: 407–418. https://doi.org/10.1016/j.jenvman.2010.11.011.

53 Barakat, M.A. (2011). New trends in removing heavy metals from industrial wastewater. *Arab. J. Chem.* 4: 361–377. https://doi.org/10.1016/j.arabjc.2010.07.019.

54 Bodnár, M., Hajdu, I., Rőthi, E. et al. (2013). Biopolymer-based nanosystem for ferric ion removal from water. *Sep. Purif. Technol.* 112: 26–33. https://doi.org/10.1016/j.seppur.2013.03.043.

55 Zhang, L., Zeng, Y., and Cheng, Z. (2016). Removal of heavy metal ions using chitosan and modified chitosan: a review. *J. Mol. Liq.* 214: 175–191. https://doi.org/10.1016/j.molliq.2015.12.013.

56 Altintas, Z., Chianella, I., Da Ponte, G. et al. (2016). Development of functionalized nanostructured polymeric membranes for water purification. *Chem. Eng. J.* 300: 358–366. https://doi.org/10.1016/j.cej.2016.04.121.

57 Zhou, K., Zhang, Q.G., Han, G.L. et al. (2013). Pervaporation of water–ethanol and methanol–MTBE mixtures using poly (vinyl alcohol)/cellulose acetate blended membranes. *J. Membr. Sci.* 448: 93–101. https://doi.org/10.1016/j.memsci.2013.08.005.

58 Mohamed, M.A., Salleh, W.N.W., Jaafar, J. et al. (2015). Incorporation of N-doped TiO$_2$ nanorods in regenerated cellulose thin films fabricated from recycled newspaper as a green portable photocatalyst. *Carbohydr. Polym.* 133: 429–437. https://doi.org/10.1016/j.carbpol.2015.07.057.

59 Kim, K., Ingole, P.G., Yun, S. et al. (2015). Water vapor removal using CA/PEG blending materials coated hollow fiber membrane. *J. Chem. Technol. Biotechnol.* 90: 1117–1123. https://doi.org/10.1002/jctb.4421.

60 Crini, G. and Badot, P.-M. (2008). Application of chitosan, a natural aminopolysaccharide, for dye removal from aqueous solutions by adsorption processes using batch studies: a review of recent literature. *Prog. Polym. Sci.* 33: 399–447. https://doi.org/10.1016/j.progpolymsci.2007.11.001.

61 Vakili, M., Rafatullah, M., Salamatinia, B. et al. (2014). Application of chitosan and its derivatives as adsorbents for dye removal from water and wastewater: a review. *Carbohydr. Polym.* 113: 115–130. https://doi.org/10.1016/j.carbpol.2014.07.007.

62 Salehi, E., Daraei, P., and Arabi Shamsabadi, A. (2016). A review on chitosan-based adsorptive membranes. *Carbohydr. Polym.* 152: 419–432. https://doi.org/10.1016/j.carbpol.2016.07.033.

63 de Castro Silva, F., da Silva, M.M.F., Lima, L.C.B. et al. (2016). Integrating chloroethyl phosphate with biopolymer cellulose and assessing their potential for absorbing brilliant green dye. *J. Environ. Chem. Eng.* 4: 3348–3356. https://doi.org/10.1016/j.jece.2016.07.010.

64 Tsurumi, T., Osawa, N., Hitaka, H. et al. (1990). Structure of cuprammonium regenerated cellulose hollow fiber (BMM hollow fiber) for virus removal. *Polym. J.* 22: 751–758. https://doi.org/10.1295/polymj.22.751.

65 Jiang, G., Zhang, J., Qiao, J. et al. (2015). Bacterial nanocellulose/Nafion composite membranes for low temperature polymer electrolyte fuel cells. *J. Power Sources* 273: 697–706. https://doi.org/10.1016/j.jpowsour.2014.09.145.

6

Properties and Applications of Gelatin, Pectin, and Carrageenan Gels

Dipali R. Bagal-Kestwal, M.H. Pan and Been-Huang Chiang

Institute of Food Science and Technology, National Taiwan University, No.1, Roosevelt Road, section 4, Taipei, Taiwan, ROC

6.1 Introduction

A gel is a solid jelly-like material that can have texture ranging from soft and weak to hard and tough. The word "gel" was coined by the 19th century Scottish chemist Thomas Graham by shortening the term "gelatin" [1]. Gels are defined as non-fluid colloidal networks or polymer networks that are expanded throughout their whole volume by a fluid [2, 3]. They are liquid by weight, but due to a three-dimensional crosslinked network within the liquid, they behave like solids. Crosslinking within the fluid results in a gel structure due to the combination of hydrogen bonds, helix formation, and complexation at the network junction points. In this way, gels are a dispersion of molecules of a liquid within a solid framework in which the solid is the continuous phase and the liquid is the discontinuous phase [3]. Typical types of gels are hydrogels, organogels, xerogels, nanocomposite hydrogels, etc. A hydrogel is a network of polymer chains that are hydrophilic, and sometimes exhibits as a colloidal gel in which water is the dispersion medium. Hydrogels are highly absorbent and they can contain over 90% water. Hydrogels also possess a degree of flexibility very similar to natural tissue due to their significant water content. The first appearance of the term "hydrogel" in the literature was in 1894 [3, 4]. Hydrogels can be classified by various methods based on source (natural or synthetic) [5], configuration (crystalline, semi-crystalline, or non-crystalline), crosslinking type (chemical or physical), and polymeric composition (homopolymeric, copolymeric, or multipolymer interpenetrating polymeric hydrogel [IPN]). Hydrogels play an important role in wound care and dressings, contact lenses, hygiene products, implant materials, and drug delivery. They are an important and integral part of tissue engineering, cosmetics, dentistry, and the food industry. Readers can get more detailed information from review articles [6–9]. In this chapter we will focus on biopolymeric hydrogels, especially gelatin, pectin, and carrageenan, which are widely used in the food industry.

6.2 Gelatin

Gelatin is a product of the structural and chemical degradation of collagen [10]. It is defined as "a product obtained by the partial hydrolysis of collagen derived from the skin,

Bio Monomers for Green Polymeric Composite Materials, First Edition.
Edited by P.M. Visakh, Oguz Bayraktar and Gopalakrishnan Menon.
© 2020 John Wiley & Sons Ltd. Published 2020 by John Wiley & Sons Ltd.

white connective tissue and bones of animals" Gelatin derived from an acid-treated procedure is known as Type A and gelatin derived from an alkali-treated process is known as Type B [11–13]. In the Food Chemicals Codex gelatin is defined as the product obtained from the acid, alkaline, or enzymatic hydrolysis of collagen, the chief protein component of the of the skin, bones, and connective tissue of animals, including fish and poultry [14]. The principal raw materials used in gelatin production are cattle bones, cattle hides, pork skins, poultry, and fish [15, 16]. Fish gelatin has similar functional characteristics to mammalian gelatin and has received considerable attention in recent years [17, 18]. Molecular weight distribution and amino acid composition influence the physical and structural properties of gelatins from various sources [19–21]. Extraneous substances, such as minerals (in the case of bone), fats, and albuminoids (found in skin), are removed by chemical and physical treatment to give purified collagen. These pretreated materials are then hydrolyzed to gelatin, which is soluble in hot water [14]. Due to similar amino acid composition and, to some extent, structure, collagen and gelatin show similarities in properties [22–30]. However, gelatin as a polymeric product exhibits its own properties and these properties are important in various industrial applications [16, 31–38].

6.2.1 Structural Unit of Gelatin

Gelatin contains 18 distinct amino acids with a basic repeat unit of Gly-Pro-x- or Gly-x-Hyp, where x is either an alkaline or an acidic amino acid. It contains many glycine (almost one in three residues, arranged every third residue), proline, and 4-hydroxyproline residues [38]. A typical structure is -Ala-Gly-Pro-Arg-Gly-Glu-4Hyp-Gly-Pro-. Collagen may be considered an anhydride of gelatin. The hydrolytic conversion of collagen to gelatin yields different mass peptide chains [22]. Thus, gelatin is not a single chemical entity, but a mixture of fractions composed entirely of amino acids joined by peptide linkages to form polymers varying in molecular mass from 15 to 400 kD [10, 22, 39–46]. Normally gelatin contains 14% hydroxyproline, 16% proline, and 26% glycine [47](Figure 6.1).

6.2.2 Molecular Structure of Gelatin

Gelatin is composed of 50.5% carbon, 6.8% hydrogen, 17% nitrogen, and 25.2% oxygen [48]. It is a heterogeneous mixture of single- or multi-stranded polypeptides, each with extended left-handed proline helix conformations and containing between 50 and 1000

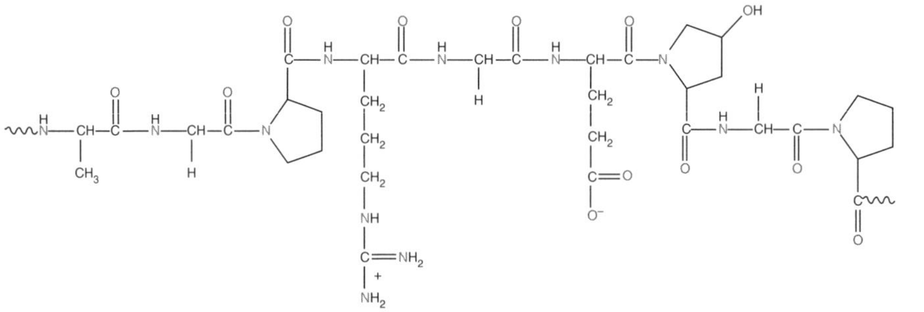

Figure 6.1 Representative gelatin structure. Source: Reproduced with permission from Elsevier [17].

amino acids. Gelatin is a mixture of α-chains (one polymer/single chain), β-chains (two α-chains covalently crosslinked), and γ-chains (three covalently crosslinked α-chains) [49, 50].

Collagen, the precursor of gelatin has been identified in many distinct forms (more than 28 forms) depending upon their structural arrangements [22, 38–42, 48]. Different forms of collagen are numbered with Roman numerals (Type I to Type XXVIII). The helical structure of type I collagen (extracted from skin and bones) is composed of two α1 (I) and one α2 (II) chains, each with molecular mass ~95 kD, width ~1.5 nm, and length ~0.3 μm [38]. It consists of mixtures of these strands together with their oligomers and breakdown (and other) polypeptides. Like collagen, the helices undergo coil transition followed by aggregation in solution due to the formation of right-handed triple-helical proline/hydroxyproline-rich junction zones. The high content of these pyrrolidines results in a stronger polymeric gel [38].

6.2.3 Properties of Gelatin

Gelatin is an amphoteric protein and its isoionic point is between pH 5.0 and 9.0, depending on raw material and preparation method. Gelatin swells in cold water and is completely soluble in hot water. A temperature of above 60 °C is necessary in order to release the ordered dry gelatin structure. Gelatin possesses specific features when compared to other biopolymers. First, gelatin macromolecules show both acidic and basic functional groups. Second, it has the capacity to form a specific triple-stranded helical structure, which is absent in most synthetic polymers at low temperatures. The rate of formation of the helical structure of gelatin depends on numerous factors such as the presence of covalent bonds [25, 49], gelatin molecular weight [50], the presence of iminoacids [51], and the gelatin concentration in the solution [11, 12]. The third peculiarity of gelatin as a biopolymer is its specific interaction with water, which is different from that observed in synthetic hydrophilic polymers [9, 25].

Gelatin is a nearly tasteless, odorless, vitreously brittle solid that is faintly yellow in color [51]. Gelatin contains 8–13% moisture with a specific gravity of 1.3–1.4. The properties of gelatin solutions are influenced by temperature, pH, ash content, method of manufacture, thermal history, and concentration [10]. Gelatin is insoluble in less polar organic solvents such as benzene, acetone, primary alcohols, and dimethylformamide [52], but soluble in aqueous solutions of polyhydric alcohols such as glycerol, propylene glycol, etc. It is also soluble in highly polar organic solvents such as acetic acid, trifluoroethanol, and formamide [52]. Gelatin has long shelf life under dry conditions, but it loses its ability to swell and dissolve when heated or stored in humid conditions (relative humidity more than 60%) [53]. Liquid gelatin is highly stable at low temperature. Gelatin solution does not contain non-colloidal ions other than H^+ and OH^-. When the pH is at the isoionic point it is known as isoionic gelatin. These solutions may be prepared by an ion-exchange process [10].

6.2.3.1 Thickening Ability

Gel strength and viscosity are the most important physical properties of gelatin [48]. Gelatin is hygroscopic, i.e. it readily absorbs and retains water. Gelatin has positive and negative charges along with uncharged hydrophilic and hydrophobic groups, which make it a polyampholyte [54]. The process of thickening involves the specific and non-specific entanglement of polymeric chains which are conformationally disordered in the solvent. Thickening happens when the concentration is high enough. The temperature, pH of the system, concentration, and type of material have a large impact on the thickening ability of gelatin.

6.2.3.2 Gelling Ability

Gelatin is partially soluble in cold water with approximate solubility of 34 g in 100 ml of water. Gelatin's solubility increases with increasing temperature. When gelatin is dissolved in water at around 35–45 °C and cooled slowly it will form gel [41]. The gel obtained from gelatin is thermoreversible. This is certainly its most interesting property. One part of gelatin can trap 99 parts of water. The gel structure formation always involves the association and formation of junction zones in the network. Unlike most hydrocolloids of polysaccharide origin, gelatin gels independently of pH and without the need for other reactive agents [43]. When a gelatin solution is cooled, its viscosity increases progressively and converts from a sol to a gel. On the other hand, if the gel is heated, it melts and once again becomes a solution. The sol/gel conversion is reversible and depends on temperature, so it is recognized as a thermoreversible gel. However, a succession of heating and cooling processes may cause gelatin to deteriorate to some extent.

6.2.3.3 Film-Forming Property

When a gelatin solution is spread in a thin layer on a surface and dried, it forms a film. This property has been exploited for the manufacture of hard and soft capsules and in microencapsulation. Gelatin films can be used to immobilize cells, proteins, and enzymes to fabricate biosensors.

6.2.3.4 Other Properties

Emulsification with gelatin is possible through a mashing and mixing diffusion process. Due to its aerating ability, microbead gelatins are created by beating the mixture in the gaseous phase. Gelatin has numerous other functions, such as coating, texture improvement, protecting, syneresis prevention, and gluing. Very often it performs more than one function at a time. Certain types of gelatin, obtained through a special manufacturing process, are cold soluble. Two types of gelatins are on the market: the gelling gelatins, which have similar properties to native gelatins and are heat-soluble (Type A), and hydrolyzed gelatins, which have no gelation capacity (Type B).

6.2.3.5 Microbiological Properties

Gelatin is an excellent nutrient for most bacteria, hence contamination must be avoided when handling it. Most countries have microbiological specifications for gelatin. Total mesophyllic plate counts of 1000 CFU are generally accepted. Some countries limit the presence of Coliforms, *E. coli*, Salmonella, Clostridial spores, Staphylococci, and sometimes even Pseudomonades for safety.

6.2.4 Gelatin Applications

The most important properties of gelatin are its melt-in-the-mouth characteristics and its ability to form thermoreversible gels. Gelatin is also stable at a wide range of pH and remains unaffected by ionic strength. It is preferred in many applications because of its clarity and bland flavor [10]. In most cases, except for the food industry, gelatin is used in the solid state [9].

6.2.4.1 Food Applications

Gelatin has unique characteristics depending on its origin and sources [17]. Gelatin is regarded as a food ingredient rather than an additive and it is Generally Regarded as Safe (GRAS) [47]. An important use of gelatin in the food industry is in desserts and confectionery applications such as marshmallows and gummi-candies. It is also used as a binding and/or glazing agent in meat and aspics. In confectionary products, gelatin is a key ingredient. Marshmallows, gummy candy, and fruit snacks are composed of gelatin. The dairy industry also uses gelatin extensively in the manufacture of sour cream, mousses, puddings, yogurt, ice cream, cheese, and specialty desserts, while meat processors depend on the properties of gelatin in the production of head cheese, pates, and luncheon meats [47, 55]. Gelatin is sometimes even used in bakery products. Other than as a food ingredient, gelatin is also used as a food additive and acts as a thickening agent, gelling agent, stabilizer, and emulsifier. Gelatin is an ingredient that is high in protein, fat-free, cholesterol-free, and low in calories. This makes it important in the processing of many food products [56]. In reduced-fat margarines and butters, gelatin is used to replace part of the fats, thus making it possible to bind the product and reduce its calorific value with controllable viscosity, texture, and taste. Gelatin in the form of films has been used to coat the surface of some food and agricultural products to prolong their shelf life. Gelatin has an animal origin, therefore it is perfectly compatible with meat-based products. Gelatin has also been reported to possess antioxidant activity. A study by Gomez-Guillén et al. [21] revealed antimicrobial activity associated with gelatin. Gelatin films have commercial application as food packaging films [17, 57–66] because they are edible and biodegradable with an antibacterial function, heat sealing ability, and moisture and oxygen barrier properties.

Gelatin is also a good candidate as carrier material for bioactive components such as natural antioxidants and antimicrobial substances [67]. Gelatin-based composite films can incorporate ingredients such as corn oil, sunflower oil [68], chitosan, etc. to achieve the desired quality of the films. Gelatin nanocomposite is also used to improve the mechanical and barrier properties of gelatin-based films. Such nanocomposite films have wide application in food packaging where antibacterial activity against foodborne pathogens is the primary concern [69]. Reinforcement and strengthening of gelatin films can be obtained by addition of carbon nanotubes without affecting film barrier properties [70].

6.2.4.2 Cosmetics and Pharmaceutical Applications

There are many different grades of gelatin for a variety of uses in the pharmaceutical industry. It is a key ingredient in the production of soft and hard shell gelatin capsules. Gelatin is used as a binder in tablet formulations and as a coating material to ease swallowing or mask unpleasant tastes. Cosmetics companies are also important users. Gelatin's properties are particularly well suited to the encapsulation of bath oils, and use in moisturizing lotions and skin creams, making it an important contributor to these products. Fish-derived gelatin film has attracted tremendous attention in the cosmetic industry in recent years. Citrus-based essential oils, including bergamot, kaffir lime, lemon, and lime, were incorporated in fish-derived gelatin film to increase the antioxidant activity [71]. Manufacture of active packaging films by incorporating bioactive materials such as root essential oils, namely, ginger, turmeric, and plai [72], citrus essential oil [71], green tea [73, 74], grape seed extract [75], gingko leaf extract, butylated-hydroxyl-toulene (BHT) [76], silver [74], and α-tocopherol [76] in fish skin-derived gelatin films showed remarkable results [70].

Being digestible and easily assimilated by humans, gelatin has a great advantage in the pharmaceutical industry. It is used mainly for its film-forming, gelling, and binding properties, and for its solubility in hot water. A sterile, non-antigenic, water-insoluble, and absorbable gelatin film (Gelfilm™) is obtained from a specially prepared gelatin/formaldehyde solution [77]. Being non-toxic, readily soluble in biological fluids at body temperature, a good film-forming material, and producing a strong flexible film with homogeneous structure, gelatin is often used in drug formulation. The commonly recognized dosage forms using gelatin are two-piece hard capsules, soft elastic gelatin capsules (Softgels), tablet, tablet coating, granulation, encapsulation, and micro-encapsulation [78–81].

6.2.4.3 Other Applications

Gelatin is used in a wide range of technical applications. Different grades of gelatin are used as a clarifying agent. In the beverage industry, gelatin is used as a clarifying agent in products like beer, juice, cider, and wine, in fact whenever high clarity is needed. Gelatin is also used in the production of photographic film and paper. Other commercial applications include the manufacture of adhesives, matches, and specialty micro-encapsulated products. Coacervation is useful in the photographic industry [82, 83]. A common application of coacervation is the use of gelatin and gum arabic to produce oil-containing microcapsules for carbonless paper manufacture [84–86]. Gelatin is still the best medium known for making photographic emulsions [87–89]. Technical gelatins are different from edible and pharmaceutical gelatins and are not required to meet the rigid specifications for human consumption. Such gelatins are used for paper manufacture, printing processes, formation of match heads, coated abrasives, book binding, corrugated cardboard sealing [10], etc.

6.3 Pectins

Pectin is a structural hetero-polysaccharide contained in the primary cell walls of terrestrial plants [89]. It makes up 2–35% of plant cell walls. The biological function of pectin is to crosslink cellulose and hemicellulose fibers, and provide rigidity in the cell wall, like "cement." Pectin is also a major component of the middle lamella, where it helps to bind cells together [90, 91]. Pectin was discovered by Vauquelin in 1790 [92] and later isolated and described by Henri Braconnot in 1825 [93, 94]. As a water-soluble gel, the texture of pectin is similar to that of gelatin, but slightly stickier [95, 96].

6.3.1 Natural Sources of Pectin

Pectin is present in all plants, but the content and composition varies depending on the species, variety, maturity of the plant, plant part, and growing condition. Pectins are high in legumes and citrus fruits (60–70%) than in cereals. Pears, apples, guavas, quince, plums, gooseberries, oranges, and other citrus fruits contain large amounts of pectin, while soft fruits like berries, peaches, apricots, cherries, grapes, blackberries, raspberries, dewberries, and strawberries contain small amounts of pectin. Pineapples, bananas, damsons, greengages, and redcurrants also show low level of pectin. Other sources of pectin include beets, cabbage, carrots, and potato [97].

6.3.2 Structural Unit of Pectin

Pectin does not have exact structures, and may contain approximately 17 different monosaccharides and over 20 types of linkages [98]. The pectin structure mostly depends on the source, and even on the extraction methods used as the neutral sugar-containing side chains may be degraded during the extraction process. Pectin is composed of at least three polysaccharide domains: homogalacturonan (HGA), rhamnogalacturonan-I (RG-I), and rhamnogalacturonan-II (RG-II) [99–101] (see Figure 6.2).

Pectin has an α-(1→4)-linked D-galacturonic acid polysaccharide backbone. The acid groups along the chain are largely esterified with methoxy groups. There

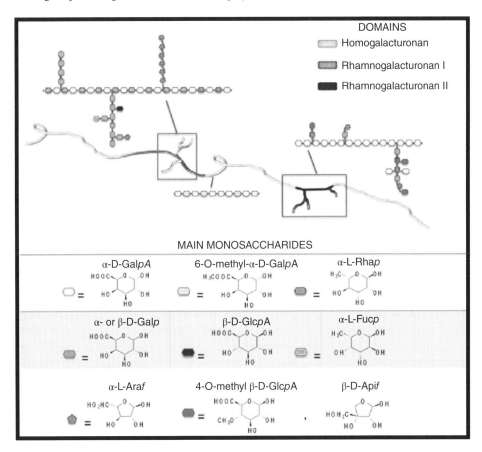

Figure 6.2 Schematic of pectin structure with its monomers. Source: Reproduced with permission from Elsevier [101].

can also be acetyl groups present on the free hydroxy groups. The majority of the structure consists of homopolymeric partially 6-methylated, and 2- and/or 3-acetylated poly-α-(1→4)-D-galacturonic acid (1–4%) residues. However, there are substantial "smooth and flexible hairy" non-gelling areas of alternating α-(1→2)-L-rhamnosyl-α-(1→4)-D-galacturonosyl sections containing branch points with mostly neutral side chains (1–20 residues) of mainly α-L-arabinofuranose and α-D-galactopyranose (rhamnogalacturonan I). The molecule does not adopt a straight conformation in solution, but is extended and curved ("worm like") with a large amount of flexibility. Its structure and biosynthesis have been recently reviewed. At least 67 transferases are involved in its biosynthesis [102, 103]. Pectins from some sources are xylogalacturonan blocks of α-(1→4)-D-galacturonic acid units, partially substituted at the O-3 position with a single non-reducing β-D-xylopyranose and/or with longer (dimer to octamer) β-D-xylopyranose chains [104]. The pectin's rhamnogalacturonan II side chains may also contain other residues, such as D-xylose, L-fucose, D-glucuronic acid, D-apiose, 3-deoxy-D-manno-2-octulosonic acid (Kdo), and 3-deoxy-D-*lyxo*-2-heptulosonic acid (Dha) attached to poly-α-(1→4)-D-galacturonic acid regions [105]. Neutral sugars such as L-rhamnose, xylose, galactose, and arabinose are also present in side chains. However, the total content of neutral sugars varies with the source, the extraction conditions, and subsequent treatments [106]. They also carry non-sugar substituents, essentially methanol, acetic acid, phenolic acids, and occasionally amide groups.

The esterification of galacturonic acid residues with methanol is a very important structural characteristic of pectic substances. Commercial pectin products are categorized according to their methoxy content and their gel-forming ability. The degree of methylation (DM) is defined as the percentage of carbonyl groups esterified with methanol. If more than 50% of the carboxyl groups are methylated the pectin is called high-methoxy pectin (HMP), and less than 50% of degree of methylation gives low methoxy pectin (LMP) [105, 106]. The properties of pectin depend on the degree of esterification.

6.3.3 Low Methoxyl Pectins

LMPs can gel in the presence of divalent cations, usually calcium. The gelation is the result of the formation of intermolecular junction zones between homogalacturonic smooth regions of different chains. The structure of such a junction zone is generally ascribed to the so-called "egg box" binding process. Initial strong association of two polymers into a dimer is followed by the formation of weak interdimer aggregation, mainly governed by electrostatic interactions [107]. The gel-forming ability of LMPs increases with decreasing degree of methylation. LMPs with a blockwise distribution of free carboxyl groups are very sensitive to low calcium levels. The presence of acetyl groups prevents gel formation with calcium ions but gives the pectin emulsion stabilizing properties.

6.3.4 High Methoxyl Pectins

HMPs show insignificant dimerization upon binding with calcium due to the lack of sufficient carboxylate groups. If the methoxyl esterified content is greater than about

50%, calcium ions show some interaction but do not form gel. The controlled removal of methoxyl groups converts HMPs to LMPs.

HMPs (> 43% esterified, usually ~67%) form gels because of hydrophobic interactions and hydrogen bonding around pH ~3.0 to reduce electrostatic repulsions in acids and sugars [108]. LMPs (~35% esterified), in the absence of added cations, form transparent gels by the formation of cooperative "zipped" associations at low temperatures (~10 °C) [109]. The pattern of hydrogen-bonded association is highly similar to that of alginate. Thus, the rheological properties of LMPs are dependent on the type of cation, salt concentration, and pH.

6.3.5 Gelation of Pectins

Pectins are soluble in water. Gels are formed when polymer molecules interact to a certain extent to form a network that entraps solvent and solutes. The junction zones resulting from polymer molecule interactions must be of limited size. Due to large chain size, precipitation may happen instead of gel formation [106]. Coordinate bonding with Ca^{2+} ions, hydrogen bonding, and hydrophobic interactions may be involved in gel formation. The gelation mechanism of pectins is mainly governed by their degree of esterification. For LMPs gelation results from specific non-covalent ionic interactions between blocks of galacturonic acid residues of the pectin backbone and with divalent ions such as calcium. The affinity of pectin chains toward calcium increases with decreasing degree of esterification or ionic strength, and with increasing polymer concentration. The charge density of the polygalacturonate chain, temperature, and the distribution pattern of free and esterified carboxyl groups have high impact on the calcium binding capacity.

6.3.6 Pectin Extraction

The main raw materials for pectin production are dried citrus peel or apple pomace, both of which are by-products of juice production. Pomace from sugar beet, banana, and pea hulls are also used to a small extent. The extraction conditions can lead to large variations in the chemical structure of the final product. Pectins are industrially extracted from citrus peel or apple pomace by hot acidified water. Extraction conditions in the range of pH 1.5–3.5, temperature 60–100 °C, and extraction time of 0.5–6 hours are used. Optimization of these conditions can lead to products with anticipated gelling capacity and degree of methylation. The hot and dilute acid is used in long extraction processes to break down protopectin into small lengths. After extraction, the extract with small segmented branching and side chains is filtered and then concentrated using vacuum evaporation. The pectin is then precipitated by adding ethanol or isopropanol. Originally, pectin precipitation was carried out with aluminum salts, but this method is no longer used. Alcohol treatment results in HMPs while ammonia treatment produces LMPs. After alcohol precipitation, pectin is separated, washed, and dried. However, amidated pectins are obtained by using ammonium hydroxide in this process [110]. After drying and milling, pectin is usually standardized with sugar and sometimes calcium salts or organic acids to have optimum performance in particular applications [111]. Enzymatic degradation of the pectin is prevented by addition of surfactant such as sodium dodecyl sulfate (SDS). Dilute sodium deoxycholate (SDC) is used to remove pigments and lipids, and 90% dimethyl sulfoxide (DMSO) will remove starch. Enzymatic extraction always results in short branched segments. In order to extract intact pectins, arabinase and galactanase could be used to avoid degradation [104].

6.3.7 Pectin Functionality and Applications

Gelation is probably the most important property of pectin. Pectin is also used as a thickener in liquid foods, where it adds body with little cling. In recent years, pectin has often been used as an effective emulsifier in numerous food applications. It has been suggested that the protein moiety, feruloyl, and acetyl groups play major roles in pectin emulsifying activities, but the emulsion-stabilizing properties of pectins are controlled by the HGA domain and the neutral sugar side chains of rhamnogalacturonan-I. Pectins extracted from citrus fruits are produced commercially as a white to light brown powder. The classical application is to provide the jelly-like consistency to jams or marmalades [111]. Pectin reduces syneresis in jams and marmalades and increases the gel strength. For household use, pectin is an ingredient in gelling sugar (also known as "jam sugar") where it is diluted to the right concentration and mixed with sugar and citric acid [112]. For conventional jams and marmalades that contain above 60% sugar and soluble fruit solids, HMPs are used, while LMPs and amidated pectins are used for reduced-sugar products. Pectin is used in confectionery jellies to give the gel structure, a clean bite, and confer a good flavor release. It can also be used to stabilize acidic protein drinks (such as drinking yogurt), to improve the mouth-feel and pulp stability in juice-based drinks, and as a fat replacer in baked goods [113]. Pectin can keep casein colloids stable and enhance the functional properties of whey proteins [114]. Pectin is also used in dessert fillings, sweets, and as a source of dietary fiber. Typical levels of pectin used as a food additive are between 0.5% and 1.0%, about the same amount of pectin as in fresh fruit [115]. There is increasing evidence that pectin may have some health benefits beyond its role as a dietary fiber. Small pectin fragments have anti-cancer function as they bind to and inhibit various actions of the pro-metastatic protein galectin [116].

Pectins are found to be useful in many other industrial applications. They can be used as a stabilizer for water and oil emulsions [114]. Incorporation of pectin in films acquires increasing interest because it makes the films biodegradable and may even be used in some *in vivo* pharmaceutical applications [117, 118]. Because of its film-forming properties, pectin is useful as a sizing agent for paper and textiles [119–121]. It is useful for the preparation of membranes for ultrafiltration and electrodialysis. Strong and self-supporting films are prepared from blends of pectin and starch. Biodegradable straws are also prepared using pectin. The combinations of pectin with other polymers and their applications are summarized in Table 6.1.

In medicine, pectin increases the volume of stool so it is used against constipation. It is also the main ingredient in Kaopectate, an over-the-counter medicine to treat mild diarrhea, along with kaolinite. Pectin also can be used to reduce blood cholesterol levels and gastrointestinal disorders, and remove heavy metals from the human body [125]. Pectin is also used in throat lozenges as a demulcent, and in wound-healing preparations and specialty medical adhesives, such as colostomy devices. Pectin could be used in various oral drug-delivery platforms, e.g. controlled-release systems, gastro-retentive systems, colon-specific delivery systems, and muco-adhesive delivery systems, because of its non-toxicity and low cost [129]. However, pectins from different sources have

Table 6.1 Applications of pectins and other polymer blends.

Pectin composite	Conditions and properties	Applications
Pectin–gelatin	7–10% gelatin, elastic gel Gels with long, tough texture, high thermal and storage stability	Hybrid hydrogels for confectionery products, jelly beans, sour cream [122]
Pectin–proteins	LMP, poly-L-lysine DE, 36%; pH near neutrality Gel formation by crosslinking and physical bonds Clear elastic gel with controlled gel strength and network swelling	To develop satisfactory food texture in dairy gel products, sauces, dressings, etc. [122, 123]
Pectin–polysaccharides (alginate)	Neutralized pH with cold setting conditions for LMPs Reversible gel forms at low solids and below pH 3.8	Confectionery products, jelly beans, jelly toppings [122]
Pectins with starch/oxidized starch/potato maltodextrin	Long and viscous gelling agents Can be used to replace/ reduce sucrose	Jelly beans [123], paper making, oil extraction, coffee and tea fermentation [124–126]
Pectins with agar-agar	Composite, 50–50%, substitute powdered for gelatin	Food products with better preservation and longer storage stability of gelled dairy desserts, low calorie for vegan and vegetarian consumers [101]
Pectins with guar gum	Emulsifying agents to thicken liquid, suspend particles, retain water/moisture Adhesive gum gels	To produce low viscosity gels for cholesterol-lowering diet [127], low-fat cheese products, bakery goods, etc.
	Reduction of cholesterol, for the treatment of obesity, lowering body mass index, body fat, etc.	Supplementation in dietary food for special purpose [128]
	Emulsifying agents, controlled texture and shelf life	Dairy food products, chewing gum [124]
Nopal mucilage extract (LMP)	Moisturizing agent	Non-food applications such as cosmetics, soaps, creams, and shampoo [113]
	Used to produce mortar Mixed with lime in the protection and restoration of historical monuments	For protection and restoration of historical monuments [122, 123]

different gelling abilities due to variations in molecular size and chemical composition. The inconsistency may result in poor reproducibility in drug-delivery characteristics.

Other applications of pectin include edible films, paper substitute, foams, and plasticizers, etc. In addition to pectolytic degradation, pectins are susceptible to heat degradation during processing, and the degradation is influenced by the nature of the ions and salts present in the system [130]. Pectin gel containing sulfuric acid is used in lead accumulators [131]. In cosmetic products, pectin acts as emulsion stabilizer and is used in the formulation of body and hand lotions, make-up foundations, shampoos, hair conditioners, and permanent waves and personal cleanliness products [132, 133].

6.4 Carrageenans

Carrageenans, also known as carrageenins, form a family of polysaccharides obtained by extraction of various species of red seaweeds (Rhodophyceae) [111, 134]. They are gel-forming and thickening mucopolysaccharides from the cell wall of red algae. According to Tseng (1945), the name "carrageenin" was taken from the village of Carraghen in Ireland, where it was first used [135]. Furthermore, as recorded by Smith et al., carrageenan production increased after World War II and fractionation of the crude extract started in the 1950s [136].

6.4.1 Sources

Different seaweed species produce various types of carrageenan. *Betaphycus gelatinum* and *Eucheuma gelatinae* are mainly found in China and Taiwan. *Fucus crispus* (Irish Moss) and *Chondrus crispus*, also known as rock moss, are harvested in Ireland, Canada, the United States, France, Iridaea, and North Atlantic [137]. *Eucheuma striatum* and *Kappaphycus alvarezii* (cottonii/*Eucheuma cottonii*) are found in the Philippines, Indonesia, and the east coast of Africa. *Gigartina canaliculate* and *Gigartina skottsbergii* (*Mastocarpus stellatus*) are from France, Morocco Argentina, and Chile.

6.4.2 Carrageenan Structure

Carrageenan is a mixture of water-soluble, linear, sulfated galactans. It is formed by alternate units of D-galactose and 3,6-anhydro-galactose (3,6-AG) linked by α-1,3- and β-1,4-glycosidic linkages. It is an anionic linear sulfated polygalactan polymer with 15–40% ester sulfate content. The molecular weight of carrageenan is usually high but depends on many factors, such as type of seaweed species, age of seaweed, harvesting season, extraction method, and condition. Based on the position and number of sulfate groups and the number of 3,6-anhydrogalactosyl rings of each disaccharide unit, carrageenans can be classified into different types. The various types of carrageenan with different characteristic structures are identified by Greek prefixes such as λ, κ, ι, ε, and μ, which all contain 22–35% sulfate groups [136]. This classification is based on solubility in potassium chloride, therefore does not reflect definitive chemical structures but only general differences in the composition and degree of sulfation at specific locations in the polymer [136]. ι- and κ-carrageenan are gel-forming carrageenans, whereas λ-carrageenan is a thickener/viscosity builder [135]. In ionic solutions, κ- and ι-carrageenan self-associate into helical structures to form rigid or flexible gels, respectively. Commercial enzymatic production of ι- and κ-carrageenan is carried out using sulfohydrolase. The natural precursors of ι- and κ-carrageenans are called ν- and μ-carrageenan and are non-gelling carrageenans [137, 138]. Higher levels of ester sulfate are responsible for low solubility temperature and low gel strength.

6.4.3 Properties of Carrageenans

Carrageenans are linear, water-soluble hydrophilic colloid polymers that became commercially important after World War II. They are highly flexible with helical structures. They have gel-forming ability even at room temperature but have poor acid solubility. These high-molecular-weight polysaccharides contain sulfate half-ester, which makes carrageenan strongly anionic. They form gels in the presence of cations such as sodium, potassium, and calcium. Electrostatic interactions of carrageenan with protein may result in gelling as well as non-gelling products. Reactivity with proteins depends upon the protein/carrageenan net charge ratio, the function of the isoelectric point of the protein, the system pH and the ratio of carrageenan to protein [139]. The viscosity of carrageenan solution depends on concentration, temperature, the presence of other solutes, salts, the type of carrageenan, and its molecular weight. Viscosity increases nearly exponentially with concentration. κ-carrageenan gel is turbid (cloudy) while ι-carrageenan forms clear gels. The original classification of carrageenan was determined by the fractionation of the polysaccharide with potassium chloride.

In the past, the fraction that was soluble in 0.25 M KCl was called lambda (λ) carrageenan and the fraction that was insoluble was called kappa (κ, furcellaran) carrageenan. Rees and his co-workers later altered this so that κ- and λ-carrageenan referred to idealized specific disaccharides [140]. In 1967, the third carrageenan's structure was elucidated [141].

The commercial classification of carrageenan is as follows:

- κ-carrageenan forms a strong, rigid gel in the presence of potassium ions with high syneresis. It is produced mainly from *Kappaphycus alvarezii* and reacts with dairy proteins due to the presence of calcium. κ-carrageenan with sodium salt is water soluble, but potassium and calcium salts of κ-carrageenan are insoluble at room temperature. κ-carrageenan forms thermo-reversible gels upon cooling.
- ι-carrageenan forms soft and elastic gels in the presence of calcium ions. It is produced mainly from *Eucheuma denticulatum*. Sodium salts of ι-carrageenan are water and milk soluble, forming gels, but ι-carrageenan is insoluble with potassium in water at room temperature.
- λ-carrageenan is milk soluble and does not gel. It is often used to thicken dairy products. λ-carrageenan is soluble in cold water.

6.4.4 Extraction of Carrageenans

Different species of red seaweed are cleaned, dried and baled, and supplied to manufacturers for further processing. At the manufacturing plant the dried seaweed is ground, sifted to remove impurities such as salt, sand, and marine organisms, washed thoroughly, and dried again. After treatment with hot alkali solution (e.g. 5–8% potassium hydroxide), the carrageenan solution is separated from weed solids using centrifugation followed by filtration, and then concentrated by evaporation and pH adjusted prior to the recovery of the carrageenan from solution [142]. Carrageenan recovery can be achieved by several methods. Alcohol precipitation or potassium chloride followed by steam heating is commonly used to get refined carrageenan (RC) [143]. Another process involves immersion and boiling in hot aqueous potassium hydroxide. The resultant extract is then soaked in fresh water to remove alkali. The final dried product is subjected to milling to obtain fine powder [144]. This product is known as semi-refined carrageenan (SRC). Both RC and SRC are basic grades in the market. RC has 2% maximum acid insoluble material while SRC contains a higher level of cellulosic material.

6.4.5 Applications of Carrageenans

Carrageenans are recognized as GRAS by most countries. They can be classified as low molecular weight, degraded carrageenan, or high molecular weight, undegraded carrageenan. Degraded, low molecular weight carrageenans have been demonstrated to be a carcinogen in laboratory animals, and are therefore classified as a "possible human carcinogen" by the International Agency for Research on Cancer [145]. Food-grade carrageenans are defined as carrageenans with a viscosity of no less than 5 mPa.s at 1.5% concentration and 75 °C [146]. Carrageenans are used in a variety of commercial applications as gelling, thickening, and stabilizing agents, especially in food products and sauces. Aside from these functions, carrageenans are also used in experimental medicine, pharmaceutical formulations, cosmetics, and industrial applications (Table 6.2).

Table 6.2 Varieties and properties of carrageenan and its applications.

Product application		Carrageenan type	Role/function
Food application			
Dairy products	Desserts, ice cream, cream, milkshakes, yogurts, salad dressings, sweetened condensed milks, skimmed milk, filled milk, milk gel, whipped cream and products, shakes	κ, κ + tara composite	Whey prevention, control meltdown, bodying, thickening and gelation [142, 143], prevents crystallization of saturated water or sugar solutions
	chocolate and chocolate products	Whey/vegetable oil-based κ/λ	Binder [134]
Bakery products	Cake icings, tart fillings, bread dough	ι + starch, κ + starch	Increases high viscosity and thermal resistance of food products [147]
			Agent for thickening and gelation [142, 148]
Infant formulations	Soy-based infant formulas	ι, κ	Emulsion, fat, and protein stabilization [149]

(Continued)

Table 6.2 (Continued)

Product application		Carrageenan type	Role/function
Diet-related food products	Diet sodas	κ, λ + locust bean gum	Protein stabilizer [134, 149]
	Low-fat cottage cheese	κ, ι + locust bean gum composite	Gelation [150]
	Low-calorie jellies	κ + galactomannan composite	Texture enhancer and flavor suspensor [151]
Soy and vegetarian/vegan food products	Soy milk and milk products	κ, ι	Improves textural characteristics [151]
	Soy creamer		Prevents crystallization of saturated water or sugar solutions [134]
Beverages	Alcoholic beverages (beer, wine, etc.), fruit juice, vinegars, chocolate milk, syrups, powdered fruit juices, diet shakes	κ	Clarification and refining to remove haze causing proteins [150]
Pet foods	Paté-type (loaf) canned pet foods	κ + locust bean gum composite, guar galactomannans	Gelling agent and fat stabilization, for smooth texture, emulsifier, stabilizer, and thickener [146, 147]
Meat, fish, and poultry applications	Fish gels and processed meat, cooked ham, imitation meat, sausages, canned meat, hamburgers, pureed meat, poultry, processed meat	κ, ι + locust bean gum composite	Substitute for fat, increases water retention, increases volume, or improves slicing by gelation [2, 148, 151]
Other applications			
Cosmetic	Toothpastes	κ, λ, and ι with sodium	Binder and emulsion stabilizer, prevents constituent separation
	Shampoo, body lotion, and other cosmetic creams		Bodying, emulsion stabilizer, thickener, dispersion medium
	Personal lubricants		Ingredient binder [148]

(Continued)

Table 6.2 (Continued)

Product application		Carrageenan type	Role/function
Pharmaceutical	Pharmaceutical products, soft and hard gel capsules	κ, ι	Inactive excipient in pills and tablets
	Dietary supplements		Substitute for animal-based specific gelling, thickening, and stabilizing applications
	Topical lotions/ointments, eye drops, suppositories		Increases formulation stability
Foam/gel products	Solid air freshener Composite gel-foam air freshener	ι blends with other gums + oil	Improves syneresis and freeze–thaw stability [148]
	Firefighting foam	κ	Foam thickener and stabilizer
Water-based products	Paints, inks, and shoe polish	κ + oil	Suspension, flow control, emulsion stabilization
Biotechnology	Cell and enzyme immobilization	κ	Supports hydrogel matrix [146]
	Antibiotics and aspartic acid production		Large-scale antibiotic and semi-antibiotic production using different types of immobilized strains [148]

6.5 Future Prospects

Pectin, carragennans, and gelatin are commercially important hydrophilic polysaccharides and have wide applications in the food, pharmaceutical, cosmetic, and biotechnology industries as gelling, thickening, stabilizing, and emulsifying agents. Investigation of their blends or composites with various bio- and synthetic polymers will give high-quality functional ingredients for use in food, cosmetics, and pharmaceuticals.

However, the structural and molecular properties of these composites need detailed investigation, as do facile and cost-effective methods for their production. These high-quality hydrogel composites will open up new application opportunities in many areas and increase the world's technological advancements in this field.

Acknowledgments

We want to acknowledge the financial assistance from the Ministry of Science and Technology (MOST), Taipei, Taiwan [NSC-100-2221-E-002-032-MY3, MOST-105-2311-B-002-037], and the National Taiwan University [Project No. 104R4000] for this research. We also want to thank Dr. Rakesh Mohan Kestwal for his valuable suggestions during his stay in Taiwan.

References

1 Harper, D., Online Etymology Dictionary: gel. Online Etymology Dictionary. Retrieved 9 December 2013.

2 Ferry, J.D. (1980). *Viscoelastic Properties of Polymers*, 3e. New York: Wiley. ISBN: 978-0-471-04894-7.

3 https://en.wikipedia.org/wiki/Gel#cite_note-1 saved on 17 February 2017.

4 Van Bemmelen, J.M. (1894). The hydrogel and the crystalline hydrate of copper oxide. *Z. Anorg. Chem.* 5: 466.

5 Wen, Z., Xing, J., Yang, C. et al. (2013). Degradable natural polymer hydrogels for articular cartilage tissue engineering. *J. Chem. Technol. Biotechnol.* 88: 327.

6 Ahmed, E.M. (2015). Hydrogel: Preparation, characterization, and applications: A review. *J. Adv. Res.* (6): 105.

7 Calo, E. and Khutoryanskiy, V.V. (2015). Biomedical applications of hydrogels: A review of patents and commercial products. *Eur. Polym. J.* 65: 252.

8 Pal, K., Bhatinda, A.K., and Mujumdar, D.K. (2009). *Polymeric hydrogels: characterization and biomedical applications. Des. Monomers Polym.* 12 (3): 197–220, 1568–5551.

9 Kozlov, P.V. (1983). The structure and properties of solid gelatin and the principles of their modification. *Polymer* 24: 651.

10 http://www.gelatin-gmia.com/images/GMIA_Gelatin_Manual_2012.pdf web search on 17 February 2017.

11 Guerrero, P., Stefani, P.M., Ruseckaite, R.A., and de la Caba, K. (2011). Functional properties of films based on soy protein isolate and gelatin processed by compression molding. *J. Food Eng.* 105: 65.

12 Stainsby, G., Pearson, A.M., and Dutson, T.R. (1987). *Collagen as Food, in Advances in Meat Research*, vol. 4 (ed. A.J. Bailey), 209. New York: Van Nostrand Reinhold Company.

13 Eysturskaro, J., Haug, I.J., Ulset, A., and Draget, K.I. (2009). Structural and functional properties of soy protein isolate and cod gelatin blend films. *Food Hydrocolloids* 23 (8): 2094.

14 US Pharmacopoeia 34/National Formulary 29 (2011). *Food Chemicals Codex 7*. Rockville, MD: United States Pharmacopeial Convention, Inc.

15 Hill, T.T. (1965). *Literature on Gelatin. in Literature Resources of the Chemical Process Industries*, 2e, vol. 1. Washington: American Chem. Soc.

16 Tucker, H.A. (1965). Absorbable gelatin (gelform) sponge. In: *An Annotated Bibliography, 1945–1965*, 111. Thomas, C.C., Springfield.

17 Nur Hanani, Z.A., Roos, Y.H., and Kerry, J.P. (2014). Use and application of gelatin as potential biodegradable packaging materials for food products. *Int. J. Biol. Macromol.* 71: 94.

18 Chiou, B.S., Avena-Bustillos, R.J., Bechtel, P.J. et al. (2008). Cold water fish gelatin films: Effects of cross-linking on thermal, mechanical, barrier, and biodegradation properties. *Eur. Polym. J.* 44: 3748.

19 Nur Hanani, Z.A., Roos, Y.H., and Kerry, J.P. (2012). Use of beef, pork and fish gelatin sources in the manufacture of films and assessment of their composition and mechanical properties. *Food Hydrocolloids* 29: 144.

20 Mhd Sarbon, N., Badii, F., and Howell, N.K. (2013). Preparation and characterisation of chicken skin gelatin as an alternative to mammalian gelatin. *Food Hydrocolloids* 30: 143.

21 Gómez-Guillén, M.C., Giménez, B., López-Caballero, M.E., and Montero, M.P. (2011). Functional and bioactive properties of collagen and gelatin from alternative sources: A review. *Food Hydrocolloids* 25: 1813.

22 Sokolov, S.I. (1937). *Physical Chemistry of Collagen and its Derivatives*. Moscow and Leningrad: Gizlegprom.

23 Zaides, A.L., The Structure of Collagen and its Changes in Processing, Rostechizdat, Moscow (1960).

24 Mikhailov, A.N. (1971). *Collagen of Skin and Principles of its Processing*. Moscow: Legkaya Insustriya.

25 Raikh, G. (1969). *Kollagen (Collagen)*. Moscow: Legkaya Industriya.

26 Chien, J.C.W. (1975). Solid-state characterization of the structure and property of collagen. *Macromol, Sci. Rev. Macromol. Chem.* C12 (1): 1.

27 Veis, A. (1964). *The Macromolecular Chemistry of Gelatin, Book series: Molecular Biology: An International Series of Monographs and Textbooks*, vol. 5, 21. New York and London: Academic Press.

28 Ramachandran, G.N. (1957). *Recent Advances in Gelatin and Glue Research* (ed. G. Stainsby), 32. London and New York: Pergamon Press.

29 Yannas, I.V. (1972). Collagen and gelatin in the solid state. *J. Macromol, Sci. Rev. Macromol. Chem.* C7 (l): 49.

30 Hopp, V. (1965). *Chemiker-Ztg. Chem. Apparature* 89: 469.

31 Hill, T.T. (1968). The literature of gelatin. *Adv. Chem. N.* 78: 381.

32 Miz, K. and James, T. (1973). *Theory of Photographic Process*. Leningrad: Khimiya.

33 Brezlav Yu, A.Z. (1970). *Nauchnoi i Prikladnoi Fotografii i Kinematographii* 15: 458.

34 Bagal-Kestwal, D.R., Kestwal, R.M., and Chiang, B.H. (2016). Bio based nanomaterials and their bio- nanocomposites. In: *Book of Nanomaterials and Nanocomposites, Zero- to Three- Dimensional Materials and their Composites* (ed. M. Visakh), 255. Germany: Wiley-VCH Verlag GmbH and Co 978-3-527-33780-4.

35 Titov, A.A. (1952). *Proteins in Industry and Agriculture*, 162. Moscow: Academy of Sciences of the USSR.

36 Wood, H.W. (1961). The role of gelatin in photographic emulsions. *J. Photogr. Sci.* 9: 151.

37 Borginon, H. (1967). Photographic properties of the gelatin macromolecule. *J. Photogr. Sci.* 15: 207.

38 www1.lsbu.ac.uk/water/gelatin.html#sou saved on 17 February 2017.

39 Aoyagi, S. (1985). Photographic gelatin. In: *Proceedings of the Fourth IAG Conference (1983)* (ed. H. Ammann-Brass and J. Pouradier), 79–94.

40 Larry, D. and Vedrines, M. (1985). Photographic gelatin. In: *Proceedings of the Fourth IAG Conference (1983)* (ed. H. Ammann-Brass and J. Pouradier), 35–54.

41 Chen, X. and Peng, B. (1985). Photographic gelatin. In: *Proceedings of the Fourth IAG Conference (1983)* (ed. H. Ammann-Brass and J. Pouradier), 55–64.

42 Beutel, J. (1985). Photographic gelatin. In: *Proceedings of the Fourth IAG Conference (1983)* (ed. H. Ammann-Brass and J. Pouradier), 65–78.

43 Koepff, P. (1984). *Photographic Gelatin Reports 1970–1982* (ed. H. Ammann-Brass and J. Pouradier), 197–209.

44 Tomka, I. (1984). *Photographic Gelatin Reports 1970–1982* (ed. H. Ammann-Brass and J. Pouradier), 210.

45 Bohonek, J., Spuhler, A., Ribeaud, M., and Tomka, I. (1967). *Photographic Gelatin II* (ed. R.J. Cox), 37–55. London: Academic Press.

46 Courts, A. and Stainsby, C. (1958). *Recent Advances in Gelatin and Glue Research*, 100. New York: Pergamon Press.

47 Cole, C.G.B. (2000). Gelatin. In: *Encyclopedia of Food Science and Technology*, 2e, vol. 4 (ed. F.J. Francis), 1183–1188. New York: Wiley.

48 Smith, C.R. (1921). Osmosis and swelling of gelatin. *J. Am. Chem. Soc.* 43: 1350.

49 Papon, P., Leblon, J., and Meijer, P.H.E. (2007). *The Physics of Phase Transitions*, 22. Berlin: Springer.

50 Ross, P.I. (1987). Encyclopedia of Polymer Science and Engineering. In: *Fibers, Optical to Hydrogenation*, vol. 7 (ed. H.F. Mark, N.M. Bikales, C.G. Overberger, et al.), 488. New York: Wiley-Interscience.

51 Chaplin, M., Water structure and science: Gelatin. www.lsbu.ac.uk/water/hygel.html.

52 Vold, R.D. and Vold, M.J. (1983). *Colloid and Interface Chemistry*. Reading, MA: Addison-Wesley.

53 Finch, C.A. and Jobling, A. (1977). The science and technology of gelatin. In: *The Science and Technology of Gelatin* (ed. A.G. Ward and A. Courts), 258–260. London: Academic Press.

54 Howe, A.M. (2000). Some aspects of colloids in photography. *Curr. Opin. Colloid Interface Sci.* 5: 288.

55 Mariod, A.A. and Adam, H.F. (2013). Review: gelatin, source, extraction and industrial applications, Acta Sci. Pol. Technol. Aliment. 12 (2): 135.

56 Karim, A.A. and Bhat, R. (2008). Gelatin alternatives for the food industry: recent developments, challenges and prospects. *Trend Food Sci. Technol.* 19: 644.

57 Villegas, R., O'Connor, T.P., Kerry, J.P., and Buckley, D.J. (1999). Effect of gelatin dip on the oxidative and colour stability of cooked ham and bacon pieces during frozen storage. *Int. J. Food Sci. Technol.* 34: 385.

58 Villegas, R., O'Connor, T.P., Kerry, J.P. et al. (2000). Effect of dietary alpha-tocopherol acetate supplementation and gelatine dip on the oxidative and colour stability of bacon and pepperoni during frozen storage. *Fleischwirtschaft Int.* 79 (4): 86.

59 Liu, L., Kerry, J.F., and Kerry, J.P. (2006). Effect of food ingredients and selected lipids on the physical properties of extruded edible films/casings. *Int. J. Food Sci. Technol.* 41: 295.

60 Piotrowska, B., Sztuka, K., Kolodziejska, I., and Dobrosielska, E. (2008). Influence of transglutaminase or 1-ethyl-3-(3-dimethylaminopropyl) carbodiimide (EDC) on the properties of fish-skin gelatin films. *Food Hydrocolloids* 22: 1362.

61 Jiang, M., Liu, S., Du, X., and Wang, Y. (2010). Physical properties and internal microstructures of films made from catfish skin gelatin and triacetin mixtures. *Food Hydrocolloids* 24 (1): 105.

62 Ahmad, M., Benjakul, S., Prodpran, T., and Agustini, T.W. (2012). Physico-mechanical and antimicrobial properties of gelatin film from the skin of unicorn leatherjacket incorporated with essential oils. *Food Hydrocolloids* 28 (1): 189.

63 Ahmad, M., Benjakul, S., Sumpavapol, P., and Nirmal, N.P. (2012). Quality changes of sea bass slices wrapped with gelatin film incorporated with lemongrass essential oil. *Int. J. Food Microbiol.* 155: 171.

64 Nowzari, F., Shábanpour, B., and Ojagh, S.M. (2013). Comparison of chitosan-gelatin composite and bilayer coating and film effect on the quality of refrigerated rainbow trout. *Food Chem.* 141: 1667.

65 Pereda, M., Ponce, A.G., Marcovich, N.E. et al. (2011). Chitosan-gelatin composites and bi-layer films with potential antimicrobial activity. *Food Hydrocolloids* 25: 1372.

66 Hoque, M.S., Benjakul, S., and Prodpran, T. (2011). Cuttlefish *Sepia pharaonis* skin gelatin-based film: storage stability and its effectiveness for shelf-life extension of chicken meat powder. *Int. Aquat. Res.* 3: 165.

67 Gennadios, A., McHugh, T.H., Weller, C.L., and Krochta, J.M. (1994). *Edible Coatings and Films to Improve Food Quality* (ed. J.M. Krochta, E.A. Baldwin and M.O. Nisperos-Carriedo), 201–277. Lancaster: Technomic Publishing Co., Inc.

68 Pérez-Mateos, M., Montero, P., and Gómez-Guillén, M.C. (2009). Formulation and stability of biodegradable films made from cod gelatin and sunflower oil blends. *Food Hydrocolloids* 23: 53.

69 Kanmani, P. and Rhim, J.W. (2014). Physical, mechanical and antimicrobial properties of gelatin based active nanocomposite films containing agnps and nanoclay. *Food Hydrocolloids* 35: 644.

70 Ortiz-Zarama, M.A., Jiménez-Aparicio, A., Perea-Flores, M.J., and Solorza-Feria, J. (2014). Barrier, mechanical and morpho-structural properties of gelatin films with carbon nanotubes addition. *J. Food Eng.* 120: 223.

71 Tongnuanchan, P., Benjakul, S., and Prodpran, T. (2012). Properties and antioxidant activity of fish skin gelatin film incorporated with citrus essential oils. *Food Chem.* 134: 1571.

72 Tongnuanchan, P., Benjakul, S., and Prodpran, T. (2013). Physico-chemical properties, morphology and antioxidant activity of film from fish skin gelatin incorporated with root essential oils. *J. Food Eng.* 117: 350.

73 Giménez, B., López de lacey, A., Pérez-Santín, E. et al. (2013). Release of active compounds from agar and agar–gelatin films with green tea extract. *Food Hydrocolloids* 30: 264.

74 Wu, J., Chen, S., Ge, S. et al. (2013). Preparation, properties and antioxidant activity of an active film from silver carp (*Hypophthalmichthys molitrix*) skin gelatin incorporated with green tea extract. *Food Hydrocolloids* 32: 42.

75 Li, J.H., Miao, J., Wu, J.L. et al. (2014). Preparation and characterization of active gelatin-based films incorporated with natural antioxidants. *Food Hydrocolloids* 37: 166.

76 Jongjareonrak, A., Benjakul, S., Visessanguan, W., and Tanaka, M. (2008). Antioxidant activity and properties of fish skin gelatin film incarporated with BHT and a-tocophero. *Food Hydrocolloids* 22: 449.

77 Correll, J.T., and Kalamazoo, M., U.S. Patent No. 2,507,244, Surgical Gelatin Dusting Powder and Process for Preparing Same, issued 9 May 1950. https://patentimages.storage.googleapis.com/43/40/85/de475416a3ac24/US2507244.pdf

78 O'Neill, J.F., (1990) Gelatin for the Pharmaceutical Industry, Pharmtech Conference 1990, Pharmaceutical Raw Materials-Standards, Applications, and Vendor Considerations, Moderator Dr. Garnet Peck, Purdue University.

79 Eaton, F., (1989) Manufacture of Hard Shell Gelatin Capsules and Associated Problems, Presented at Hard Shell and Softgel Capsule Manufacturing Conference, Medical Manufacturing Techsource.

80 Withered, D.F., (1986) Advances in capsule sealing technology, Presented at SME Conference on Proven Methods of Capsule Manufacturing Technology.

81 Stringer, W., (1989) SoftGel (Soft Gelatin Capsule) Update, Presented at Hard Shell and SoftGel Capsule Manufacturing Conference, Medical Manufacturing Techsource.

82 Kuznicka, B. and Kuznicki, J. (1987). Photographic gelatin. In: *Photographic gelatin, Proceedings of the Fifth RPS Symposium (1985)* (ed. S.J. Bond), 206–212.

83 Croome, R.J. (1985). Photographic gelatin. In: *Proceedings of the Fourth IAG Conference (1983)* (ed. H. Ammann-Brass and J. Pouradier), 267–282.

84 Green, B.K. and Lowell, S., US Patent No. 2,800,457, Oil-containing microscopic capsules and method of making them, issued 23 July 1957. https://patents.google.com/patent/US2800457A/en.

85 Gutcho, M.W. (1976). *Microcapsules and Microencapsulation*, 230–236. Park Ridge, NJ: Noyes Data Corporation.

86 Wood, P.D. (1977). *The Science and Technology of Gelatin* (ed. A.G. Ward and A. Courts), 419–422. London: Academic Press.

87 Mees, C.E.K. (1959). *The Theory of the Photographic Process*. New York: MacMillan Publishing.

88 Neblette, C.B. (1958). *Photography – Its Materials and Processes*. New York: D. Van Nostrand.

89 Calixto, S., Ganzherli, N., Gulyaev, S., and Figueroa-Gerstenmaier, S. (2018). Gelatin as a photosensitive material. *Molecules* (Basel, Switzerland) 23 (8): 2064. https://doi.org/10.3390/molecules23082064.

90 Coffin, D.R. and Fishman, M.L. (1993). Viscoelastic properties of pectin/starch blends. *J. Agric. Food Chem.* 41 (8): 1192.

91 Pérez, S., Rodríguez-Carvajal, M.A., and Doco, T. (2003). A complex plant cell wall polysaccharide: rhamnogalacturonan II. A structure in quest of a function. *Biochimie* 85: 109.

92 Vauquelin, M. (1790). Analyse du tamarin. *Ann. Chim.* 5: 92.

93 Braconnot, H. (1825). Nouvelles observations sur l'acide pectique. *Ann. Chim. Phys.* 30: 96.

94 Braconnot, H. (1825). Recherches sur un nouvel acide universellement répandu dans tous les vegetaux (Investigations into a new acid spread throughout all plants). *Ann. Chim. Phys.*, Series 2 28: 173.

95 Grierson, D., Maunders, M.J., Slater, A. et al. (1986). Gene expression during tomato ripening. *Phil. Trans. R. Soc. B* 314: 399.

96 Voragen, A.G.J., Coenen, G.J., Verhoef, R.P., and Schols, H.A. (2009). Pectin, a versatile polysaccharide present in plant cell walls. *J. Struct. Chem.* 20: 263.

97 Woo, K.K., Chong, Y.Y., Li, H., and Tang, P.Y. (2010). Pectin extraction and characterization from Red Dragon Fruit (*Hylocereus polyrhizus*): A preliminary study. *J. Biol. Sci.* 10: 631.

98 Pérez, S., Mazeau, K., and Hervé du Penhoat, C. (2000). The three-dimensional structures of the pectic polysaccharides. *Plant Physiol. Biochem.* 38: 37.

99 Mohnen, D. (2008). Pectin structure and biosynthesis. *Curr. Opin. Plant Biol.* 11: 266.

100 Yapo, B.M. (2011). Pectic substances: from simple pectic polysaccharides to complex pectins: a new hypothetical model. *Carbohydr. Polym.* 86: 373.

101 Munarin, F., Tanzi, M.C., and Petrini, P. (2012). Advances in biomedical applications of pectin gels. *Int. J. Biol. Macromol.* 51 (4): 681.

102 Round, A.N., Rigby, N.M., MacDougall, A.J., and Morris, V.J. (2011). A new view of pectin structure revealed by acid hydrolysis and atomic force microscopy. *Carbohydr. Res.* 345: 487.

103 Ridley, B.L., O'Neill, M.A., Mohenen, D. et al. (2001). Pectins: structure, biosynthesis, and oligogalacturonide-related signaling. *Phytochemistry* 57: 929.

104 Sharma, B.R., Naresh, L., Dhuldhoya, N.C. et al. (2006). An overview on pectins. *Times Food Process. J.* 44: 1.

105 Atmodjo, M.A., Hao, Z., and Mohnen, D. (2013). Evolving views of pectin biosynthesis. *Annu. Rev. Plant Biol.* 64: –747.

106 BeMiller, J.N., (1986) An introduction to pectins: structure and properties, ACS Symposium Series; American Chemical Society: Washington, DC.

107 Ralet, M.C., Dronnet, V., Buchholt, H.C., and Thibault, J.F. (2001). Enzymatically and chemically de-esterified lime pectins: characterisation, polyelectrolyte behaviour and calcium binding properties. *Carbohydr. Res.* 336: 117.

108 Tsoga, A., Richardson, R.K., and Morris, E.R. (2004). Role of cosolutes in gelation of high-methoxy pectin. Part 1. Comparison of sugars and polyols. *Food Hydrocolloids* 18: 907.

109 Kjøniksen, A.L., Hiorth, M., and Nyström, B. (2004). Temperature-induced association and gelation of aqueous solutions of pectin. A dynamic light scattering study. *Eur. Polym. J.* 40: 2427.

110 Eisenbrand, G. and Schreier, P. (2006). *RÖMPP Lexikon Lebensmittelchemie*; Thieme, Stuttgart; Mai, Broschiert, 2e. ISBN: 13: 978-3137366010.

111 Berg van den, L. (2008). *Dissertation: Texture of Food Gels Explained by Combining Structure and Large Deformation Properties*, 26. Wageningen University. ISBN: 9789085049432.

112 Mortensen, A., Aguilar, F., Crebelli, R. et al. (2017). Re-evaluation of potassium nitrite (E 249) and sodium nitrite (E 250) as food additives. *EFSA J* 15: e04786. https://doi.org/10.2903/j.efsa.2017.4866.

113 Francisco, G.M. and Cárdenas, A. (2003). Pectins from Opuntia spp.: a short review. *J. Prof. Assoc. Cactus Dev.* 5: 17.

114 Thakur, B.R., Singh, R.K., Handa, A.K., and Rao, M.A. (1997). Chemistry and uses of pectin – a review. *Crit. Rev. Food Sci. Nutr.* 37: 47.

115 May, C.D. (1990). Industrial pectins: Sources, production and applications. *Carbohydr. Polym.* 12 (1): 79.

116 Maxwell, E.G., Belshaw, N.J., Waldron, K.W., and Morris, V.J. (2012). Pectin – An emerging new bioactive food polysaccharide. *Trends Food Sci. Technol.* 24: 64.

117 Bagal-Kestwal, D.R., Kestwal, R.M., and Chiang, B.H. (2015). Invertase-nanogold clusters decorated onion membranes for fluorescence-based sucrose sensor. *J. Nanotechnol.* 13: 1.

118 May, C.D. (1997). Pectin. In: *Thickening and Gelling Agents for Food* (ed. A. Imeson), 230–261. London: Blackie Academic and Professional.

119 Miers, J.C., Swenson, H.A., Schultz, T.H., and Owens, H.S. (1953). Pectinate and pectate coating I. gerenal requirements and procedures. *Food Technol.* 7: 229.

120 Schultz, T.H., Miers, J.C., Owens, H.S., and MaClay, W.D. (1949). Permeability of pectinate film to water vapor. *J. Phys. Colloid Chem.* 53: 320.

121 Swenson, H.A., Miers, J.C., Schultz, T.H., and Owens, H.S. (1953). Pectinate and pectate coatings. II. Application to nut and fruit products. *Food Technol.* 7: 232.

122 Cárdenas, A., Higuera-Ciapara, I., and Goycoolea, F.M. (1997). Rheology and aggregation of cactus (*Opuntia ficus-indica*) mucilage in solution. *J. Prof. Assoc. Cactus Dev* 2: 152.

123 Cárdenas, A., Arguelles, W.M., and Goycoolea, F.M. (1998). On the possible role of *Opuntia ficus-indica* mucilage in lime mortar performance in the protection of historical buildings. *J. Prof. Assoc. Cactus Dev* 3: 64.

124 Essays, UK. (November 2013). Food product analysis: Dutch lady full cream milk. Retrieved from https://www.ukessays.com/essays/sciences/food-product-analysis-dutch-lady-cream-2575.php?vref=1. Web search saved on 18 February 2017.

125 Zy, Z., Liang, L., Fan, X. et al. (2008). The role of modified citrus pectin as an effective chelator of lead in children hospitalized with toxic lead levels. *Altern. Ther. Health Med.* 14 (4): 34.

126 Evageliou, V., Richardson, R.K., and Morris, E.R. (2000). Co-gelation of high methoxy pectin with oxidised starch or potato maltodextrin. *Carbohydr. Polym.* 42 (3): 233.

127 Trautwein, E.A., Kunath-Rau, A., and Erberdsobler, H.F. (1998). Effect of different varieties of pectin and guar gum on plasma, hepatic and biliary lipids and cholesterol gallstone formation in hamsters fed on high-cholesterol diets. *Br. J. Nutr.* 79: 463.

128 Babiker, R., Merghani, T.H., Elmusharaf, K. et al. (2012). Effects of gum arabic ingestion on body mass index and body fat percentage in healthy adult females: two-arm randomized, placebo controlled, double-blind trial. *Nutr. J.* 11: 111.

129 Sriamornsak, P. (2011). Application of pectin in oral drug delivery. *Expert Opin. Drug Deliv.* 8 (8): 1009.

130 Thiele, H. (1967). German Patent 1,249,517. *Chem. Abstr.* 109390m: 67.

131 Pertix, V. (1966). Union Gmbh, British Patent 1,025,975. *Chem. Abstr.* 64: 18969.

132 Evageliou, V., Richardson, R.K., and Morris, E.R. (2000). Effect of sucrose, glucose and fructose on gelation of oxidised starch. *Carbohydr. Polym.* 42 (3): 261.

133 Kashyap, D.R., Vohra, P.K., Chopra, S., and Tewari, R. (2001). Applications of pectinases in the commercial sector: a review. *Bioresour. Technol.* 77: 215.

134 Yermak, I.M., Mischchenko, N.P., Davydova, V.N. et al. Carrageenan-sulfated polysaccharides from red seaweeds as matrices for the inclusion of echinochrome. *Mar. Drugs* 15 (11): 337. https://doi.org/10.3390/md15110337.

135 Velde, F. and Ruiter, G.A. (2002). Carrageenan. In: *Biopolymers Vol. 6: Polysaccharides II: Polysaccharides from Eukaryotes* (ed. A. Steinbüchel, S. De Baets and E.J. VanDamme), 245–274. Weinheim: Wiley-VCH. ISBN: 3-527-30227-1.

136 Smith, D.B. and Cook, W.H. (1953). Fractionation of carrageenan. *Arch. Biochem. Biophys.* 232: 232.

137 Laurienzo, P. (2010). Marine polysaccharides in pharmaceutical applications: An overview. *Mar. Drugs* 8: 2435. https://doi.org/10.3390/md8092435.

138 Sun, Y., Liu, Y., Jiang, K. et al. (2014). Electrospray ionization mass spectrometric analysis of κ-carrageenan oligosaccharides obtained by degradation with κ-carrageenase from *Pedobacter hainanensis*. *J. Agric. Food Chem.* 62 (11): 2398.

139 MacMullan, E.A. and Eirich, F. (1963). The precipitation reaction of carrageenan with gelatin. *J. Colloid Sci.* 18: 526.

140 Dolan, T.C.S. and Rees, D.A. (1965). 649 The carrageenans. Part II. The positions of the glycosidic linkages and sulphate esters in λ-carrageenan. *J. Chem. Soc.* 3534.

141 Muller, G.P. and Rees, R.A. (1968). Current structural views of red seaweed polysaccharides. In: *Proceedings of Drugs from the Sea* (ed. H.D. Freudental), 241–255. Washington, DC: Marine Technology Society.

142 Stanley, N. (1987). Production, properties and uses of carrageenan. In: *Production and Utilization of Products from Commercial Seaweeds* (ed. D.J. McHugh), 288. Rome: FAO, Fisheries Technical Paper.

143 Bourgade, G., 1871. Improvement in treating marine plants to obtain gelatin. US Patent 112, 535.

144 Istinii, S., Ohno, M., and Kusunos, H. (1994). *Methods of analysis for agar, carrageenan and alginate in seaweed, Bull. Mar. Sci. Fish. Kochi Univ.* 14: 49.

145 World Health Organization (1983). *IARC Monographs on the Evaluation of the Carcinogenic Risk of Chemicals to Humans, Some Food Additives, Feed Additives and Naturally Occurring Substances*, vol. 31, 79. International Agency for Research on Cancer. ISBN: 92832-1231-2.

146 Necas, J. and Bartosikova, L. (2013). *Carrageenan: a review, Vet. Med. Czech* 58 (4): 187.

147 De Ruiter, G.A. and Rudolph, B. (1997). *Carrageenan biotechnology, Trends Food Sci. Technol.* 8: 389.

148 Vilela, A., Cosme, F., and Pinto, T. (2018). Emulsions, foams, and suspensions: The microscience of the beverage industry. *Beverages* 4 (2): 25. https://doi.org/10.3390/beverages4020025.

149 Ruttenberg, M.W. (1980). *Handbook of Water-Soluble Gums and Resins*, vol. 22 (ed. R.L. Davidson), 108. New York: McGraw-Hill Book Company.

150 Iborra, J.L., Manjón, A., and Cánovas, M. Immobilization in carrageenans. In: *Protocol: Immobilization of Enzymes and Cells*, vol. 1 (ed. G.F. Bfckerstaff), 53–60. Totowa, NJ: Methods in Biotechnology, Humana Press Inc.

151 Lin, C.F. (1977). Interaction of sulfated polysaccharides with proteins in food colloids. In: *Food Colloids* (ed. H.D. Graham), 320–346. Westport, CT: Avi Publishing Co. Inc.

7

Biodegradation of Green Polymeric Composites Materials

Karthika M.[1], Nitheesha Shaji[1], Athira Johnson[1], Neelakandan M. Santhosh[1], Deepu A. Gopakumar[1,2] and Sabu Thomas[1]

[1] *International and Inter-University Center for Nanoscience and Nanotechnology, Mahatma Gandhi University, Kottayam, Kerala, India*
[2] *School of Industrial Technology, Universiti Sains Malaysia, Penang, Malaysia*

7.1   Introduction

Sustainable progress in the research and development (R&D) area is intended to design materials that are biodegradable in nature and make a significant contribution to the protection of the environment. Researchers and technologists are focused on the production of highly eco-friendly products to conserve our natural and ecological systems. In recent decades, synthetic polymers have been manufactured from petrochemicals and studies show that these are highly dangerous to nature because of their non-degradability and emission of toxins [1]. To overcome these problems, natural polymers (green polymers) are employed because of their biodegradability and sustainable nature. The term "biodegradability" means the decomposition of materials by microorganisms into methane, carbon dioxide, inorganic compounds, water, and biomass [2]. Use of biopolymers helps to preserve fossil resources, save energy, and reduce pollution. The activity of abiotic factors transforms the properties of polymeric molecules and thereby results in the weakening of their structure and initiating degradation [3]. Along with abiotic factors mechanical, light, and thermal degradation also occur. Shear forces and compression are the primary reasons for mechanical degradation [4]. Photosensitive materials and thermal-responsive materials undergo light and thermal degradation, respectively.

A huge number of biopolymers (collagen, cellulose, starch, chitin, etc.) have been identified and extracted from various biological sources. Based on their source, they are classified into biomass products, microorganism-derived products, biotechnologically obtained products, and oil products [2]. Figure 7.1 shows the classifications of biodegradable polymers from different sources.

Green polymers refer to the biomass products obtained from renewable sources that are environmentally compatible in nature. Biopolymers like starch, cellulose, and chitin are the major examples of agro-based polymers or biopolymers and have a wide range of applications because of their high availability and biodegradability, and low toxicity. Starch is a linear polymer obtained from cereals and tubers composed of

Bio Monomers for Green Polymeric Composite Materials, First Edition.
Edited by P.M. Visakh, Oguz Bayraktar and Gopalakrishnan Menon.
© 2020 John Wiley & Sons Ltd. Published 2020 by John Wiley & Sons Ltd.

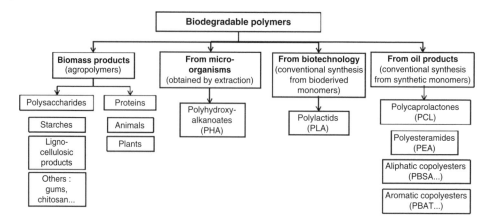

Figure 7.1 Classification of biodegradable polymers based on their source.

two polymer chains of D-glucose units called amylase and amyl pectin. Chitin is the second most abundant polymer after cellulose and is obtained from the exoskeleton of arthropods and crustaceans. It is composed of N-acetyl-D-glucosamine units linked by β-1,4-linkages. Cellulose is known as the most abundant polymer in nature and is obtained from plants, bacteria, and algae. Cellulose is a polymer composed of linear chains of D-glucose units linked by β-1,4-linkages.

Polymers and polymeric composite materials have large-scale applications in the aerospace, military, sports, industrial, and biological sectors [5–7]. Green polymeric composites are prepared by using biopolymers as matrix and solid particle as fillers. This combination improves the properties of both polymers and fillers. The degradation of polymeric materials is followed through various stages, including bio-deterioration, depolymerization, assimilation, and mineralization. One major feature is that progress can stop at each stage. During the initial stage of decomposition, microbes and other factors decompose the materials into small fragments [8]. Depolymerization happens with the support of enzymes and free radicals produced from microbes which cleave the polymers and generate oligomers, dimers, and monomers. As a result of the incorporation, molecular energy is converted to new biomass by the activity of microbial metabolism. As a result, new biomasses and energy are produced (assimilation). This process is called assimilation. Finally, during mineralization simple molecules like carbon dioxide (CO_2), nitrogen (N_2), and methane (CH_4) are formed and released into the environment [8]. A lot of literature work has already been carried out on the properties and applications of green polymeric composites. In this chapter, we emphasize the mechanisms of biodegradation of green polymers and green polymeric composites.

7.2 Biodegradation of Green Polymers

7.2.1 Green Polymers: Definition and Properties

Biocompatibility, sustainability, and eco-efficiency are new concerns that are vital for the development of next-generation materials and technologies. The term "green" is used to describe materials that are biodegradable and renewable. Green polymers, in other words, biodegradable polymers, are those produced partially or entirely from natural resources. Thus, they can reduce and prevent environmental pollution, which makes them beneficial over other synthetic polymers. A wide variety of green polymers like cellulose, starch, lignin, and chitin are derived from biological sources such as microorganisms, plants, trees, etc. [9]. They can also be produced by synthetic routes from biological resources such as vegetable oils, sugars, fats, resins, proteins, amino acids, etc.

The structure of biodegradable polymers is instrumental in their properties. They tend to consist of ester, amide, or ether bonds. Non-toxic behavior, an ability to maintain great mechanical integrity until degradation, and capable of controlling rates of degradation are major advantages of these types of polymers. In general, green polymers can be divided into two groups based on their source [10]. One of these groups are agro-polymers, which are derived from biomass. The other consists of biopolyesters, which are derived from microorganisms or synthetically made from either natural or synthetic monomers. Agro-based polymers include polysaccharides like cellulose which consist of glycosidic bonds and proteins obtained from amino acids (Figure 7.2). Polymers based on natural products are both biocompatible and biodegradable, although they should be technologically acceptable. There is a worldwide research effort to develop biodegradable polymers as potential candidates for a wide variety of applications as a waste management option.

With developments in technology and ideas, more biodegradable polymers are employed in agricultural mulching films [11], packaging [12], and controlled-release fertilizers [13, 14]. All biodegradable polymers ought to be stable enough to be used in their specific application, but, upon disposal, they must simply break down. Polymers, specifically biodegradable polymers, have very strong carbon backbones that are tough to break; such degradation usually starts from the end groups [15]. Biodegradable polymers additionally tend to possess minimal chain branching as this crosslinking usually decreases the number of end groups per unit weight [9]. The major advantages of biodegradable polymers are that they can be composted with organic wastes and returned to enrich the soil, so their use will not only reduce environmental pollution but will also lessen the labor cost for the removal of plastic waste in the environment

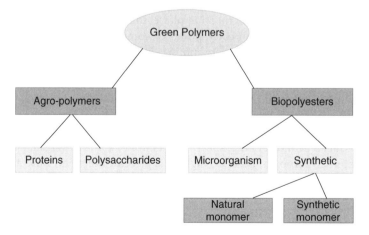

Figure 7.2 General classification of green polymers based on their origin.

because they degrade naturally. Their decomposition will help to increase the longevity and stability of landfills by reducing the volume of garbage [16]. They could also be recycled to make useful monomers and oligomers by microbial and enzymatic treatment. Recently, the development of the manufacture of such polymers has been substantially increased. The term "biodegradable" is critical in determining the commercialization of a product because of the environmental concerns of most developing countries.

7.2.2 Mechanism of Biodegradation

In simple terms, biodegradation can be defined as the disintegration of materials by bacteria, fungi, or other biological means [17]. Biodegradation of a polymeric material is chemical degradation in the natural environment by the action of microorganisms via enzymatic action that includes changes in the chemical structure, resulting in loss of mechanical and structural properties and ultimately leading to the formation of metabolic products like water, methane, carbon dioxide, and biomass that are beneficial to the environment [18]. Apart from biotic environmental factors like microorganisms, abiotic factors like photo-degradation, hydrolysis, and oxidation, add a lot to the biodegradation process [19–21]. The process of biodegradation can therefore be further understood as the deterioration of a material's physical as well as chemical properties that results in its diminishing molecular mass associated with the formation of carbon dioxide, water, and methane by the action of microbes in both aerobic and anaerobic conditions, along with the influence of some abiotic factors like photo-degradation, oxidation, and hydrolysis. The rate of biodegradation can therefore be affected by studying the conversion of carbon to carbon dioxide. There are a variety of internationally accepted standards for the measurement of biodegradability and use of microbes for the test polymer. The rate and extent of biodegradation or the microbial use of a test plastic material can be measured by using it as the sole added carbon source in a test system containing a microbially rich matrix-like compost in the presence of air, and under optimal temperature conditions (preferably 588 °C, representing the thermophilic phase) [22]. Figure 7.3 shows typical data obtained when the percent of carbon released (as carbon dioxide) from a bioplastic exposed in a composting environment is plotted as a function of time. The lag phase in the figure corresponds to the period at which the microbes are adapted to the test polymer. In the biodegradation phase, the polymer is degraded to evolve carbon dioxide. The plateau region corresponds to complete biodegradation [22]. There can be various other methods for determining the rate of biodegradation of polymers. de Campos and coworkers studied polyvinyl alcohol (PVA) degradation by measuring PVA concentration after it had undergone the action of the soil microorganisms, leachate microorganisms, and the soil with leachate microorganisms in the mineral medium (MM) [23] (Figure 7.4).

Generally, biodegradation of polymers occurs in two steps (Figure 7.5). In the first step, the polymers are reduced to shorter chains by the influence of some extracellular enzymes and abiotic factors. Microorganisms are highly adaptive to the environment and secrete both endoenzymes and exoenzymes, which attack the substrate and cleave the molecular chains into segments [24]. The next stage includes the biomineralization process. After the depolymerization steps leading to the formation of oligomers, the oligomers are bio-assimilated by the microorganisms and mineralized. Conversion of biodegradable materials or biomass to gases, water, salts, minerals, and residual

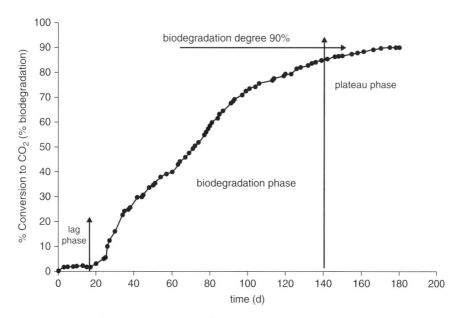

Figure 7.3 Example data from a biodegradation test of a biodegradable biopolymer assessed as CO_2 release over 180 days. The CO_2 release curve shows the typical lag phase, the biodegradation phase, and a plateau phase [22] (http://rstb.royalsocietypublishing.org/content/364/1526/2127.short).

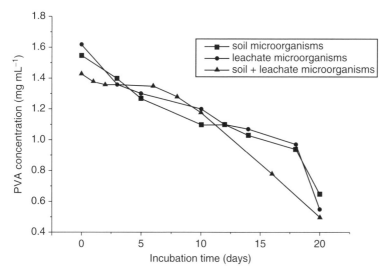

Figure 7.4 Variation of PVA concentration recorded in the presence of an acclimated mixed culture [23] (http://www.scielo.br/scielo.php?pid=S1516-89132011000600024&script=sci_arttext).

biomass is called mineralization [24]. The degradation process can be either aerobic or anaerobic [25]. If the degradation takes place in the presence of oxygen, it is said to be aerobic biodegradation, whereas in anaerobic degradation oxygen will not be present. Carbon dioxide is produced in aerobic degradation whereas methane is produced instead in anaerobic degradation. Equation (7.1) corresponds to aerobic degradation

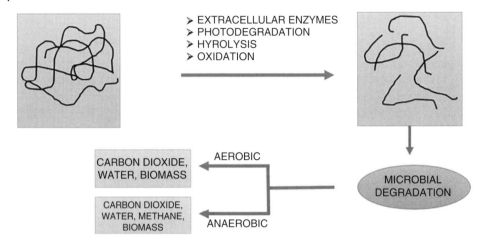

Figure 7.5 Schematic illustration of the biodegradation of polymers.

whereas Eqn (7.2) represents anaerobic degradation:

$$\text{polymer} + O_2 \rightarrow CO_2 + H_2O + \text{biomass} + \text{residue} \tag{7.1}$$

$$\text{polymer} \rightarrow CO_2 + CH_4 + H_2O + \text{biomass} + \text{residue} \tag{7.2}$$

If oxygen (O_2) is present, aerobic biodegradation occurs and carbon dioxide (CO_2) is produced. If there is no oxygen, anaerobic degradation occurs and methane (CH_4) is produced instead of carbon dioxide [26]. The biodegradation process is illustrated in Figure 7.5.

The structure of biodegradable polymers is instrumental in their properties [26]. Natural polymers (i.e. proteins, polysaccharides, nucleic acids) are degraded in biological systems by oxidation and hydrolysis. Biodegradation includes a change in the properties, such as tensile strength, color, shape, molar mass, etc., of a polymer or polymer-based product under the action of environmental agents like heat, light, or chemicals [27]. These types of changes in properties are often termed "ageing." The biodegradation process is associated with some morphological changes in the polymer structure that can be well understood with the help of scanning electron microscopy (SEM) micrographs. Figure 7.6 shows the morphological changes associated with the biodegradation of a composite made of a poly(hydroxybutyrate-co-hydroxy valerate) matrix reinforced with curaua fibers (with and without alkaline treatment) in simulated soil [28].

According to the micrographs, the degradation process expands all over the sample with time with an increase in the number of holes across the surface along with a gradual reduction in the sample size. This may be attributed to the multiplication of the microbial population and their activity with time. The micrographs show that the poly(3-hydroxybutyrate-co-valerate) (PHBV) sample surfaces have spherical cavities, whereas PHBV/CF and PHBV/CF/NaOH samples exhibit longitudinal surface erosion [28]. Figure 7.7 shows the mechanical behavior of the same composites before and after biodegradation. Higher decreases in tensile strength and elastic modulus occurred for the PHBV/CF and PHBV/CF/NaOH composites, indicating that the presence of CF

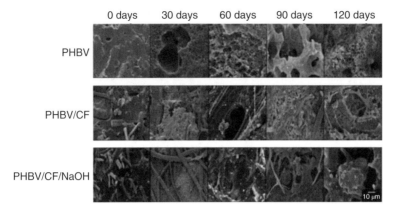

Figure 7.6 SEM micrographs of the polymer and composites after different degradation periods [28] (http://onlinelibrary.wiley.com/doi/10.1002/app.40712/abstract).

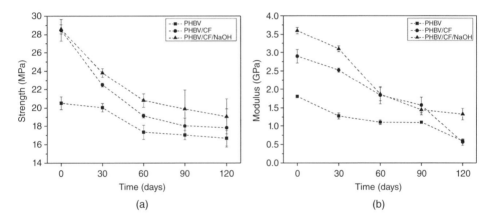

Figure 7.7 (a) Tensile strength and (b) elastic modulus of samples before and after biodegradation [28] (http://onlinelibrary.wiley.com/doi/10.1002/app.40712/abstract).

favors biodegradation, with a particularly large effect observed after 60 days of exposure [28]. This shows that there is a serious deterioration in mechanical properties of the polymer after biodegradation.

Further clarity on the morphological changes of polymers can be understood from Figure 7.8. The figure corresponds to the degradation of vegetable cellulose and poly(3-hydroxybutyrate) (PHB) that were buried in pots containing soil [29]. Before biodegradation PHB showed a smooth surface, which changed to a spongy and porous surface after degradation that accounts for the degradation of the polymer due to microbial attack, whereas in the case of cellulose the agglomerated structure before degradation changed with the separation of fibers due to microbial degradation [29].

In a finished product, such a change is to be prevented or delayed. The breakdown of these polymers relies on a variety of elements as well as the polymer. The properties of the polymer that critically influence the degradation are the type of bonds, solubility, etc. along with the pH, temperature, and microbes present in the surroundings [30]. The polymer material's complexity, structures, and compositions are the foremost

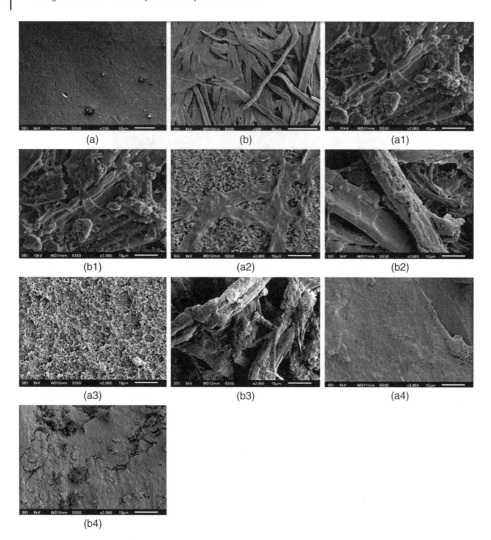

Figure 7.8 PHB and standard cellulose: (a) PHB 0 days (350X) and (b) standard cellulose 0 days (500X). (a1) and (b1) after 30 days of degradation (2000X), (a2) and (b2) after 60 days of degradation (2000X), (a3) and (b3) after 90 days of degradation (2000X), (a4) and (b4) after 150 days of degradation (2000X) [29] (http://www.scielo.br/scielo.php?pid=S0104-14282015000200154&script=sci_arttext).

vital aspects that govern polymer biodegradability. Another structural characteristic of the polymer is the possible branching of chains or the formation of networks (crosslinked polymers) [17]. Despite having the same overall composition these different structures of a polymer can directly influence the accessibility of the material to enzyme-catalyzed polymer chain cleavage, and also have a crucial impact on the higher ordered structures of the polymers. In recent years a considerable amount of qualitative and semi-quantitative information has been accumulated to draw some conclusions about the following important factors that affect the rate of degradation of synthetic polymers in a biological environment [31–33]:

- structure of the polymer
- molecular mass
- wettability
- preparation methods
- environmental conditions (temperature, pH, etc.).

As previously discussed, the process of degradation generally starts with the breakdown of the polymer to form oligomers. This is because almost all polymers are too large to pass through the cell membrane. After depolymerization, these small fragments of polymers can be easily absorbed and degraded with the microbial cells.

7.2.3 Biodegradation of Green Polymers

Polymers that are obtained from nature (such as starch or flax fiber) are liable to degradation by microorganisms [9]. The polymer may or may not decompose sooner in aerobic conditions, according to the formulation and microbes used. In the case of materials where starch is employed as an additive to a plastic matrix, the polymer in touch with the soil and/or water is attacked by the microbes [34–36]. The microbes digest the starch, yielding a porous, sponge-like structure with a high surface area and low structural strength. Once the starch element has been depleted, the compound matrix begins to be degraded by an enzymatic attack. Every reaction leads to the cleavage of a molecule, slowly reducing the load of the matrix until all the material has been digested. Another approach to microorganism degradation of biopolymers involves growing microorganisms for the particular purpose of digesting polymeric materials. Though this technique reduces the amount of waste, it does not account for the preservation of non-renewable resources. In the case of photodegradable polymers like polyolefins, degradation takes place as a result of the action of sunlight [37]. Several other polymers are degraded by exposure to chemicals and broken into small particles, which are further degraded upon microbial degradation. It has been reported that soil microbes are able to start the depolymerization of many natural polymers such as cellulose and hemicellulose [38]. They are able to secrete enzymes in the soil that then depolymerize the polymers. da Luz and coworkers have reported that the fungus Pleurotusostreatus PLO6 can degrade green polyethylene [39]. A decrease in half-life can be obtained if the polymer is exposed to sunlight before the action of the fungus.

Microorganisms utilize PHB and PHBV as energy sources, thus allowing these polymers to be biodegraded in microbial active environments [26]. Microbes colonize the polymer surface and secrete enzymes that degrade PHB into HB (hydroxybutyrate) and PHBV into HB and HV (hydroxy valerate) segments [26]. These fragments are used as a carbon source by the cells for the purpose of growth. Lee et al. investigated the degradation of PHB by fungi samples collected from various environments [40]. PHB depolymerization was tested in vials filled with a PHB-containing medium, which was inoculated with isolates from the samples. The degradation activity was detected by the formation of a clear zone below and around the fungal colony. In total, 105 fungi were isolated from 15 natural habitats and 8 lichens, among which 41 strains showed PHB degradation. Ratajska and Boryniec investigated the biodegradation of new polymeric materials of natural origin and of their mixtures with other natural and synthetic polymers [17]. The processes of biodegradation were carried out in an aqueous medium and a soil medium in the presence of air. Experiments with the biodegradation of cellulose carbamate showed that the product was more susceptible to biological decomposition than cellulose or its viscose derivatives [17]. Experiments with the biodegradation of chitosan showed that the process was dependent on both the origin and properties of the samples, as well as on the degree of their deacetylation and conditions of the biodegradation process. Enzymatic degradation of cellulose is generally carried out by hydrolases. Chitin is found in crab shells, the cell wall of bacteria and fungi. Chitosan and partially deacetylated chitins, prepared via alkaline deacetylation of chitin, are mainly composed of N-acetyl glucosamine (GlcNAc) and o-glucosamine (GlcN) residues [9].

Various glycosidases, such as chitinase and lysozyme, can degrade chitin via hydrolysis, whereas chitosan is hydrolyzed by chitosanase [9]. Studies on the enzymatic degradation of chitin by Shigemasa et al. show that the degradation of chitin and partially deacetylated chitin by the chitinase from Bacillus sp. PI-7S was much higher than that of lysozyme [41]. The authors also reported "smoother" degradation of β-chitin than α-chitin. Alginate, a copolymer of α-L-guluronate (G) and its C5 epimer β-D-mannuronate (M), is an important biopolymer that has been successfully applied in economically important enzyme and whole-cell immobilization, which have a significant impact in biochemical as well as biomedical research [41]. Alginate, in nature, is degraded by alginases or alginate lyases through α,β-elimination mechanism. Dextran, a homoglycan of α-D-glucopyranose linked with α-1,6 linkages, is a natural biopolymer that has a diverse range of applications in the food industry, biomedicals, cosmetics, etc. The most important type of enzymatic polymer cleavage reaction is hydrolysis. Glycosidic bonds, but also ester and peptide linkages, are subject to hydrolysis through nucleophilic attack on the carbonyl carbon atom. Polysaccharides, fats, PHB, gelatin, keratin, and also the synthetic polymers such as polylactate or polymalate are all degraded through such reactions [40]. Generally speaking, all biopolymers that are extracted from natural resources are biodegradable in nature, although there are some polymers that are synthetically prepared but still show a biodegradable nature. An enormous effort to employ these two types of polymers in place of conventionally used polymers may build the foundation for a greener future.

7.3   Biodegradation of Composite Materials

The chemical dissolution of composite materials by microbes or other biological means is referred to as biodegradation. Biodegradation of composites means that at least one step is mediated by biological agents. Depending on the aerobic or anaerobic environment, polymeric composites can degraded in different ways. During degradation, any attack on the composite material may lead to the loss of strength due to fracture, disbonding, or delamination, and ultimate failure. A composite material that is biodegradable and is formed by a matrix (resin) and reinforcement of natural fibers is called a biocomposite. Biodegradable composites have great importance because of the problem of the solid waste generated by plastic materials after their use. The factors that are necessary for the development of biodegradable composites include proper mixing of lignocellulosic selection of appropriate biopolymer matrix, suitable surface treatments if required, and low-cost but high-speed fabrication techniques. Some of the benefits of the processing of LC-based biodegradable composites are less abrasiveness for tooling, unlike glass fiber composites, and the absence of airborne particles, reducing respiratory problems for workers.

There are different mechanisms for the microbial degradation of polymeric composites:

- direct attack of the resin by acids or enzymes
- blistering due to gas evolution enhances cracking due to calcareous deposits and gas evolution
- polymer destabilization by concentrated chlorides and sulfides [42].

Generally, the biodegradability of composites is determined by their weight loss in soil burial. Some of the reported standards for the determination of biodegradability include ASTM D6400-99, which includes D6002-96, D5338-98, and D6340-98, European standards (CEN/TC 261/SC 4N 99 and ISO 14855), DIN-Standard draft 54 900, and ISO/CD-standard 15986.2. The compatibility of biodegradable polymeric materials is evaluated in the standard for burial test. According to these standards, chemical examination and ultimate degradation of polymer (mineralization) and disintegration under real conditions in a composting plant or laboratory conditions are to be followed [43]. The degradation of thermoplastic nanocomposites in the environment can be assessed by irradiating with UV in biotic environments [44]. Nanocomposites have higher degradability than control samples. It was assumed that, once oxygen reached the matrix, it will remain there for a while because the clay will interfere in the path, thus O_2 is easily available for longer to initiate the degradation faster than for the neat polymer. By blending natural fibers with polymers like poly(lactic acid) (PLA), starch blends, cellulose acetates, and poly(hydroxy alkanoates) (PHA), a fully biodegradable material can be manufactured. Materials of this kind are known to decay under defined conditions, which are usually different from the ambient conditions under which they are used. PLA is one of the most common environmentally friendly composite materials. Different studies have reported on the decay of PLA under real-life conditions [45]. The presence of fillers and fibers has crucial role in the biopolymer degradation process [46]. Large amounts of the hydrophilic filler starch can increase the biodegradation rate of PLA.

As far as biodegradability is concerned, it has been confirmed that PLA is naturally degraded in soil or compost. Bayerl et al. reported the degradation of the PLA-flax fiber composite as the result of some microbial action as well as hydrolysis [47]. The amount of fiber content and time contributed to greater degradation. The biggest weight loss observed was about 2.75% for UD-30 after six weeks. However, the weight loss is very much less than what was observed with the compost samples. The short and long fiber samples also displayed similar trends, with the maximum recorded weight loss values after six weeks being about 1.6% for SF-30 and 1.3% for LF-30. Bayerl et al. found that the degradation (weight loss) increases with both fiber content and time in the compost. Neat PLA samples, having no fibers, undergo only marginal degradation on the outside surfaces, even over eight weeks. The presence of fibers and the architecture of fiber structure in a PLA composite can intensely affect the degradation behavior of the composite.

As far as biodegradability of PLA-basednano composites is concerned, Nieddu et al. investigated the biodegradation of PLA-clay composites [48]. They observed higher rates of PLA biodegradation in compost with the addition of nano-clays, which was attributed to the high relative hydrophilicity of the clay, allowing an easier permeability of water into the polymer matrix and activating the hydrolytic degradation process. Indeed, the presence of clays can increase the hydrophilicity of a polymer matrix and thus increase the amount of water at equilibrium in the material (solubility). Nevertheless, the water diffusion rate into the polymer can be decreased (diffusion coefficient) due to the clay platelet-like morphology which maximizes the permeate path length through the composite [49]. The water permeability in the polymer is the product of both the solubility and the diffusion coefficient, and, in consequence, this can be increased or decreased depending on the variation of these two parameters.

Plastics are a major constituent of municipal solid waste that poses a growing disposal and environmental pollution problem due to their recalcitrant nature. Common plastics are resistant to biological degradation [50]. The biodegradation of plastics is limited by their molecular weight, chemical structure [51], water solubility, and the fact that most plastics are xenobiotic. To reduce their harmful impact on the environment and allow them to be altered during organic waste recycling processes, various materials have been added to improve the biodegradability of plastics. Additives are used to modify conventional plastics to enhance their biodegradability to form bio-based plastics and natural fiber composites. Biodegradable composite materials are used in a wide range of applications. Different types of biodegradable plastics are used for food packaging and for waste containment. Biodegradable composites have also been developed for medical applications, including medical devices and drug delivery [52]. In the case of biodegradable plastics made from PLA and natural fibers, physical properties and performance have been found to be similar to conventional plastics for greenhouse crop production [53]. The degradability of biodegradable plastics can be initiated in non-biodegradable plastics by incorporation of special additives which become responsible for degradation of plastics. Since biodegradable polymers are relatively costly, their use is restricted to specific high-value added applications only. Researchers have developed composites in which different plant fibers serve both as reinforcement and matrices. Natural fiber reinforced composites have advantages like renewability, environmentally friendly, low cost, lightweight, and high specific performance. The chemical structure and constitution of the composites determine the biodegradability of plastics. Biodegradation is brought about by biological activity predominantly by the enzymatic action of microorganisms and can be measured by standard tests for a specified period of time.

Gómez et al. determined the rate and extent of mineralization of a wide range of commercially available plastic alternative materials during composting, anaerobic digestion, and soil incubation [54]. These workers assessed biodegradability by measuring the amount of carbon mineralized from these materials during incubation under conditions that simulate these three environments and by examination of the materials by SEM. The results showed that during 660-day soil incubation, substantial mineralization was observed for polyhydroxyalkanoate plastics, starch-based plastics, and materials made from compost, but only one type of polyhydroxyalkanoate-based plastic biodegradation rate is similar to that of the positive control cellulose. No significant degradation was observed for polyethylene or polypropylene plastics or the same plastics amended with commercial additives meant to confer biodegradability. During anaerobic digestion for 50 days, 20–25% of the bio-based materials but less than 2% of the additive-containing plastics were converted to biogas (CH_4 and CO_2). SEM analysis showed a substantial disintegration of polyhydroxyalkanoate-based plastic, and some surface changes for other bio-based plastics and coconut coir materials, but no evidence of degradation of polypropylene or polypropylene-containing additives. While some bio-based plastics and natural fibers show degradation to an appreciable extent in the three environments, only a polyhydroxyalkanoate-based resin biodegraded to any significant extent during the timescale of composting and anaerobic digestion processes used for solid waste management.

Priyanka and Palsule developed various composites of polypropylene (PP) using natural fibers such as pineapple leaf fiber, banana fiber, and bamboo fiber, and studied their degree and rate of aerobic biodegradation [55]. Composites contained 10%, 15%, and 50% volume fractions of pineapple leaf fiber, banana fiber, and bamboo fiber, respectively, which are the optimum fiber percentages of the respective composites. All the composites showed partial biodegradation in the range of 5–15% depending on the fiber content. Degradation did not take place in the covalent ester linkages between the natural fiber and the MA-g-PP compatibilizer but in those areas of the fibers which remained physically embedded in the resin matrix. Thus, although natural fiber reinforced PP composites are not excellent biodegradable materials, they can address the management of waste plastics by reducing the amount of polymer content used, which, in turn, will reduce the generation of non-biodegradable polymeric wastes [56].

Wagner and coworkers examined fiber-reinforced polymer composites for susceptibility to microbiologically influenced degradation [57]. They exposed composites, resins, and fibers to sulfur/iron-oxidizing, calcareous-depositing, ammonium-producing, hydrogen-producing, and sulfate-reducing bacteria (SRB) in batch culture to evaluate the degradation. They found that physiological types of bacteria were populated evenly on the surfaces. Epoxy and vinyl ester neat resins, carbon fibers, and epoxy composites were not adversely affected by the microbial species. SRB degraded the organic surfactant on glass fibers. Hydrogen-producing bacteria disrupted bonding between fibers and vinyl ester resin, and penetrated the resin at the interface. Acoustic emission testing demonstrated reduction of tensile strength in a stressed carbon fiber-reinforced epoxy composite after exposure to SRB [58].

The soil burial test under the effect of cow dung can be used to test the biodegradation of coir fiber-reinforced epoxy composites. Biodegradation of the composites was established by carbon dioxide evolution and analysis of the morphology and chemical structure of the degraded composite by SEM and Fourier transform infrared spectroscopy. Verma and coworkers reported that the biodegradability of coir/epoxy composites was increased to 30% by mixing the cow dung with the composite [59].

In polymeric material, because of the presence of naturally occurring microorganisms, biodegradation is like chemical degradation through enzymatic action into the metabolic products of microorganisms such as bacteria and fungi, which is the chemical dissolution or breakdown of materials. Molecular degradation in aerobic and anaerobic conditions was triggered by enzymes, leading to complete or partial removal of the residue from the environment. The rate of biodegradation of composites of natural polymers studied in various environments such as different soils, compost, and weather conditions. Biodegradable natural fiber polymer composites could be used as an alternative to synthetic fiber polymer composites. These polymer composites could be disposed of in a safe and environmentally friendly way using processes such as making compost, soil application, and biological wastewater treatment. Because of their importance, the application of biodegradable natural fibers in polymer reinforced composites is attracting more research attention all the time. In order to prove the nature of their biodegradability, biocomposites were buried in garbage dump land, which contains cellulolytic bacteria. Recently, bamboo fibers have been hybridized with more corrosion-resistant synthetic fibers such as glass and carbon fibers in order to tailor the properties of the composite according to the desired structure under consideration. As the degradation of synthetic fibers occurs at a slow rate or not at all, inclusion with natural fibers may lead to improved biodegradable performance. Natural fiber reinforced PLA composites are fully renewable materials with effective degradability as compared to non-renewable petroleum-based products.

Melt intercalation of polymers into layered silicates of clay has been proven to be an excellent technique to prepare polymer-layered silicate (PLS) nanocomposites [60]. With only a small percentage of clay PLS exhibits greatly improved thermal, mechanical, and barrier properties compared with the pristine polymers. The advantages of using clay are that it is environmentally friendly, naturally abundant, and economical. Since 1997 there have been no reports on the preparation of biodegradable PLS composites except those composites based on poly(ε-caprolactone) (PCL) and organic clays. Jimenez et al. found that PCL/clay nanocomposites showed improved biodegradability compared to pure PCL due to the catalytic role of clay in the biodegradation mechanism, but it is unclear how the clay increases the biodegradation rate of PCL [60].

Lee et al. reported the biodegradability of aliphatic polyester-based nanocomposites under compost in 2002 [61]. In that study APES/organic montmorillonite (MMT) nanocomposites decreased with an increasing number of organoclays except for the APES/Cloisite 10A hybrid. In addition, APES/Cloisite 10A hybrids exhibited higher biodegradability than the APES/Cloisite 30B hybrid. Lowering of the biodegradability of APES/organic MMT may be due to the presence of dispersed silicate layers with a large aspect ratio in the APES matrix, which forces the micro-organism diffusing in the bulk of the film through more tortuous paths. Therefore the biodegradation of APES was hindered because the effective path length and time for the micro-organism diffusion increased. It assumed that the retardation in biodegradation is due to the improvement of barrier properties of the aliphatic polyester after composite preparation with clay [61].

The degradation of PLA under compost is a complex process that involves four major steps: water absorption, ester cleavage and formation of oligomer fragments, solubilization of oligomer fragments, and finally diffusion of soluble oligomers by bacteria. The tendency for hydrolysis of neat PLA mainly controls the PLA degradation. Biodegradability studies on PLA-layered silicate nanocomposite showed a similar tendency to hydrolysis of PLA and PLA-layered silicate composites. Ray et al. expected that the presence of terminal hydroxylated edge groups in the silicate layers might be one of the factors responsible for this [49]. Wang et al. investigated the biodegradability of poly(3-hydroxybutyrate-co-3-hydroxy valerate)/organophilic montmorillonite (PHBV/OMMT) nanocomposites by a degrading cultivation method in soil suspension [62]. The experimental conditions were similar to those in the natural soil environment, using soil suspension as the biodegradable environment to investigate whether biodegradation can accelerate the process and shorten the experimental time compared to using natural soil. The investigators found 295 kinds of PHBV degraders from different kinds of soil, mostly soil bacteria and moulds. Wang et al. assumed that the biodegradability of PHBV/OMMT in soil suspension decreased with an increasing amount of organophilic montmorillonite [62], as for the biodegradability for aliphatic polyester/OMMT in activated soil reported by Lee et al. [61]. The biodegradable PHBV/OMMT nanocomposites with low OMMT content offer the opportunity to reduce the cost of PHBV and show better application potential of biodegradable PHBV in the future [62].

Research has been carried out to study the biodegradability of starch–PVA blends and their nanocomposites. Chen et al. found that the biodegradation rate in starch–PVA cast films decreases as the PVA content in compost increases, but films containing both starch and PVA showed much faster degradation than pure PVA [64]. Another study reported that starch–glycerol samples incubated in compost lost up to 70% of their dry weight within 22 days. Yet, there was only 59% weight loss in starch–glycerol formulations with added PVA, which suggests that the addition of PVA slows the process of degradation [63]. The degradability of thermoplastic starch and PVA blends under anaerobic conditions simulates the most common disposal environment for household waste. It was mostly PVA that remained at the end of the digestion and the starch was almost entirely degraded. However, the PVA content significantly impacted the rate of starch solubilization [65]. The reactivity kinetic model for the modified starch–PVOH blends shows that the biodegradability of PVOH was enhanced by the addition of starch. Based on the kinetic models, the growth rate of the microorganism was found to increase with increasing starch content in the PVA–starch blend as a first-order reaction [66]. Both neat PVA and blends that had been crosslinked exhibited comparatively slow degradation and a stimulating effect of lignocellulosic fillers on the biodegradation of PVA in blends. Tang et al. studied the biodegradability of nano-SiO_2-reinforced starch/PVA nanocomposite films in 2008 [67]. With 5% nano-SiO_2 (SPS5), the weight loss could be up to 60%, which is very close to the weight loss of starch–PVA without nano-SiO_2 [67]. This might indicate that the nanoparticles had no significant influence on the biodegradability of these films. However, the biodegradation of clay–starch–PVA nanocomposite films depended on both the type and content of the nanoparticles, and the nanoparticles hindered the rate of biodegradation [68].

7.4 Conclusion

Synthetic polymers are vital in several branches of industry, particularly in the packaging industry. However, they have an undesirable effect on the environment and cause problems with waste deposition and utilization. There is therefore increasing interest in using polymers incorporating natural materials like starch. The negative influence of synthetic polymers on the natural environment creates several issues with waste deposition and utilization. Nowadays, a few perishable water-soluble polymers are still available commercially, in spite of the obvious and pressing need for environmental protection. Biodegradable polymers have been researched, but polymers based on renewable sources (especially starch) are of most interest. The key advantages of perishable polymers are that they can be composted with organic wastes and can enhance the soil, their use will not only reduce injuries to wild animals caused by dumping of standard plastics, but will also reduce the labor price for the removal of plastic wastes within the atmosphere, because they degrade naturally, their decomposition can increase the longevity and stability of landfills by reducing the volume of garbage, and they may be recycled to produce useful monomers and oligomers by microorganism and enzymatic treatment. Recently, the manufacturing process of biodegradable green polymers has been determined. In this chapter, we have discussed the biomechanical pathways for the degradation of green polymers and green polymer composites. Several studies have been carried out to design polymers with biodegradable properties to keep the environment safe and clean.

References

1 Griffin, G. (1980). Synthetic polymers and the living environment. *Pure Appl. Chem.* 52: 399–407.

2 Avérous, L. and Pollet, E. (2012). Biodegradable polymers. In: *Environmental silicate nano-biocomposites* (ed. L. Avérous and E. Pollet), 13–39. Springer.

3 Helbling, C., Abanilla, M., Lee, L., and Karbhari, V. (2006). Issues of variability and durability under synergistic exposure conditions related to advanced polymer composites in the civil infrastructure. *Compos. A Appl. Sci. Manuf.* 37: 1102–1110.

4 Briassoulis, D. (2004). Mechanical design requirements for low tunnel biodegradable and conventional films. *Biosyst. Eng.* 87: 209–223.

5 Sun, X.S. (2005). 1 – Overview of plant polymers: resources, demands, and sustainability. In: *Bio-based Polymers and Composites* (ed. R. Wool and S.S. Xiuzhi), 1–14. Burlington: Academic Press.

6 Wool, R.P. (2005). 15 – Nanoclay biocomposites. In: *Bio-based Polymers and Composites* (ed. R. Wool and S.S. Xiuzhi), 523–550. Burlington: Academic Press.

7 Wool, R.P. (2005). 16 – Lignin polymers and composites. In: *Bio-based Polymers and Composites* (ed. R. Wool and S.S. Xiuzhi), 551–598. Burlington: Academic Press.

8 Lucas, N., Bienaime, C., Belloy, C. et al. (2008). Polymer biodegradation: mechanisms and estimation techniques – a review. *Chemosphere* 73: 429–442.

9 Ghanbarzadeh, B. and Almasi, H. (2013). Biodegradable polymers. In: *Biodegradation – Life of Science* (ed. R. Chamy and F. Rosenkranz), 141–185. Rijeka: InTech.

10 Clarinval, A.M. and Halleux, J. (2005). 1 – Classification of biodegradable polymers. In: *Biodegradable Polymers for Industrial Applications* (ed. R. Smith), 3–31. Woodhead Publishing.

11 Touchaleaume, F., Martin-Closas, L., Angellier-Coussy, H. et al. (2016). Performance and environmental impact of biodegradable polymers as agricultural mulching films. *Chemosphere* 144: 433–439.

12 Siracusa, V., Rocculi, P., Romani, S., and Rosa, M.D. (2008). Biodegradable polymers for food packaging: a review. *Trends Food Sci. Technol.* 19: 634–643.

13 González, M.E., Cea, M., Medina, J. et al. (2015). Evaluation of biodegradable polymers as encapsulating agents for the development of a urea controlled-release fertilizer using biochar as support material. *Sci. Total Environ.* 505: 446–453.

14 Sempeho, S.I., Kim, H.T., Mubofu, E., and Hilonga, A. (2014). Meticulous overview on the controlled release fertilizers. *Adv. Chem.* 2014: 1–16.

15 Leja, K. and Lewandowicz, G. (2010). Polymer biodegradation and biodegradable polymers – a review. *Pol. J. Environ. Stud.* 19: 255–266.

16 Hernandez, N., Williams, R.C., and Cochran, E.W. (2014). The battle for the "green" polymer. Different approaches for biopolymer synthesis: bio advantaged vs. bio-replacement. *Org. Biomol. Chem.* 12: 2834–2849.

17 Ratajska, M. and Boryniec, S. (1998). Physical and chemical aspects of biodegradation of natural polymers. *React. Funct. Polym.* 38: 35–49.

18 Göpferich, A. (1996). Mechanisms of polymer degradation and erosion. *Biomaterials* 17: 103–114.

19 Engineer, C., Parikh, J., and Raval, A. (2011). Review on hydrolytic degradation behavior of biodegradable polymers from controlled drug delivery system. *Trends Biomater. Artif. Organs* 25: 79–85.

20 Hawkins, W.L. (1964). Thermal and oxidative degradation of polymers. *Polym. Eng. Sci.* 4: 187–192.

21 Yousif, E. and Haddad, R. (2013). Photodegradation and photostabilization of polymers, especially polystyrene: review. *Springerplus* 2: 1–32.

22 Song, J., Murphy, R., Narayan, R., and Davies, G. (2009). Biodegradable and compostable alternatives to conventional plastics. *Philos. Trans. R. Soc. London B Biol. Sci.* 364: 2127–2139.

23 Campos, A., Marconato, J.C., and Martins-Franchetti, S.M. (2011). Biodegradation of blend films PVA/PVC, PVA/PCL in soil and soil with landfill leachate. *Braz. Arch. Biol. Technol.* 54: 1367–1378.

24 Mohan, S.K. and Srivastava, T. (2010). Microbial deterioration and degradation of polymeric materials. *J. Biochem. Technol.* 2: 210–215.

25 Doble, M. (2005). Biodegradation of polymers. *Indian J. Biotechnol.* 4 (2): 186–193.

26 Alshehrei, F. (2017). Biodegradation of synthetic and natural plastic by microorganisms. *J. Appl. Environ. Microbiol.* 5: 8–19.

27 Porter, R.S., Cantow, M.J.R., and Johnson, J.F. (1967). Polymer degradation. V. Changes in molecular weight distributions during sonic irradiation of polyisobutene. *J. Appl. Polym. Sci.* 11: 335–340.

28 Beltrami, L.V., Bandeira, J.A., Scienza, L.C., and Zattera, A.J. (2014). Biodegradable composites: morphological, chemical, thermal, and mechanical properties of composites of poly(hydroxybutyrate-co-hydroxy valerate) with curaua fibers after exposure to simulated soil. *J. Appl. Polym. Sci.* 131 (40712): 1–8.

29 Schröpfer, S.B., Bottene, M.K., Bianchin, L. et al. (2015). Biodegradation evaluation of bacterial cellulose, vegetable cellulose and poly(3-hydroxybutyrate) in soil. *Polímeros* 25: 154–160.

30 Tokiwa, Y., Calabia, B.P., Ugwu, C.U., and Aiba, S. (2009). Biodegradability of plastics. *Int. J. Mol. Sci.* 10: 3722–3742.

31 Kale, S.K., Deshmukh, A.G., Dudhare, M.S., and Patil, V.B. (2015). Microbial degradation of plastic: a review. *J. Biochem. Technol.* 6: 952–961.

32 Dussud C and Ghiglione J-F. (2014). Bacterial Degradation of Synthetic Plastics In CIESM Workshop Monograph No. 46. The Tara Expeditions Foundation.

33 Singh, B. and Sharma, N. (2008). Mechanistic implications of plastic degradation. *Polym. Degrad. Stab.* 93: 561–584.

34 Nehra, K., Jamdagni, P., and Lathwal, P. (2017). Bioplastics: a sustainable approach toward healthier environment. In: *Plant Biotechnology: Recent Advancements and Developments*, 297–314. Springer.

35 Adhikari, D., Mukai, M., Kubota, K. et al. (2016). Degradation of bioplastics in soil and their degradation effects on environmental microorganisms. *J. Agric. Chem. Environ.* 5: 23.

36 Emadian, S.M., Onay, T.T., and Demirel, B. (2017). Biodegradation of bioplastics in natural environments. *Waste Manag.* 59: 526–536.

37 Arkatkar, A., Arutchelvi, J., Sudhakar, M. et al. (2009). Approaches to enhance the biodegradation of polyolefins. *Open Environ. Eng. J.* 2: 68–80.

38 López-Mondéjar, R., Zühlke, D., Becher, D. et al. (2016). Cellulose and hemicellulose decomposition by forest soil bacteria proceeds by the action of structurally variable enzymatic systems. *Sci. Rep.* 6: 25279.

39 da Luz, J.M.R., Paes, S.A., Ribeiro, K.V.G. et al. (2015). Degradation of green polyethylene by *Pleurotus ostreatus. PLoS One* 10: e0126047.

40 Lee, K.-M., Gimore, D., and Huss, M. (2005). Fungal degradation of the bioplastic PHB (poly-3-hydroxybutyric acid). *J. Polym. Environ.* 13: 213–219.

41 Banerjee, A., Chatterjee, K., and Madras, G. (2014). Enzymatic degradation of polymers: a brief review. *Mater. Sci. Technol.* 30: 567–573.

42 Abbasi, Z. (2012). Water resistance, weight loss and enzymatic degradation of blends starch/polyvinyl alcohol containing SiO_2 nanoparticle. *J. Taiwan Inst. Chem. Eng.* 43: 264–268.

43 Kaiser JP. Testing the Performance and the Disintegration of Biodegradable Bags for the Collection of Organic Wastes. Macromolecular Symposia: Wiley Online Library; 2001. p. 115–122.

44 Pandey, J.K. and Singh, R.P. (2004). On the durability of low-density polyethylene nanocomposites. *e-Polymers* 4: 566–578.

45 Leejarkpai, T., Suwanmanee, U., Rudeekit, Y., and Mungcharoen, T. (2011). Biodegradable kinetics of plastics under controlled composting conditions. *Waste Manag.* 31: 1153–1161.

46 Liu, X., Wang, T., Chow, L.C. et al. (2014). Effects of inorganic fillers on the thermal and mechanical properties of poly(lactic acid). *Int. J. Polym. Sci.* 2014: 8.

47 Bayerl, T., Geith, M., Somashekar, A.A., and Bhattacharyya, D. (2014). Influence of fibre architecture on the biodegradability of FLAX/PLA composites. *Int. Biodeterior. Biodegrad.* 96: 18–25.

48 Nieddu, E., Mazzucco, L., Gentile, P. et al. (2009). Preparation and biodegradation of clay composites of PLA. *React. Funct. Polym.* https://doi.org/10.1016/j .reactfunctpolym.2009.03.002.

49 Ray, S.S. and Bousmina, M. (2005). Biodegradable polymers and their layered silicate nanocomposites: in greening the 21st century materials world. *Prog. Mater. Sci.* 50: 962–1079.

50 Chum, H.L. (ed.) (1991). Polymers from bio-based materials. In: *Process Chemistry and Technology.* William Andrew.

51 Vroman, I. and Tighzert, L. (2009). Biodegradable polymers. *Materials* 2: 307–344.

52 Shalaby, S., Ikada, Y., Langer, R. et al. (1995). Polymers of biological and biomedical significance. *Ann. Biomed. Eng.* 23: 333.

53 Evans, M.R., Taylor, M., and Kuehny, J. (2010). Physical properties of biocontainers for greenhouse crops production. *HortTechnology* 20: 549–555.

54 Gómez, E.F. and Michel, F.C. (2013). Biodegradability of conventional and bio-based plastics and natural fiber composites during composting, anaerobic digestion

and long-term soil incubation. *Polym. Degrad. Stab.* https://doi.org/10.1016/j
.polymdegradstab.2013.09.018.

55 Priyanka, P.S. (2013). Banana fiber/chemically functionalized polypropylene composites with in-situ fiber/matrix interfacial adhesion by Palsule process. *Compos. Interfaces* https://doi.org/10.1080/15685543.2013.799012.

56 Chattopadhyay, S.K., Singh, S., Pramanik, N. et al. (2011). Biodegradability studies on natural fibers reinforced polypropylene composites. *J. Appl. Polym. Sci.* 121: 2226–2232.

57 Wagner, P.A., Little, B.J., Hart, K.R., and Ray, R.I. (1996). Biodegradation of composite materials. *Int. Biodeter. Biodegr.* 38: 125–132.

58 Verma, P., Dixit, S., and Asokan, P. (2013). Biodegradation of coir/epoxy composites. *Appl. Polym. Compos.* 1: 75–84.

59 Li, X. and Ha, C.S. (2003). Nanostructure of EVA/organoclay nanocomposites: effects of kinds of organoclays and grafting of maleic anhydride onto EVA. *J. Appl. Polym. Sci.* 87: 1901–1909.

60 Jimenez, G., Ogata, N., Kawai, H., and Ogihara, T. (1997). Structure and thermal/mechanical properties of poly(ε-caprolactone)-clay blend. *J. Appl. Polym. Sci.* 64: 2211–2220.

61 Lee, S.-R., Park, H.-M., Lim, H. et al. (2002). Microstructure, tensile properties, and biodegradability of aliphatic polyester/clay nanocomposites. *Polymer* 43: 2495–2500.

62 Wang, S., Song, C., Chen, G. et al. (2005). Characteristics and biodegradation properties of poly (3-hydroxybutyrate-co-3-hydroxy valerate)/organophilic montmorillonite (PHBV/OMMT) nanocomposite. *Polym. Degrad. Stab.* 87: 69–76.

63 Mao, L., Imam, S., Gordon, S. et al. (2000). Extruded cornstarch-glycerol-polyvinyl alcohol blends: mechanical properties, morphology, and biodegradability. *J. Polym. Environ.* 8: 205–211.

64 Chen, L., Imam, S.H., Gordon, S.H., and Greene, R.V. (1997). Starch-polyvinyl alcohol crosslinked film-performance and biodegradation. *J. Environ. Polym. Degrad.* https://doi.org/10.1007/BF02763594.

65 Russo, M.A., O'Sullivan, C., Rounsefell, B. et al. (2009). The anaerobic degradability of thermoplastic starch: polyvinyl alcohol blends: potential biodegradable food packaging materials. *Bioresour. Technol.* 100: 1705–1710.

66 Chai, W.-L., Chow, J.-D., Chen, C.-C. et al. (2009). Evaluation of the biodegradability of polyvinyl alcohol/starch blends: a methodological comparison of environmentally-friendly materials. *J. Polym. Environ.* 17: 71.

67 Ali, S.S., Tang, X., Alavi, S., and Faubion, J. (2011). Structure and physical properties of starch/poly vinyl alcohol/sodium montmorillonite nanocomposite films. *J. Agric. Food Chem.* 59: 12384–12395.

68 Spiridon, I., Popescu, M.C., Bodârlău, R., and Vasile, C. (2008). Enzymatic degradation of some nanocomposites of poly(vinyl alcohol) with starch. *Polym. Degrad. Stab.* 93: 1884–1890.

8

Applications of Green Polymeric Composite Materials

Bilahari Aryat[1], V.K. YaduNath[1], Neelakandan M. Santhosh[1] and Deepu A. Gopakumar[1,2]

[1] *International and Inter University Center for Nanoscience and Nanotechnology, Mahatma Gandhi University, Kottayam, Kerala, India*
[2] *School of Industrial Technology, Universiti Sains Malaysia, Penang, Malaysia*

8.1 Introduction

Interest in the design of green polymeric composites that can reduce fossil fuel use and provide a platform for the protection of ecosystems and the environment has greatly increased in recent years. Currently, the use of synthetic polymers and their derivatives contributes to the decline of soil quality due to the presence of eliminative hazardous chemicals and their non-biodegradability. To overcome this, natural polymers are preferred [1]. Biopolymers like cellulose, starch, collagen, keratin, and chitin etc. are obtained from various biological sources such as plants, animals, bacteria, and algae. During their degradation, the chemical structure and mechanical properties of these biopolymers are completely lost and they converted into other compounds that are useful to the environment. Polymeric matrices (continuous phase) and reinforcing materials (discontinuous/dispersed phase) are the main constituents of polymeric composites. Green polymeric composite materials are formed by the association of a biodegradable polymer as matrix and natural fibres as reinforcing materials [2]. These ecofriendly composites have good mechanical (e.g. strength and elastic modulus), thermal, electrical, and chemical properties [3]. Based on renewability, green polymeric composites can be divided into two categories: renewable composites and partially renewable composites. Renewable composites contain a matrix and reinforcing materials that are both renewable. In the case of partially renewable composites, either the matrix or the reinforced material is renewable [4].

Biodegradable polymers have huge applications in the clinical instrument, packaging, and agricultural industries. Thin films of biodegradable polymers are used to facilitate early cropping and interrupt early weed formation [5]. Modified biopolymer nanocomposites have excellent electrical and magnetic properties so are employed in the manufacturing of solar cells, light emitting diodes (LEDs), medical devices, sensors, and display panels. Green polymer nanocomposites are widely used to build hurricane-resistant houses and automotive machinery. Hollow keratin fibres and chemically modified soybean oil are used for electronic applications due to their

Bio Monomers for Green Polymeric Composite Materials, First Edition.
Edited by P.M. Visakh, Oguz Bayraktar and Gopalakrishnan Menon.
© 2020 John Wiley & Sons Ltd. Published 2020 by John Wiley & Sons Ltd.

low dielectric constants [6]. Great attention needs to be paid to the packaging of materials, especially food items. Green polymers like cellulose have excellent barrier properties and keep materials airtight [7]. The toxicity and non-biodegradability of synthetic polymers mean they are not suitable for applications in the clinical instrument industry. Currently, technologies are focused on formulating engineered materials that are ecofriendly, biocompatible, and non-toxic to nature and the environment. Chitin-based polymer composites provide an alternative platform for the preparation of biodegradable scaffolds and regenerative tissues [8]. This chapter mainly focuses on the different green composite materials and their major applications.

8.2 Biotechnological and Biomedical Applications of PEG

8.2.1 Biological Separations

The "phase partitioning" method was first established by Albertsson [9]. The production of aqueous polymer two-phase systems has been known since the last century, but Albertsson took advantage of the phase system for biological purification. This method involves purification of biological components (cells, viruses, nucleic acids, proteins) by partitioning them between the two phases acquired by a solution of a combined set of polymers (typically polyethylene glycol [PEG] and dextran) in aqueous buffer. It is fascinating that immiscible lamina can be made, although each lamina consists of more than 90% water. As indicated earlier, this incompatibility of PEG with new polymers is one of its main disadvantages.

The physical mechanism for this incompatibility is not evident (it has been considered in terms of Flory–Huggins theory) [10–12] but the incompatibility of PEG with neutral polymers, such as dextran, is obviously connected to its incompatibility with proteins and nucleic acids, which represent possible non-immunogenic enzymes and non-fouling surfaces. It should be noticed that PEG is not uncommon in the development of two-phase systems; many pairs of polymers exhibit this characteristic, and it is indeed possible to create multilayered systems using various polymer composites. Salt solutions also establish two-phase systems with PEG and water (the salt essentially decreases the lower consolute temperature), and PEG will serve a two-phase system with water above its lower consolute temperature. Separation of substances between the two phases, or between the phase and the phase interface in the event of particles, can be handled by changing phase components, including salt concentration, salt identity, pH, polymer identity, polymer molecular weight (MW), and phase-volume ratio. It is especially remarkable that increasing PEG MW acts to make proteins in the dextran-rich phase while increasing dextran MW acts to drive proteins to the PEG-rich phase. Also, various salts give "neutral" phase systems in which purifications are not susceptible to the charge of the required protein or cell but depend somewhat on complex factors such as disparity in membrane lipids.

Other salts, however, are not partitioned equally between the phases and therefore allow "charged" phase systems in which purifications are reliant on the charge of the target entity. Affinity partitioning can be implemented by covalent connection of affinity ligands to PEG. Ligands that have been used in this way include dyes, antibodies, enzyme inhibitors, metals, and hydrophobic alkyl chains. The target object can be divided from the affinity ligand by conventional pH and salt variations or, if an ethylene oxide–propylene oxide copolymer is adopted, by warming above the lower consolute point of the polymer. Affinity partitioning separations are an important alternative to other affinity methods such as cell sorting and chromatography. Frequently, the approach has taken as the treatment of multiple separations in the form of countercurrent distribution or countercurrent chromatography (in which the two immiscible liquids flow past each alternative). Finally, it is essential to mention that phase partitioning offers many advantages for massive-scale, industrial purifications. The approach can be scaled up quickly, and many procedures are available for dealing with liquid–liquid two-phase systems. The inexpensive polymers can be retrieved and recycled after the procedure. A significant advantage in using cell homogenates (e.g. in recombinant DNA technology) is that cell debris collects at the phase interface while nucleic acids can be gathered in the bottom phase. This separation is attractive as an initial step in the large-scale recovery of proteins. There are several modifications to this two-phase partitioning theme. For example, PEG bis-copper-chelates can be exploited for metal-affinity precipitation of proteins [13]. The PEG-derived detergent Triton X-1l4 can be dispersed in the buffer, stirred with a crude protein mixture, and heated above the cloud point to form a two-phase system and partition the proteins. Also, a chromatographic form of partitioning can be derived by immobilizing the dextran–water phase on a chromatographic support and eluting with the PEG–water phase.

8.2.2 PEG Proteins and PEG Peptides for Medical Applications

In 1977, Davis and Abuchowski, and their co-workers established that covalent connection of PEG to a protein gives minimal disruption of activity and makes the protein non-immunogenic and non-antigenic, therefore imparting highly increased serum lifetimes [14, 15]. This knowledge is remarkably important because it appropriates the path to intravenous administration of proteins for employing the intrinsic errors of metabolism and diverse complications. Additional benefits of protein "pegylation" are reduced percentage of kidney clearance and increasing cooperation with cells. Examples of medically useful PEG-altered proteins include PEG-asparaginase for analysis of acute leukemias [14], PEG-adenosine deaminase for therapy for acute combined immunodeficiency disease [16], and PEG-superoxide dismutase and PEG-catalase for reducing tissue damage emanating from reactive oxygen species correlated with ischemia and associated pathological circumstances [17]. Lightly pegylated protein allergens can be exploited in hyposensitization of allergic subjects [18]. There is likewise much pharmaceutical interest in the PEG modification of nano drug molecules (penicillin, procaine, aspirin, etc.) because of their increased water solubility, altered pharmacokinetics, and much-enlarged size and diminished percentage of renal clearance [19]. For example, there is interest in using PEG-peptides, such as PEG interleukin, for stimulation of the immune system [20].

8.2.3 Poly(lactic acid): Properties and Applications

In addition to the therapeutic and separation studies of PEG-proteins outlined above, poly(lactic acid) (PLA) conjugates have diverse demands in biotechnology. In particular, it is significant that PEG attachment induces solubility of PEG-enzymes in dry organic solvents, such as chloroform, where they display attractive functions [21].

Another application of PEGs is in synthesizing artificial enzymes by connecting binding sites and catalytic sites via PEG linkers. The PEG linker is an important feature of this application because of its length, hydrophilicity, and flexibility. Yomo et al. [22] illustrate this application in developing semi-synthetic oxidases. In a similar way, [23] connected catalytically active metal-porphyrin compounds to PEG-modified bovine serum albumin, which operates as a binding spot, to achieve hybrid catalysts that are active in organic media.

8.2.3.1 Activity of PEG on Non-fouling Surfaces

There is considerable opportunity to apply surface-attached PEG for to the inhibition of protein adsorption. Attaching a monolayer of PEG to a surface can reduce the occurrence of many medically unwanted processes. Drug delivery from PEG-based gels, which also feature good protein-rejecting properties, displays considerable potential as effectively. The incorporation of PEG lipids into the surface of liposomes caused increased serum lifetime and altered pharmacokinetics, as displayed in a similar study [24, 25]. Therefore, it emerges that PEG on the surface of a liposome, as for PEG-proteins, shields small particles from the immune system.

8.2.3.2 Tether between Molecules and Surfaces

PEG displays useful characteristics when applied as a tether or linker to couple an active molecule to a surface. In this treatment, the PEG operates to suppress non-specific protein adsorption on the surface. Additionally, the tethered molecules have been shown to be remarkably active, performing virtually as loose molecules in solution. This knowledge can be justified by understanding that the remarkably long, fully-hydrated PEG linker moves the active molecule further out into solution, some distance from the surface [26].

8.2.3.3 Control of Electro-osmosis

Electro-osmosis is the fluid flow adjoining a charged surface that appears (e.g. in capillary electrophoresis) when an electrical potential is used, leading soluble counter-ions to shift toward the oppositely charged electrode. Studies in capillary electrophoresis have long been involved in getting rid of or managing electro-osmosis. One example of this process includes adsorbing polysaccharides onto the charged surface to develop a viscous layer that impedes counter-ion movement. One drawback of this method is the desorption of polysaccharides in the presence of proteins. In 1986 it was demonstrated that covalently bound PEG-5000 essentially eliminates the electro-osmotic stream, while lower MW PEGs reduce such flow. Recent research work has demonstrated capillary electrophoresis as an analytical procedure, and other investigators have studied PEG coatings [27]. An additional advantage of PEG coatings is that protein adsorption to the capillary surface is also noticeably decreased.

8.2.3.4 PLA as a Viable Biodegradable Polymer

PLA is a biodegradable polymer that has many applications. It has been extensively employed in the biomedical and pharmaceutical fields for several decades due to its biocompatibility and biodegradability concerning mammalian bodies. For many generations, however, the demand for PLA was extremely low because of the high cost of synthesis in the laboratory. For the most part, direct polycondensation (Figure 8.1) was

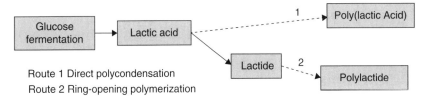

Figure 8.1 PLA production (http://dx.doi.org/10.1016/B978-1-4377-4459-0.00008-1).

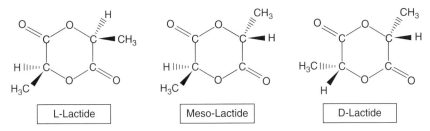

Figure 8.2 L-lactide, D-lactide (meso-lactide), and D-lactide (http://dx.doi.org/10.1016/B978-1-4377-4459-0.00008-1).

utilized to produce PLA from lactic acid. The resultant PLA had a low MW and poor mechanical properties. The properties of PLA increased with the increase in the rate of manufacture using ring-opening polymerization, which involves a transitional material recognized as lactide. Lactide is the cyclic dimer of lactic acid and can exist in the form of L-lactide, D-lactide (meso-lactide), and D-lactide stereo complex (see Figure 8.2). Nowadays, the synthesis of PLA from lactic acid is seldom from chemically synthesized lactic acid. The lactic acid used is generated from the fermentation of carbohydrates such as starch and cellulose. An enormous percentage is derived from corn and cassava. Microorganism-based fermentation yields mainly L-lactic acid.

Most of the PLA produced worldwide is for domestic applications, such as testiles, bottles, cups, food serviceware, etc. (Table 8.1). These PLA products are intended to replace existing petrochemical polymers, with the advantage that their production is environmentally friendly and they are biodegradable.

8.3 Industrial Applications

Green polymeric composite materials are utilized in a variety of industries as well as in several other applications. Due to the sudden spike in interest in green chemistry and green product development, green composites are also a large area of interest for making appliances as well as sporting goods and electronic parts. Table 8.2 summarizes the application of green composites in various industries.

Green polymer-based nanocomposites are ideal for making packaging materials and product casings such as cell phone covers and biodegradable bags and wraps. Their biocompatibility along with the antimicrobial and UV absorption properties of the fillers gives these composites an advantage over traditional packaging and casing materials. The composites can also be used to make a circuit board for electronics, wires, and

Table 8.1 Applications of PLA. Adapted from [28].

Application	Manufacturer/user (product)	Description
Apparel	Mill Direct Apparel Codiceasbarre (shirts), Gattinoni (wedding dresses), Descente (sportswear), etc.	PLA fibre is one of the best candidate for the garments. PLA based apparels shows excellent wicking properties with low moisture and odour retention. These are hypoallergenic, eliciting no skin irritation. For apparel, Ingot can be blended with a maximum of 67% natural, cellulosic or manmade fibre to achieve a variety of properties.
Bottles	Shiseido-Urara (shampoo bottles), Polenghi LAS (lemon juice bottles), Sant'Anna (mineral water bottles), etc.	PLA is well known for making bottles suitable at or slightly above room temperature since PLA bottles tend to deform at temperatures of 50–60 °C, i.e. the glass transition temperature (T_g) of PLA. When the temperature reaches T_g, the amorphous chain mobility of the plastic starts to increase significantly. The PLA material, which is glassy and rigid at room temperature, gradually turns mobile and rubbery at T_g. However, PLA bottles have excellent gloss, transparency, and clarity equal to polyethene terephthalate (PET). The PLA also has exceptional flavor and aroma barrier properties. The substitution of 100 000 of 32 oz. juice bottles can save fossil fuels equating to 1160 gal of greenhouse gases or a car traveling for 23 800 miles.
Food packaging	Lindar (thermoform containers), InnoWare Plastics (thermoform containers), Carrefour Belgium (grocery bags), etc.	PLA is suitable to be used for lightweight and transparent food packaging containers. It is highly glossy and can be easily printed like existing materials such as polystyrene, polyethylene, and polyethylene terephthalate. Container lidding made from PLA is compostable and renewable; typical lidding applications include yoghurt pots, sandwich containers, and fresh food trays for fruits, pasta, cheeses, and other delicatessen products. The design of a solution of compostable delicatessen lidding of NatureWorks PLA is shown. The advantages of this lidding design are superior flavor and aroma barrier up to 47 °C, with strong resistance to most oils and fats in contact with food products. The heating sealing can be done at temperatures as low as 80 °C with the heat seal strength 0.1.5 lb./In. PLA has good compatibility with many ink formulations with natural surface energy of 38 dyne/cm². Additional treatment with both corona and flame can further enhance surface energy to over 50 dyne/cm². The conversion of 250 000 medium-sized delicatessen containers to PLA can save 3000 gal of gas/greenhouse gas emissions progressively.

Films	PLA films are made for bakery goods, confectionery, salads, shrink wrap, envelope windows, laminated coatings, multi-layer performance packaging, etc. PLA can be made into a biaxially oriented plastic film for packaging bags. PLA plastic bags take a few months to fully degrade when buried in compost. The thickness of the film affects the rate of degradation and mass losses. PLA marketed by NatureWorks is specially made for processing using the blown film equipment for low-density polyethylene film. It can also be processed using the oriented polypropylene facility with minor modifications to the setting. Every year, millions of plastic bags are disposed of, causing white pollution to the ground and water. The substitution of petroleum-based plastic bags for PLA bags can make significant environmental savings. The replacement of 20 million medium salad package bags can help to save fossil fuel equal to 29 200 gal of greenhouse gas emissions.
Frito-Lay SunChip, Walmart (salad packaging), Naturally Iowa (EarthFirsts shrink sleeve label), etc.	
Fashion products	Environmentally friendly PLA can be used to produce typical parts of motobike helmets. This is only limited by the artistic design; the outer part of the helmet is covered with PLA-calendered cloth. Similarly, the fashion brand Rizieri, of Milan, Italy, has created an innovation known as "zero impact," involving models of "handmade" products based on PLA or Ingres fabric. These products have all the delicacy of silk to the touch.
Fashion helmet (designer helmet), Rizieri (ladies' shoes), etc.	
Expanded foam	This technology was developed by Biopolymers Network, and received the Best Innovations in Bioplastics Award at the annual European Bioplastics Conference. It relies on the use of CO_2 as the expansion agent, which is a safer substance than expandable polystyrene using pentane as the expansion agent. The compostability of the expanded PLA foam provides an environmentally friendly solution for the electrical and electronics industry, which uses expanded foam as a cushioning material during shipping.
Foam Fabricator, Inc. (expandable foam for cushioning)	
Foam trays	Foam trays are important in packaging, especially for fresh food. Styrofoam is a well-known foam tray made from polystyrene. This type of polystyrene is cheap but non-degradable. Recycling of foam trays is not a profitable business because a large collection volume is needed to rework into a small amount of dense resin. The density of styrofoam is 0.025 g/cm^3 compared to virgin polystyrene resin, which is 1.05 g/cm^3. This means that 42 foam trays are needed to revert to the original dense polystyrene at a similar volume. PLA is a good replacement because the disposed of PLA foam trays can be composted easily without causing adverse effects to the environment. Moreover, the composting of PLA provides enriching nutrients when buried in the soil.
Sealed Air (Cryovacs NatureTRAY food tray), Dyne a-Pak Inc. (Dyne-apak Nature meat foam tray), etc.	

(Continued)

Table 8.1 (Continued)

Application	Manufacturer/user (product)	Description
Surgical implants	Zimmer (Bio-steaks suture anchor and bone cement plug), Ethicon (Vicryl suture and Vicryl mesh) and Sulzer (Sysorbs screw), etc.	PLA and its copolymer polylactide-co-glycolide (PLGA) are compatible with living tissue. However, this is limited to the L-stereoisomer of PLA because mammalian bodies only produce an enzyme that breaks down the L-stereoisomer. PLA and PLGA are used to fabricate screws, pins, scaffolds, etc. to provide a temporary structure for the growth of tissue, eventually breaking down after a certain period. The purpose of copolymerizing with comonomer glycolide is to control the rate of degradation through the modification of crystallization. Sometimes, L- and D-isomers of lactides are copolymerized for this purpose. Although poly(D-lactic acid) cannot be consumed by the body's enzymes, prolonged exposure to body fluid tends to initiate hydrolysis, which eventually breaks down the macromolecules. PLA and copolymers are often used in orthopaedic surgery to fabricate artificial bones and joints. PLA has been used to make surgical sutures for decades. In short, PLA is an important material for biomedical surgical applications.
Drug carrier	Abbott (Lupron Depots for palliative treatment of advanced prostate cancer), Janssen Pharmaceuticals (Risperdal Consta, for the treatment of schizophrenia), etc.	Most of the PLA drug carriers on the market are available in copolymer form. This is because high-purity PLA possesses high crystallinity and takes a longer time to degrade while releasing active drugs. The majority of PLA drug carriers are copolymerized with different percentages of polyglycolic acid (PGA). Normally such drug carriers slowly release the medication for long-term treatments. For instance, leuprolide acetate applied with a microsphere delivery system of PLA and PLGA is used for the treatment of cancer and fibroids. PLGA can be used in the form of implants and gels with the therapeutics goserelin acetate and paclitaxel, for the treatment of prostate/breast cancer, or other anticancer drugs.

Table 8.2 Applications of green polymer composites.

Areas	Products
Packaging and product casings	Cell phones, computers, and other consumer items
Microelectronics	Chip packaging and in circuit boards
Automobiles	Panels, various vehicle parts
Medical industry (non-invasive)	Casts for broken limbs
Electrical	Wires and switches
Constructions	Doors and panels
Sports goods	Balls, skateboards etc.

switches according to whether the filer is a conducting or non-conducting material. Green composites are also used in wooden panels to give them a better finish and are used in construction to improve the finish of a number of products.

In the automobile industry, brake cables, panels and various other vehicle parts can be made from green polymer composites. The use of polymer composites reinforced with lingo-cellulosic fibres is already replacing the use of synthetic fibre composites. Automobile companies in Europe, such as Mercedes, are using natural fibre composites to make "green" cars. Figure 8.3 shows the different components made from green polymer composites in a Mercedes Benz sedan.

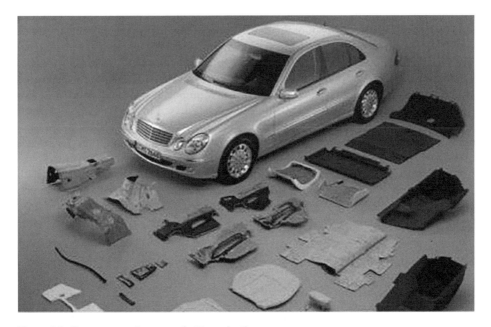

Figure 8.3 Green composite parts of a Mercedes Benz car.

A German company called Greenline produces laptop covers made from organic bioplastics and natural fiber-reinforced plastics. In short, green polymer composites can replace petroleum-based polymeric materials in many applications in our day-to-day life.

8.3.1 Biological Applications

Lately, interest has been growing in the biomedical applications of green polymeric composites, mainly due to their biodegradability. Some of these applications are described below.

8.3.2 Biosensors

Biosensors made from aluminosilicate/polymeric composites are used as sensors to detect diabetics or urea in humans. Zeolite infused in a polymeric matrix such as polysulfone or polydimethylsiloxane has been used to detect urea [29]. The use of nanocomposites can enhance interactions with specific proteins and can enhance catalytic activity, as well as minimizing structural change.

8.3.3 Tissue Engineering

Biodegradable scaffolds can be used to support and guide the in-growth of cells. Tissue scaffolds must be biodegradable, biocompatible, and sterilizable. An ideal scaffold should have enough mechanical strength to withstand physiological strains and must provide a suitable environment for cellular growth. Scaffolds should ideally have properties such as wettability, stiffness, and compliance to support cell/tissue attachment and proliferation [30–32]. In addition, their degradation products should be biocompatible, non-toxic, and transportable out of the body. There are two tissue-engineering methodologies currently in use: unseeded and seeded. In the unseeded technique, a biodegradable scaffold is implanted into the host *in vivo*, which allows the natural process of regeneration to occur. In the seeded technique, cells from the host's tissue are cultured on a biodegradable scaffold *in vitro*, creating a cell–composite grafts, followed by *in vivo* implantation of the graft in the host. Aluminosilicate-based materials can be used for scaffold preparation and tissue engineering. Biocompatible composites, using solution mixing and freeze-drying processes, can be utilized in scaffold preparation. Hydrogels can be developed from polymers and can be used in tissue engineering. Thermoresponsive polymeric nanocomposite gels can impart large elongation before breaking, with high moduli and strength. Recently, electrically conducting hydrogels have been developed for tissue engineering, which combine the advantages of conductive polymers and hydrogels, but their non-degradability hinders their application to a large extent [33].

8.3.4 Wound-healing Applications

Wounds with tissue loss, such as burns, and wounds without tissue loss, such as lacerations, can be healed using green polymeric nanocomposites. Connective tissue membranes for wound healing can be developed using a lithium chloride/dimethylacetamide mixture through solvent casting as they have good swelling and moisture transmission abilities. They also have excellent antibacterial properties [34]. Chitosan-based hydrogels can induce wound healing and will leave fewer scars. Chitosan-based hydrogels are also ideal for wound healing because they can deliver fibroblast growth factor 2 (FGF 2), which activates capillary endothelial cells and fibroblasts [35]. As well as these applications, chitosan-based hydrogels are also used for drug delivery, as contrast agents in MRI, for osteoregeneration, as hemostatic materials, and as carriers for bioactive materials for in-cell chemistry. They are also used for storing and releasing nitrous oxide into muscle cells, which can help in relaxing smooth muscle cells. Hydroxyapatite-based green polymer composite materials can be used to treat bone diseases and conditions such as fractures. They can be used to make sutures and casts, and also plates for bone repair.

8.3.5 Packaging Applications

Green polymeric composites are ideal for packaging due to their biocompatibility and biodegradability. Conventional packaging materials are not ideal as they take too long to decompose. Biopolymers such as starch, PLA, poly(ε-caprolactone) (PCL), and cellulose derivatives are being investigated to solve this problem [36–38]. PCL-based composites have already been reported to be ideal by researchers. Recent studies have shown that these types of fillers are ideal for packaging materials due to their good antibacterial and antioxidant properties [36]. Green polymer nanocomposites are considered to be a good material for food packaging because they form a gas barrier (for O_2 and CO_2) or water vapor barrier, and have high tensile strength, heat resistance, chemical and thermal stability, biodegradability, recyclability, optical clarity, configuration stability, antimicrobial and antifungal activities, and ability to sense microbiological and biochemical changes in food products [39]. Three types of technology are used for food packaging materials: nanoreinforcement packaging, nanocomposite active packaging, and nanocomposite smart packaging. Nanoreinforcement can enhance polymer flexibility, temperature/moisture stability, and gas barrier properties. PLA is usually used in this type of packaging as it has many environmental advantages and can integrate antibacterial nanoparticles such as zinc oxide (ZnO) in its matrix with stability. Polymer–clay composites made of halloysite clay mineral are already in production for packaging applications. In active packaging, the food materials can interact with nanomaterials in the polymer matrix and this can have an important role in food preservation. Smart packaging can integrate sensing (O_2 and gas sensors) and freshness indicators into the packaging material and can also act as a guard against imitation products [39].

8.4 Conclusion

Green polymeric composites are defined as biocomposites combined with natural fibres and biodegradable resins. They are called green polymeric composites due to their degradable and sustainable properties, and they can be easily degraded without affecting environment. The challenge of green composites involves the challenge of obtaining "green" polymers, which are then used as the matrix for the production of the composites. A polymer that shows environmentally favorable properties such as renewability and degradability is called a green polymer. Designing a new polymer with environmentally favorable properties such as renewability and degradability is a challenging task. Research on green polymers has produced a series of interesting green polymer composites from thermoplastic starch and its blends, PLA and its modifications, cellulose, gelatin, chitosan, etc. Natural fibers are synthesized from agricultural sources such as jute, banana, bamboo, and coconut coir, etc show more environmentally friendly properties than the synthetic fibers. There is thus a wide range of possible applications of nanocomposites from agriculture to automobiles. However, the problems of poor adhesion of matrix and fiber, the challenge of fiber orientation, the achievement of nanoscale sizes, and the evolution of truly green polymers that are environmentally friendly and renewable must first be solved.

References

1 Sahari, J. and Sapuan, S.M. (2011). Natural fibre reinforced biodegradable polymer composites. *Rev. Adv. Mater. Sci* 30: 166–174.

2 Castleton, H.F., Stovin, V., Beck, S.B.M., and Davison, J.B. (2010). Green roofs; Building energy savings and the potential for retrofit. *Energ. Buildings* 42 (10): 1582–1591.

3 Adeosun, S., Lawal, G., Balogun, S., and Akpan, E. (2012). Review of green polymer nanocomposites. *Scirp. Org* 11 (4): 483–514.

4 Thakur, V.K., Thakur, M.K., and Gupta, R.K. (2014). Review: raw natural fiber–based polymer composites. *Int. J. Polym. Anal. Charact.* 19 (3): 256–271.

5 Billingham, N.C. (1996). *Degradable Polymers: Principles and Applications*, 53 (2): 269-27010

6 Nakamura, R., Goda, K., Noda, J., and Ohgi, J. (2009). High temperature tensile properties and deep drawing of fully green composites. *Express Polym. Lett.* 3 (1): 19–24.

7 Tang, X. and Alavi, S. (2012). Structure and physical properties of starch/poly vinyl alcohol/laponite RD nanocomposite films. *J. Agric. Food Chem.* 60 (8): 1954–1962.

8 Di Martino, A., Sittinger, M., and Risbud, M.V. (2005). Chitosan: A versatile biopolymer for orthopaedic tissue-engineering. *Biomaterials* 26 (30): 5983–5990.

9 Albertsson, P. (1970). Partition of cell particles and macromolecules in polymer two-phase systems. *Adv. Protein Chem.* 24 (C): 309–341.

10 Benavides, J., Rito-Palomares, M., and Asenjo, J.A. (2011). Aqueous two-phase systems. In: *Comprehensive Biotechnology Chapter: Aqueous two-phase system Publisher: Elsevier Editors: Murray Moo-Young*, 697–713.

11 Baskir, J.N., a Hatton, T., and Suter, U.W. (1989). Protein partitioning in two-phase aqueous polymer systems. *Biotechnol. Bioeng.* 34 (1): 541–558.

12 Conover, W. (1998). Buffer Solutions: The Basics (Beynon, R. J.; Easterby, J. S.). *J. Chem. Educ.* 75 (2): 153.

13 Suh, S.S., Van Dam, M.E., Wuenschell, G.E. et al. (1990). Novel Metal-Affinity Protein Separations. *Protein Purif.* 427 (427): 139–149.

14 Abuchowski, A., Kazo, G.M., Verhoest, C.R.J. et al. (1984). Cancer therapy with chemically modified enzymes. I. Antitumor properties of polyethylene glycol-asparaginase conjugates. *Cancer Biochem. Biophys.* 7 (2): 175–186.

15 Davis, F.F., van Es, T., Palczuk, N.C. et al. (1979). Soluble, non-antigenic polyethylene glycol-bound enzymes. *Am. Chem. Soc. Div. Polym. Chem. Prepr.* 20 (1): 357–360.

16 Hershfield, M.S., Buckley, R.H., Greenberg, M.L. et al. (1987). Treatment of Adenosine Deaminase Deficiency with Polyethylene Glycol–Modified Adenosine Deaminase. *N. Engl. J. Med.* 316 (10): 589–596.

17 Beckman, J.S., Minor, R.L., White, C.W. et al. (1988). Superoxide dismutase and catalase conjugated to polyethylene glycol increases endothelial enzyme activity and oxidant resistance. *J. Biol. Chem.* 263 (14): 6884–6892.

18 Norman, P.S., Lichtenstein, L.M., Kagey-Sobotka, A., and Marsh, D.G. (1982). Controlled evaluation of allergoid in the immunotherapy of ragweed hay fever. *J. Allergy Clin. Immunol.* 70 (4): 248–260.

19 Zalipsky, S., Gilon, C., and Zilkha, A. (1983). Attachment of drugs to polyethylene glycols. *Eur. Polym. J.* 19 (2): 1177–1183.

20 Katre, N.V., Knauf, M.J., and Laird, W.J. (1987). Chemical modification of recombinant interleukin 2 by polyethylene glycol increases its potency in the murine Meth A sarcoma model. *Proc. Natl. Acad. Sci. USA.* 84 (6): 1487–1491.

21 Inada, Y., Takahashi, K., Yoshimoto, T. et al. (1988). Application of PEG-enzyme and magnetite-PEG-enzyme conjugates for biotechnological processes. *Trends Biotechnol.* 6 (6): 131–134.

22 Yomo, T., Urabe, I., and Okada, H. (1992). Principles for designing enzyme-like catalysts based on the rate-acceleration mechanisms of semisynthetic oxidases. *Eur. J. Biochem.* 203 (3): 543–550. https://doi.org/10.1111/j.1432-1033.1992.tb16581.x.

23 Yoshinaga, K., Ishida, H., Sagawa, T., and Ohkubo, K. (1992). PEG-modified protein hybrid catalyst. In: *Poly(Ethylene Glycol): Chemistry, Biotechnology and Biomedical Applications* (ed. J.M. Harris), 103–114. Boston, MA: Springer https://doi.org/10.1007/978-1-4899-0703-5_7.

24 Senior, J., Delgado, C., Fisher, D. et al. (1991). Influence of surface hydrophilicity of liposomes on their interaction with plasma protein and clearance from the circulation-studies with poly(ethylenel glycol)-coated vesicles. *Biochim. Biophys. Acta* 1062 (1): 77–82.

25 Klibanov, A.L., Maruyama, K., Torchilin, V.P., and Huang, L. (1990). Amphipathic polyethyleneglycols effectively prolong the circulation time of liposomes. *FEBS Lett.* 268 (1): 235–237.

26 Bonora, G.M., Scremin, C.L., Colonna, F.P., and Garbesi, A. (1990). HELP (high efficiency liquid phase) new oligonucleotide synthesis on soluble polymeric support. *Nucleic Acids Res.* 18 (11): 3155–3159.

27 Bruin, G.J.M., Chang, J.P., Kuhlman, R.H. et al. (1989). Capillary zone electrophoretic separations of proteins in polyethylene glycol-modified capillaries. *J. Chromatogr. A* 471 (C): 429–436.

28 Sin, L.T., Rahmat, A.R., and Rahman, W.A.W.A. (2012). Applications of Poly(lactic Acid). In: *Handbook of Biopolymers and Biodegradable Plastic Properies, Processing and Applications*, 55–69. https://doi.org/10.1016/B978-1-4557-2834-3.00003-3.

29 Lopes, A.C., Martins, P., and Lanceros-Mendez, S. (2014). Aluminosilicate and aluminosilicate based polymer composites: Present status, applications and future trends. *Prog. Surf. Sci.* 89 (3–4): 239–277.

30 Vyavahare, N., Ogle, M., Schoen, F.J. et al. (1999). Mechanisms of bioprosthetic heart valve failure: Fatigue causes collagen denaturation and glycosaminoglycan loss. *J. Biomed. Mater. Res.* 46 (1): 44–50.

31 Persidis, A. (1999). Tissue engineering. *Nat. Biotechnol.* 17 (5): 508–510.

32 Langer, R. and Vacanti, J.P. (1993). Tissue engineering. *Science (80)* 260 (5110): 920–926.

33 Guo, B., Glavas, L., and Albertsson, A.-C. (2013). Biodegradable and electrically conducting polymers for biomedical applications. *Prog. Polym. Sci.* 38 (9): 1263–1286.

34 Anitha, A., Sowmya, S., Kumar, P.T.S. et al. (2014). Chitin and chitosan in selected biomedical applications. *Prog. Polym. Sci.* 39 (9): 1644–1667.

35 Dash, M., Chiellini, F., Ottenbrite, R.M., and Chiellini, E. (2011). Chitosan – A versatile semi-synthetic polymer in biomedical applications. *Prog. Polym. Sci.* 36 (8): 981–1014.

36 Chen, G., Lu, J., Lam, C., and Yu, Y. (2014). A novel green synthesis approach for polymer nanocomposites decorated with silver nanoparticles and their antibacterial activity. *Analyst* 139 (22): 5793–5799.

37 Xie, F., Pollet, E., Halley, P.J., and Avérous, L. (2013). Starch-based nano-biocomposites. *Prog. Polym. Sci.* 38 (10–11): 1590–1628.

38 Song, Z., Xiao, H., and Zhao, Y. (2014). Hydrophobic-modified nano-cellulose fiber/PLA biodegradable composites for lowering water vapor transmission rate (WVTR) of paper. *Carbohydr. Polym.* 111: 442–448.

39 Sorrentino, A., Gorrasi, G., and Vittoria, V. (2007). Potential perspectives of bio-nanocomposites for food packaging applications. *Trends Food Sci. Technol.* 18 (2): 84–95.

9

Hydrogels used for Biomedical Applications

Nafisa Gull[1], Shahzad Maqsood Khan[1], Atif Islam[1] and Muhammad Taqi Zahid Butt[2]

[1] Department of Polymer Engineering and Technology, University of the Punjab, Quaid-e-Azam Campus, Lahore, 54590, Pakistan
[2] Faculty of Engineering and Technology, Department of Metallurgy and Materials Engineering, University of the Punjab, Quaid-e-Azam Campus, Lahore, 54590, Pakistan

9.1 Introduction

In the last few decades, many attempts have been made to replace petrochemical products by renewable, biosourced components [1]. In addition, advances towards new materials that respond to external stimuli, e.g. pH, light, temperature, etc. [2–4], have enthused the progress of innovative protocols for tissue engineering [5], drug delivery [6, 7], and nanobiotechnology [8, 9]. This environmental receptiveness has elicited major changes in both the self-assembling and physicochemical properties of such polymeric systems, which can be employed to promote the release/encapsulation of active molecules. There is a lot of work in the literature where therapeutic molecules or drug-delivery vehicles are based on synthetic polymers [10–12]. Moreover, synthetic polymers must possess a number of characteristics, such as being non-immunogenic, non-toxic, and water soluble, along with safe release properties for real applications which can be a difficult challenge. This is when biodegradable and biocompatible natural polymers demonstrate a renewable and versatile substitute of synthetic polymers. In general biopolymers mostly comprise of polysaccharides and are a basic constituent in industrial food formulations, acting as stabilizing, gelling, or emulsifying agents [13]. Moreover, they have been proven to be important materials for biomedical applications, including regenerative medicine and drug delivery [14–16].

9.2 Hydrogels

A hydrogel is a three-dimensional (3D) crosslinked network made up of a macromolecular hydrophilic polymer, crosslinked by chemical or physical interactions, with applications in advanced industry and daily life. Hydrogels remain stable upon swelling in water and are capable of absorbing a large amount of water ranging from 10% to thousands of times of their own volume. Their physical properties, including permeation, swelling, mechanical strength, and surface characteristics, can be tailored via structural

Bio Monomers for Green Polymeric Composite Materials, First Edition.
Edited by P.M. Visakh, Oguz Bayraktar and Gopalakrishnan Menon.
© 2020 John Wiley & Sons Ltd. Published 2020 by John Wiley & Sons Ltd.

modifications [17]. Moreover, hydrogels can be prepared in different physical forms, such as films, coatings, nanoparticles, microparticles, and slabs. The control of hydrogel structure has been extensively studied to attain different requirements in specialized areas, therefore the gelation phenomenon has primary importance for the fabrication of organized hydrogels with improved functional and mechanical behaviour [18–20].

Natural polymer-based hydrogels attract great interest and are suitable for drug delivery and tissue engineering of bioactive molecules [21]. In medicine, numerous hydrogel products based on natural or synthetic polymers have significant effects on patient care, for example contact lenses are normally made from poly(hydroxyethylmethacrylic)acid [P(HEMA)], biological adhesives used in surgical procedures are made from reconstituted albumin or fibrin, wound dressings are made from alginate polysaccharide, and hyaluronic acid (HA), which is used as filler, is used in several clinical indications [22, 23]. Consequently, hydrogels are used in experimental medicine and clinical practice for a broad range of applications, such as regenerative medicine and tissue engineering [24], diagnostics [25], cellular immobilization [26], barrier materials to regulate biological adhesions, separation of biomolecules or cells, etc. [27].

9.3 Short History of Hydrogels

Hydrogels have been present in the universe since life began on earth. Bacterial biofilms, which are water-swollen extracellular matrix (ECM) components, and plant structures are an omnipresent hydrated pattern in nature. Agar and gelatine are also used as hydrogels and employed in many applications early in human history. But in the advance history of hydrogels, a category of materials prepared for healthcare applications, can be precisely traced. Scientists at DuPont published a research paper on the synthesis of methacrylic polymers in 1936. In this research, poly(2-hydroxyethyl methacrylate) (polyHEMA) was also reported [17]. It was described as a brittle, glassy, hard polymer, and was not thought to be important so it was ignored until 1960. In 1960, Wichterle and Lim reported the polymerization of HEMA and a crosslinker using water and other solvents. Instead of hard and brittle, they found it to be an elastic, water-swollen, clear, soft gel. This discovering led to the development of biomedical hydrogels, as they are now known. Many formulations of hydrogels have been developed over the years [28].

9.4 Methods of Fabrication of Hydrogels

Hydrogels can be fabricated by different chemical methods. These include single-step procedures such as polymerization and simultaneous crosslinking of monomers having different functionalities as well as manifold steps including synthesis of polymers having reactive functional groups and their crosslinking, possibly also reacting the polymers with appropriate crosslinkers. The polymer engineer can fabricate and design polymer networks with molecular-scale control over structure, such as crosslinking density, and with modulated characteristics such as mechanical properties, biodegradation, and biological and chemical response to stimuli [29].

9.5 Classification of Hydrogels

Hydrogels can be classified into three types based on their source:

  i) natural polymer-based hydrogels
 ii) synthetic polymer-based hydrogels
iii) hybrid hydrogels.

Hydrogels based on synthetic or natural polymers are of great importance in cell encapsulation [30]. Over the past decade, these types of hydrogels have become especially attractive as matrices for regenerating and repairing a broad range of organs and tissues [31, 32].

Natural polymer-based hydrogels are normally biocompatible and have nominal stimulation to the immunological or inflammatory receptiveness of the host tissues. Numerous natural polymers have been extensively explored as biomaterials for reparative medicine and tissue engineering. Cartilage implants comprising biodegradable natural polymers are used to deliver cells and/or growth factor to accelerate or stimulate chondrocyte proliferation and differentiation within or on the scaffolds to attain ultimate cartilage regeneration. Additionally, excellent mechanical strength is crucial for such implants to withstand surgery and tough *in vivo* stress conditions when in use. Unfortunately, natural polymers do not have the toughness and strength required to meet pharmaceutical demands. Physical or chemical compositing and amendments are used to induce specific interactions (electrostatic interactions and hydrogen bonding) and functionalities or/and well-defined micro- or nanostructures inside the implants to enhance their toughness, strength, and bioactivity. Synthetic and natural polymers are combined to form strengthened hydrogel, which is also known as hybrid hydrogel.

9.6 Natural Polymers Used for Hydrogels

9.6.1 Protein

9.6.1.1 Collagen
Collagen is a promising candidate for pharmaceutical applications as it is the most abundant protein in mammalian tissues and is the core constituent of natural ECM. There are at least 19 types of collagen but the primary structure of all collagen comprises three polypeptide chains, which are folded to form a three-stranded rope arrangement. The strands are connected by covalent and hydrogen bonding (Figure 9.1). Collagen strands can self-assemble to form strong fibres [33]. Additionally, fibres and scaffolds can be formed from collagen and their mechanical strength can be improved by using different

Figure 9.1 Structure of collagen.

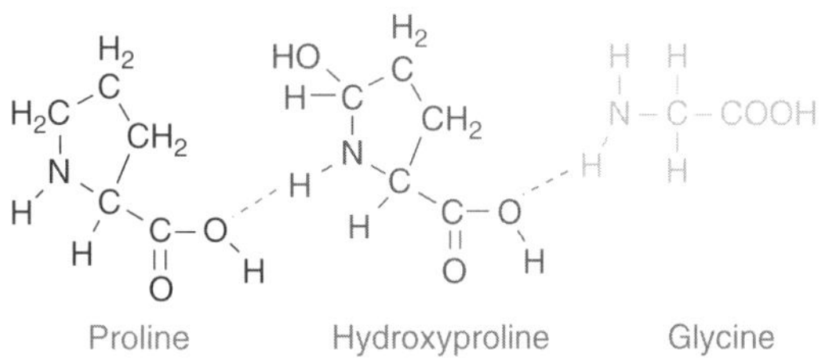

Proline   Hydroxyproline   Glycine

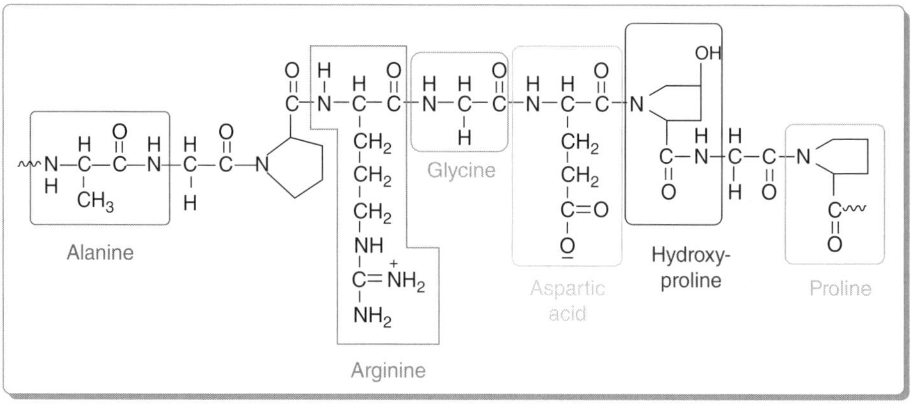

Figure 9.2 Structure of gelatin.

crosslinkers (i.e. carbodiimide, formaldehyde, glutaraldehyde) [34, 35], by crosslinking with physical treatments (i.e. heating, freeze-drying, ultraviolet (UV) irradiation) [34, 36], and by mixing with other natural and synthetic polymers, i.e. polyethylene oxide (PEO), chitosan (CS), poly(lactic-co-glycolic acid) (PLGA), poly(glycolic acid), polylactic acid (PLA), and HA [37–39]. Collagen is naturally decomposed by metallo-proteases, particularly serine proteases and collagenase, enabling its decomposition, which is monitored by the cells in engineered tissue [40].

9.6.1.2 Gelatine

Gelatine is a partial derivative of collagen prepared by hydrolysis, i.e. dissociating the natural triple helix structure of collagen into single-strand molecules, as shown in Figure 9.2 [41]. It is not as immunogenic as its precursor and retains information signals such as the arginylglycylaspartic acid (RGD) sequence, thus endorsing proliferation, differentiation, migration, and cell adhesion [42]. Hydrogels are prepared by blending gelatine and collagen, and their mechanical behaviour can be improved by using different chemical crosslinkers (i.e. carbodiimide, genipin, and glutaraldehyde) [43].

9.6.1.3 Matrigel

Matrigel is gelatinous protein mixture excreted by Engelbreth–Holm–Swarm mouse sarcoma cells mostly comprising enlactin, collagen type IV, laminin, and different growth factors [44]. This mixture bears a resemblance to the complex extracellular environment found in many tissues, and it has been extensively used as a scaffold for angiogenesis, tissue vascularization, and cell differentiation [45, 46]. Protein-based hydrogels can be prepared by thermal gelation and their mechanical characteristics can be improved via chemical crosslinking agents such as glutaraldehyde [47]. Matrigel has been used as a 3D culture model for cancer cells. Cancer cells in matrigel have morphology alteration depending on the malignancy [48, 49], and this culture model is used to differentiate malignant bronchial epithelial cells from normal cells [50]. Matrigel contains ECMs comprising approximately 30% collagen IV, 60% laminin, and 8% entactin. Matrigel also exhibits growth factor-like epidermal growth factor and tumour growth factor-platelet-derived growth factor according to its manufacturers (Corning, Tewksbury, MA). The growth factors in matrigel impart cell migration through the activation of a G-protein, modulation of cell attachment, and remodelling of the cytoskeleton [51, 52], which modifies the morphology of cells through the contribution of actin dynamics [53].

9.6.2 Polysaccharides

Polysaccharides belong to a class of degradable natural polymers that have extensive uses in tissue engineering [54]. These polymers have various sources, including animals, plants, and microbials. In spite of these multiple sources, all polysaccharides comprise repeated saccharide units which have covalent bonding with O-glycosidic bonds. Polysaccharides that have applications in cartilage tissue engineering include agarose, alginate, HA, CS, and chitin [55]. These polysaccharides are different from each other by variations in molecular weight (MW), saccharide units, and linkage types and sites. These divergences in chemical structure result in diverse physical characteristics, such as gelling behaviour, viscosity, properties at interface and surface, electrostatic behaviour, solubility, and mechanical strength [56].

9.6.2.1 Hyaluronic Acid

Hyaluronan is a liner polysaccharide that comprises alternating repeating units of disaccharides i.e., β-D-glucuronic acid, and N-acetyl-β-D-glucosamine (Figure 9.3). It is a non-sulfated glycosaminoglycan (GAG) and is present throughout the body from the vitreous of the eye to the ECM of cartilage tissue [57]. It is a vastly hydrated polyanionic macromolecule with MW ranging from $100,000$ Da in serum to $8,000,000$ Da in the vitreous. It is an indispensable constituent of the ECM, in which its biological and structural characteristics determine its functionality in matrix organization, morphogenesis, wound repair, and cellular signalling [58, 59]. Furthermore, it is quickly changed in the body by hyaluronidase, with tissues having half-lives from hours to days [60]. HA along with its derivatives has been used in clinical applications for more than three decades. Recently, it has become a premier integrant for the development of biomaterials used in regenerative medicine and tissue engineering [61–63].

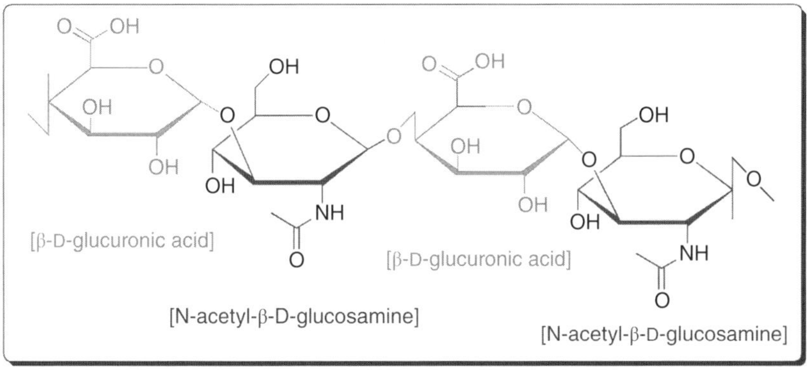

Figure 9.3 Structure of HA.

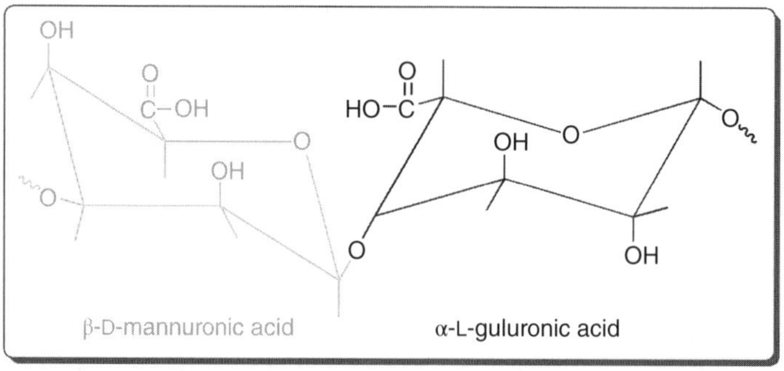

Figure 9.4 Structure of alginate.

9.6.2.2 Alginate

Alginate is brown algae-derived polysaccharide composed of R-L-guluronic acid and β-D-mannuronic acid units, as shown in Figure 9.4 [64]. The MW of alginate is in the range of 10–1000 kDa depending on the production process and source. Alginate can form an ionotropical gel with the addition of multivalent cations, which make it remarkable to be used in biomedical applications [65, 66]. The crosslinking rate should be minimized to prepare *in vivo* injection of a Ca^{+2} crosslinked alginate hydrogel. In the literature, polyols have been reported to hamper hydrogel formation. It is predicted that they obstruct the instantaneous complexation of Ca^{+2} with alginate. The polyol-containing formulation that slows down the hydrogel preparation has been registered as a patent. Alginate is non-immunogenic, biocompatible, and mucoadhesive [67]. It is frequently developed into microcarriers for cell encapsulation [68]. Alginate does not have cell interactive characteristics and several researchers have tried to deal with this problem by pairing it with growth factors, e.g. vascular endothelial growth factor or cell interactive peptides, e.g. RGD, in the backbone chain [69]. Additionally, alginate-based semi-interpenetrating polymer networks with stimuli-responsive behaviour have been prepared. Wang et al. prepared a superabsorbent, pH-sensitive hydrogel comprising polyvinylpyrolidone and sodium alginate-g-poly(sodium acrylate) via free radical polymerization using N,N-methylene-bisacrylamide as crosslinking agent and ammonium peroxidisulfate as initiator [70]. Zhao et al. developed a thermosensitive hydrogel by in situ copolymerization of poly(ethylene glycol)-co-poly(ε-caprolactone) with N-isopropylacrylamide (NiPAAM) via UV irradiation technology along with sodium alginate [71]. By grafting the polyacrylic acid (PAA) to sodium alginate using N,N−methylene-bis-acrylamide as a crosslinker and ammonium peroxidisulfate as initiator, the hydrogel was obtained [72].

9.6.2.3 Chitosan

CS is a copolymer of 2-amino-2-deoxy-D-glucopyranose and β-(1→4)-linked 2-acetamido-2-deoxy-D-glucopyranose. Its chemical structure is shown in Figure 9.5 This type of polycationic biopolymer is normally obtained by alkaline deacetylation from chitin, which is the major ingredient of the exoskeletons of crustaceans such as crabs, shrimps, etc. The major factors that affect the properties of CS are its degree of deacetylation (DDA) and MW, which represents the ratio of deacetylated units.

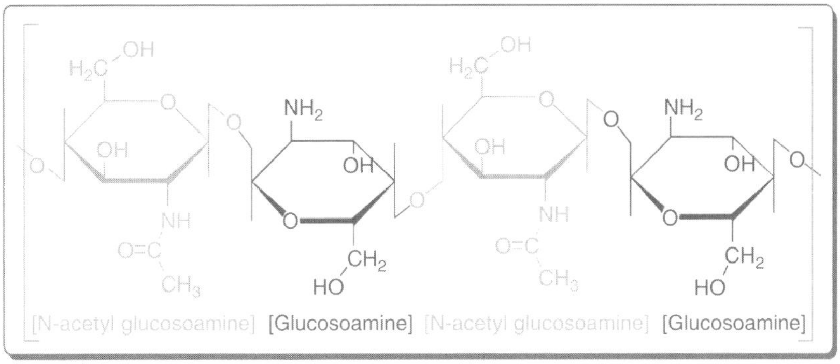

Figure 9.5 Structure of CS.

These factors are investigated by set parameters during fabrication and can be further amended. For instance, DDA can be reduced by reacetylation [73] and MW can be controlled by acidic depolymerization [74].

CS has attracted much attention for its pharmaceutical and medical applications because of its remarkable inherent properties. In fact, due to its biocompatible nature, it is used in medical applications such as implantation [75], topical ocular applications [76], injections [77], etc. Furthermore, it undergoes metabolism by certain human enzymes, particularly lysozyme, to make it biodegradable [78]. It is also documented in the literature that CS is a penetration enhancer because it can open epithelial tight-junctions [79]. It has bioadhesive properties owing to its positive charge at physiological pH, which enhances retention at the site of application [80, 81]. It endorses wound healing [82] and has bacteriostatic effects [83, 84]. Its profuse production is ecologically interesting and is of low cost [85]. It is used as an ingredient in hydrogels for pharmaceutical and medical applications [86].

9.6.2.4 Xyloglucan

Xyloglucan is a cytologically compatible polysaccharide that shows a thermally receptive nature when more than 35% of its galactose residues are removed [87]. Its stereochemical structure is shown in Figure 9.6. Its gel has been implemented as a drug-delivery carrier for diverse applications [88], but there is not sufficient literature on the morphological and rheological properties of xyloglucan-based hydrogels. Nisbet et al. [89] determined the gelation and morphological characteristics of xyloglucan hydrogels at physiological conditions. The gelation phenomenon was thought to be affected due to the presence of phosphate buffer saline (PBS) ions in comparison to deionized water. It was observed that 3 wt.% of xyloglucan in aqueous solution exhibited an elastic modulus that is considerably higher than that of other synthetic or natural hydrogels. In addition, this optimum concentration developed a gel that can be freeze-dried and investigated with scanning electron microscopy [90].

9.6.2.5 Dextran

Dextran has (1→6)-linked α-D-glucopyranosyl residues with three –OH groups per glucose ring (Figure 9.7). A tailored precursor of enzymetically degradable dextran was prepared by reaction of maleic acid with the hydroxyl group of dextran, known as

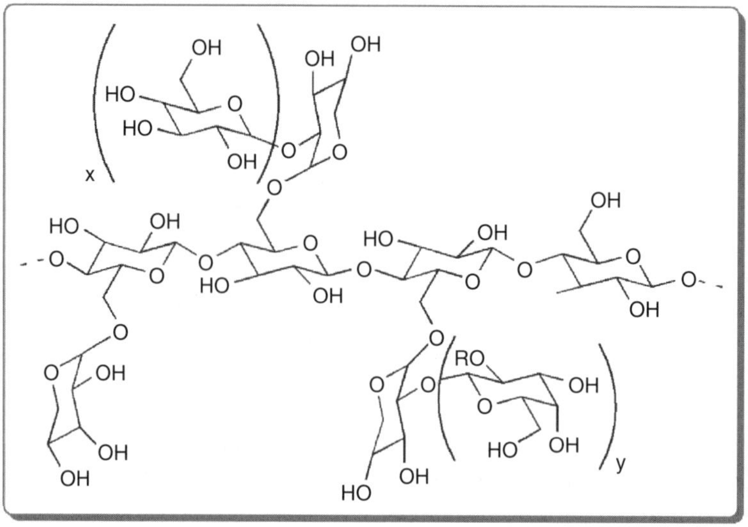

Figure 9.6 Structure of xyloglucan.

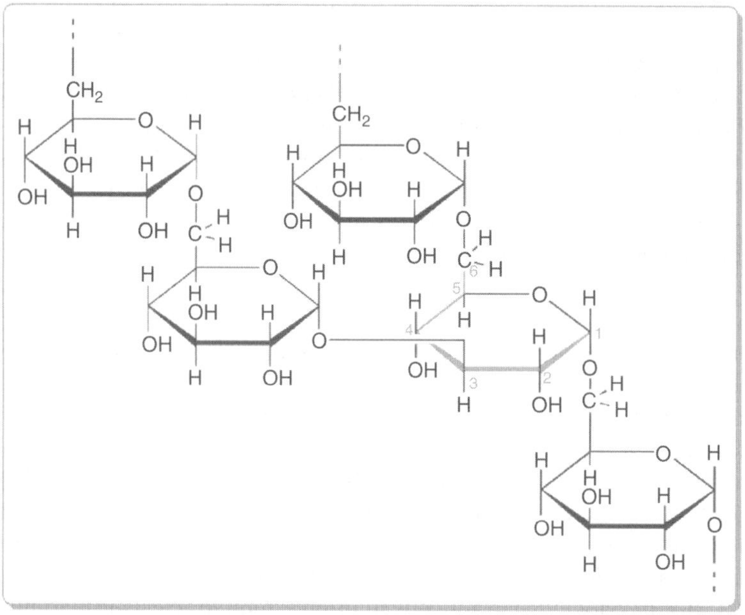

Figure 9.7 Structure of dextran.

dextran-based maleic acid. By the photocrosslinking of dextran maleic acid (Dex-MA) with NiPAAM, thermoresponsive behaviour was induced in it. This hydrogel was partly biodegradable and had a higher lower critical solution temperature (LCST) than NiPAAM owing to the biodegradable and hydrophilic nature of Dex-MA. Furthermore, the –COOH groups of Dex-MA make it a pH responsive hydrogel [91]. Huang et al. also reported a dextran-based polysaccharide [92]. A dextran macromer with

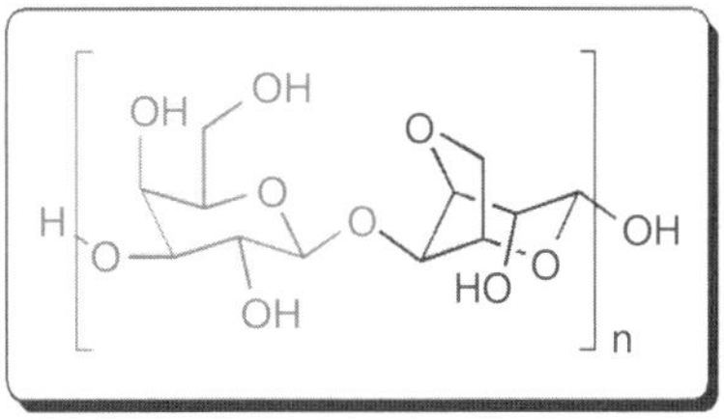

Figure 9.8 Structure of agarose.

2-hydroxyethyl methacrylate and oligolactate units, which has blocks that can degrade hydrolytically, was copolymerized with NiPAAM. This hydrogel had an LCST near to that of pNiPAAM (almost 32 °C). Its degradation and swelling in PBS solution were examined at 25 and 37 °C. At 25 °C (lower than LCST) the hydrogels degraded within two weeks, with the rate of disintegration depending on their constituents. At 37 °C, this degradation was slow due to enhanced hydrophobicity. When this hydrogel was investigated for drug delivery, it was observed that low MW drugs, e.g. methylene blue, were released at a higher rate at 37 °C than at 25 °C while the opposite effect was seen for high MW material, e.g. bovine serum albumin. The researchers concluded that the drug-release mechanism depends on several parameters, such as degradation, swelling, temperature, and interaction between hydrogel and drug [90].

9.6.2.6 Agarose

Agarose also belongs to the polysaccharide family composed of repeated units of 1,3-linked β-D-galactose and 1,4-linked 3,6-anhydro-α-L-galactose (Figure 9.8). It is water soluble above 65 °C and a gel in the temperature range of 17–40 °C, depending on the make-up of the side chain substituted hydroxyethyl groups. These gels show reversible behaviour on thermal response, i.e. they can be gels at low temperature and dissolve in water at high temperature. They are not pH sensitive, i.e. they are stable at body and room temperature. Cell-seeded agarose gels are extensively used in cartilage, as reported in the literature [93]. *In vitro* culture of chondrocyte-seeded agarose prepared for 42 days showed engineered cartilaginous tissue with GAG content up to $5.2 \pm 0.9\%$ ww (wet weight) and compressive Young's modulus up to 730 ± 65 kPa on suitably selected substrates [94]. Ephemeral contact of chondrocyte-seeded agarose hydrogel constructs to turn over transforming growth factor-β3 under serum-free conditions can even capitulate proteoglycan content of 6–7%wt of agarose constructs cultured and compressive modulus of 0.8 MPa for less than 60 days [95]. These mechanical characteristics imitate the local articular cartilage and the biosynthesis of collagen, deoxyribonucleic acid (DNA), GAGs, and other biomacromolecules that demonstrated the rejuvenation of cartilage tissue. These mechanical characteristics are closely associated with GAG content [56].

9.6.3 Heparin

Heparin is composed of α-D-glucosamine residues, β-D-glucuronic acid, and α-L-iduronic acid. Its chemical structure is shown in Figure 9.9. It has the highest negative charge density among biological macromolecules which develop ionic linkages with bioactive molecules like cytokines, growth factors, and proteins. Such non-covalent linkages in heparin not only sequester the proteins but also modulate

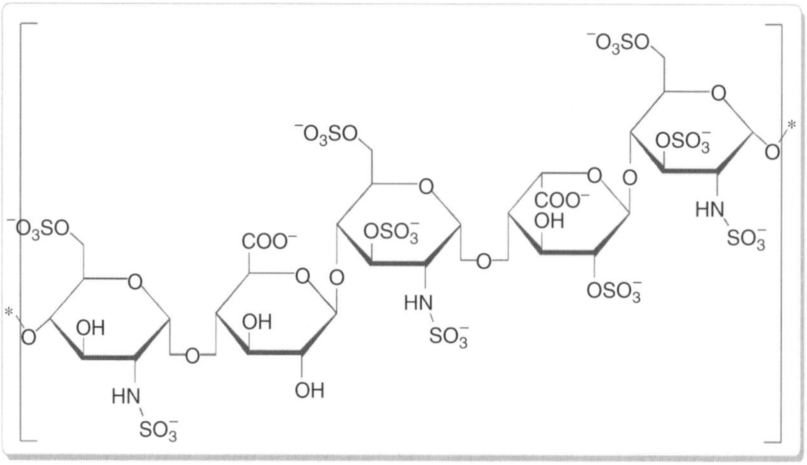

Figure 9.9 Structure of heparine.

their biological activity, such as promoting cell receptor affinity [96]. Heparin and heparin sulphate intercede in various biological interactions, such as cell proliferation, cell adhesion, and cell surface binding of lipase and other proteins, which are critical in tumour metastasis, viral invasion, angiogenesis, blood coagulation, and development processes [97]. Likewise, heparin and heparin sulphate mediate internalization of proteins, normalize protein transport through substrate membranes, and protect proteins from degradation [98]. However, significant undesirable influences of heparin, a persuasive anticoagulant, such as priapism, alopecia, osteoporosis, thrombocytopenia, and bleeding, are related to these biological activities [99]. These adverse outcomes may restrict the use of heparin in specific *in vivo* applications.

Chemically and physically crosslinked heparin-based hydrogels have been used for the assessment of controlled bioactive molecule delivery [100], cell encapsulation [101], and fate and cell function [102–104]. For example, Kiick and coworkers used heparin-based hydrogels to attune cell behaviour in a two-dimensional *in vitro* experiment [102]. To adjust cell response and adhesion, hydrogels with multiple moduli were developed via Michael addition reaction between combinations of maleimide functionalized polyethylene glycol (PEG), thiol functionalized PEG, and maleimide functionalized heparin. Such governable mechanical and biochemical properties systems render the heparin-based hydrogels important contenders for monitoring adventitial fibroblast remodelling of blood vessels. Tae and coworkers developed the heparin-based hydrogels which firmly bind the collagen type I and fibrinogen on the hydrogel surface using the heparin binding tendency by physisorption [104]. The hydrogels were developed by Michael addition using PEG diacrylate and thiolated heparin. The potent physisorption of proteins on heparin-based hydrogels, compared to the control PEG hydrogel, led to improved fibroblast proliferation and adhesion. Such protocols can be used to adhere cells on selective heparin-based surfaces of the hydrogel for applications like tissue engineering, cell culture, and biosensors. Furthermore, Werner and coworkers documented cell replacement therapies in neurodegenerative diseases using heparin-based hydrogels [105]. By tweaking the biological and mechanical behaviour of heparin-PEG hydrogels, axo-dendritic outgrowth and neural stem cell differentiation were modulated. High histocompatibility and *in vivo* stability make these hydrogels potential candidates for neuronal cell replacement therapies [106].

9.7 Synthetic Polymers Used for Hydrogels

9.7.1 Polyacrylic Acid

PAA is a water-soluble biodegradable polymer used in a variety of applications, such as water treatment, disposable diapers (as super adsorbents), etc. Its chemical structure is shown in Figure 9.10. PAA copolymers modified with block copolymers of poly(propylene oxide) and PEO have a broad range of clinical uses as these ingredients are non-toxic from a pharmaceutical point of view [107]. One of the advantages of PAA is that it exists as gel at pH 7 and a liquid at pH 5. Infiltration of cations into the gelled polymer transform the gel into liquid form [108]. As a drug-delivery transporter it is perfect for ocular delivery of ribozymes to the corneal epithelium [109]. Hydrophobically modified PAA (HMPAA) is formed by functionalization of PAA at acidic pH by alkylamines in the presence of *N,N′*-dicyclohexylcarbodiimide in an aprotic solvent. HMPAA exhibited some exceptional rheological properties in aqueous semi-dilute solutions, e.g. inter-chain agglomeration followed by an increase in shear sensitivity, viscosity, and MW. PAA-based polymers are usually used in mucosal and oral contact applications like bioadhesives, oral suspensions, and controlled-release tablets. They are also employed as emulsion stabilizers, suspending agents, and thickeners in low-viscosity materials for contemporary applications. For mucosal and bioadhesive products, high MW divinyl glycol crosslinked with PAA is widely prepared in different drug-delivery systems. All rectal, vaginal, nasal, intestinal, and buccal bioadhesive products can be prepared with this polymer [110].

9.7.2 Polyimide

Polyimide hydrogels are extensively used in reconstructive and plastic surgery. The chemical structure of polyimide is presented in Figure 9.11. Polyimide hydrogels can mimic soft tissues as they contain a large amount of water, have a high level of biocompatibility, and have the required mechanical strength. They are transparent, soft, and have excellent permeability such that oxygen in particular can easily permeate through them. One example is polyalkylimide hydrogel, which contains 96% water. It is a polyacrylate gel that has alkyl amide and imide groups in its structure. It is normally used in the form of injectable endoprosthesis, which is appropriate for the simple, non-invasive rectification of small defects [111, 112].

Figure 9.10 Structure of PAA.

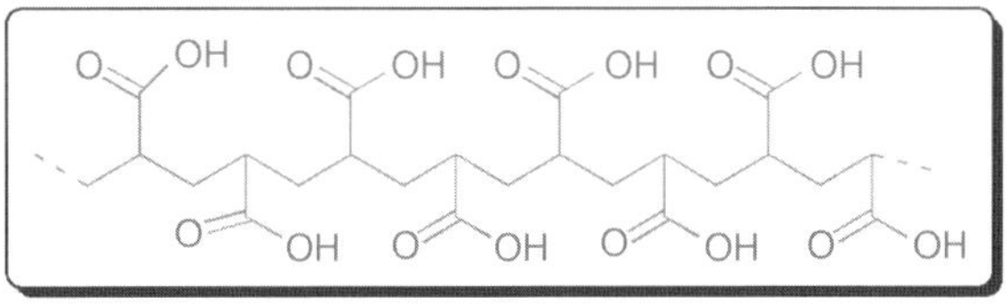

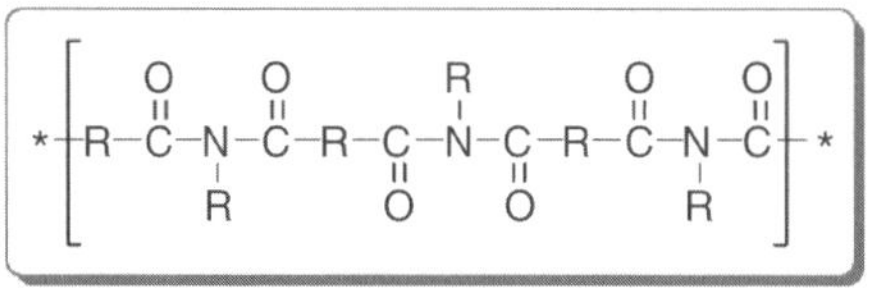

Figure 9.11 Structure of polyimide.

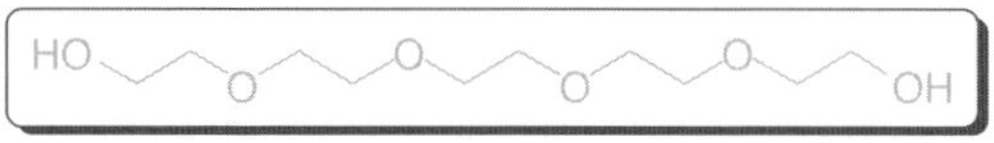

Figure 9.12 Structure of PEG.

9.7.3 Polyethylene Glycol

PEG-based hydrogels are extensively used in pharmaceutical applications. The chemical structure of PEG is shown in Figure 9.12. These hydrogels are non-immunogenic, non-toxic, and accepted by the US Food and Drug Administration for different pharmaceutical applications. They have also been used as stealth materials as they are inactive to most biological macromolecules, e.g. proteins, lipids, etc. Merrill et al. developed a hydrophilic biomaterial based on PEO and PEG [113], and presented adsorption of PEO on a glass surface which prohibited adsorption of protein. Different surface-modified PEG materials have since been used to promote surface biocompatibility and surface protein resistance. Commonly used approaches for surface modification of PEG are adsorption, covalent bonding through thiol, acrylate, and silane linkages, and covalent and ionic bonding [114]. PEG polymers can undergo covalent crosslinking by various protocols to form hydrogels. Photopolymerization of acrylate-terminated PEG monomers is the most commonly used method for the crosslinking of PEG chains [115]. PEG-based hydrogels impede protein adsorption because PEG is the passive component of the cell surroundings. For several ensuing conjugations, various approaches for tailored PEG gels have rendered them a versatile template. For instance, peptide moieties are induced in PEG gels to mutate the cell adhesion or degradation [116]. Furthermore, block copolymers of PEG, e.g. triblock copolymers of PPO and PEO, PLA, degradable PEO and similar other monomers, are used to induce special traits in PEG-based hydrogels [117, 118].

9.7.4 Polyvinyl Alcohol

Polyvinyl alcohol (PVA) is a hydrophilic synthetic polymer used as a hydrogel in tissue engineering [119]. Its stereochemical structure is shown in Figure 9.13. It can be modified with glycidyl methacrylate or acryloyl chloride to produce reactive acrylate groups through pendent –OH groups, leading to crosslinking to develop hydrogels. Additionally, it can be mixed with other water-soluble polymers to develop hydrogels. PVA-based hydrogels have properties such as high water retaining ability, good mechanical strength along with traits of flexibility, and biocompatibility, and can be employed as artificial soft tissue [120].

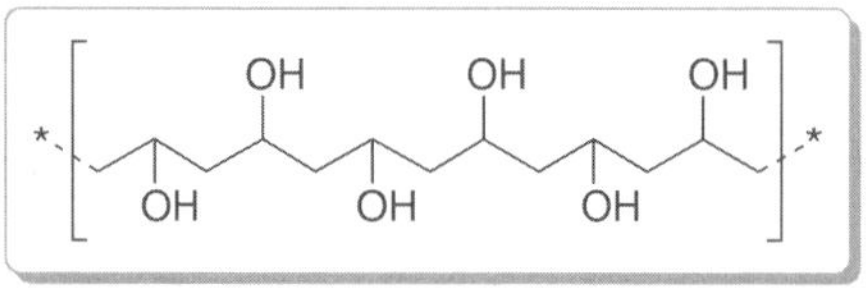

Figure 9.13 Structure of PVA.

9.8 Crosslinking of Hydrogels

Hydrogels are hydrophilic 3D polymer networks that can absorb thousand times of water of their dry weight. They may be chemically stable or may disintegrate, degrade or ultimately dissolve. Crosslinking itself is not a property of hydrogels but it induces other properties in hydrogels. The degree of crosslinking is linked with every property of hydrogels. The nature of the crosslinking is different in different modes. In fact, a network of hydrogel can be obtained in different ways. By modulating the degree of crosslinking, it is possible to optimize the properties of a material and use a single polymer in a broad spectrum of applications [121–123]. Hydrogel crosslinking can be classified as:

  (I) physical crosslinking
 (II) chemical crosslinking
(III) photo crosslinking.

9.8.1 Physical Crosslinking

When polymer networks are linked together via secondary forces, e.g. hydrophobic, hydrogen, or ionic bonding or molecular entanglements, they are called physical or reversible gels [124, 125]. Hydrogels having physical crosslinking are not homogeneous, as huddles of polymer chain entanglements and ionic and hydrophobic interactions can cause inhomogeneities. Chain loops or free chain ends represent transient network flaws in physically crosslinked hydrogels. These hydrogels are reversible as they are disrupted by changes in environmental parameters such as temperature, ionic strength of solution, addition of specific solute, pH or applied stress [126].

Physical crosslinking that occurs due to ionic linkage between a multivalent cation and polyanion forms an ionotropic hydrogel. One of these kinds of physical hydrogel is calcium alginate. When oppositely charged polyelectrolytes are combined, they may precipitate out or form a gel depending on their pH, ionic strength, and the concentration of the solutions. Such crosslinked gels are known as polyelectrolyte complexes, polyion complexes, and complex coacervates. Lim and Sun prepared a calcium alginate complex coacervate of alginate–poly(L-lysine) coated capsules to get a stable product [127]. Polyion complex and complex coacervates hydrogels are important materials for tissue engineering matrices. Sometimes physically crosslinked hydrogels can be prepared from biospecific recognition, such as polymeric sugar with conconavalin A [128] or polymeric biotin with avidin [129].

9.8.2 Chemical Crosslinking

Hydrogels formed by covalently crosslinked networks are called chemical or permanent hydrogels. In these hydrogels, a covalently bonded network of macromolecular chains is achieved by crosslinking them in solution or dry state [130, 131]. These crosslinked networks can be developed by chemical reaction between different low MW crosslinkers and activated functional groups of polymer chains that can produce a number of linkages such as amine carboxylic acid bonding and Schiff base formation via a large number of reactions, e.g. nucleophilic addition reactions, Michaelis–Arbuzov reactions, Michael's reaction, etc. [132, 133]. These hydrogels may be charged or neutral depending on the type of functional groups present in their chemical structure. The charged hydrogels mostly undergo a change in swelling when exposed to a change in pH or a change in shape with a change in electric field [134].

Crosslinked hydrogels are fabricated in two different approaches: by direct crosslinking of water-soluble polymers or by 3D polymerization in which a hydrophilic monomer is polymerized via a multifunctional crosslinker. Polymerization is normally instigated by free radical producing moieties like ammonium peroxodisulphate, 2,2-azo-isobutyronitrile, and benzoyl peroxide, or by gamma, UV or electron beam irradiation. However, 3D polymerization produces materials with residual monomers therefore purification of these materials is required. These residual monomers are also cytotoxic and constantly percolate out from hydrogel products. The decontamination of hydrogels from residual monomers is usually executed by flashing with excess water and this process can take several weeks to complete [135].

The crosslinking protocol may be covalent or ionic depending on the crosslinking agent. For instance, CS in solution form is present as polycationic, therefore it may react with polyanions to form ionic complexes. Ionic crosslinkers like sulphates or phosphates are non-toxic but their crosslinked products are unstable due to weak chemical bonds like hydrophobic and hydrogen bonding and dipole–dipole and electrostatic interactions [86].

On the other hand, covalent crosslinking agents like glutaraldehyde and formaldehyde produce chemically and physically stable hydrogels but they are minutely toxic due to unreacted molecules of by-products and crosslinkers [136–138]. Genipin is a biocompatible crosslinker that is naturally derived, i.e. extracted from gardenia [139]. It is used to bind biological tissues and crosslink biopolymers like CS and gelatine. It is less toxic than glutaraldehyde [140].

9.8.3 Photocrosslinking

In photocrosslinking, polymers that can form hydrogels can be developed by using photosensitive functional groups. The polymer can develop crosslinking with these reactive groups upon irradiation with UV light [141]. In polymer–polymer crosslinking, pre-functionalized polymer chains with reactive functional groups are required. In this type of crosslinking, covalent linkages can be formed that mainly depend on the selection of targeted reactive functional groups and the desired rate of crosslinking [142, 143].

9.9 Biomedical Applications of Hydrogels

9.9.1 Contact Lenses

Contact lenses are categorized as soft or hard on the basis of their elasticity. Although hard lenses are considered more durable, they are not as easily accepted by wearers. Hard contact lenses are prepared by hydrophobic materials like poly(hexa-fluoroisopropyl methacrylate) and poly(methyl methacrylate) while soft lenses are made from hydrogels [144].

In US Patent 3679504A, a new method of ophthalmic prostheses and preparation of coloured soft lenses is disclosed by Wichterle. By polymerizing hydrophilic monomers, a coloured compound was added between two adjacent transparent hydrogel layers. The top stratum of hydrogel could be prepared from a solution of hydrophilic macromonomer like PEG mono-methacrylate, which was prepared as explained in US Patent 3575946A. The macromonomer-based hydrogels do not require further purification due to the non-toxic behaviour of these materials [135].

Soft contact lenses are not only used for the correction of vision but also for drug delivery to the eye. Nevertheless, typical contact lenses prepared by hydrogels exhibited burst release upon ocular administration and have comparatively low drug-loading ability. Several approaches have been introduced to improve typical contact lenses to enhance their drug-delivery capacity. The modification of polymers with a controlled hydrophobic/hydrophilic copolymer ratio allows the introduction of multilayered hydrogels and impregnated drug-containing colloidal structures [145]. Venkatesh et al. [146] demonstrated the value of biomimitic hydrogels as a carrier to load significant concentration of H1-antihistamines. They also presented the drug delivery of a therapeutic dosage in five days in a controlled manner with possible expansion in the presence of protein. Xu et al. [147] observed the improved tensile strength and equilibrium swelling ratio of hydrogels used in contact lenses by incorporating β-cyclodextrin (β-CD). Puerarin was taken as a model drug and checked its release from PHEMA/β-CD hydrogels. It was observed that the *in vitro* release rate and puerarin loading depended on the concentration of β-CD in the hydrogel. PHEMA/β-CD hydrogel-based contact lenses showed a longer puerarin residence time in the tear fluid compared to typical PHEMA contact lenses and 1% puerarin eye drops in the eyes of rabbits [135].

9.9.2 Oral Drug Delivery

A hydrogel-based scaffold can be employed for drug delivery to the colon, intestine, stomach, and oral cavity. Drug delivery in the oral cavity is used to lessen diseases of the mouth without the risk of the first-pass effect. pH-sensitive hydrogels are used for targeted drug delivery like intestine, stomach, and enhanced drug bioavailability. CS-based hydrogels are fabricated to cure diseases such as inflammatory or irritable bowel diseases [148, 149].

Mucoadhesion of CS is a significant property for promoting the oral absorption of drugs. Buccal tablets of propranolol and nifedipine are prepared with CS as the mucoadhesive layer to improve the systematic bioavailability of drugs [150]. An interpenetrating network (IPN) of PEO and CS has been developed as a stomach-targeted drug-delivery material for treatment of *Helicobacter pylori*. This IPN exhibited pH-dependent drug release and swelling properties [151]. CS/PAA-based hydrogels have been developed for colon-targeted drug delivery. The pH-responsive behaviour of PAA and the biodegradability of CS by colonic normal flora make this hydrogel suitable for drug release in the colonic region [152].

9.9.3 Tissue Engineering

Hydrogels have potential applications in tissue engineering [24]. Hydrogel scaffolds prepared for tissue engineering may have a porous structure to survive in living cells or may degrade or dissolve, generating growth factor and producing pores which might be used to proliferate or infiltrate the living cells. In order to excite growth, i.e. spreading or adhesion of cells within the hydrogel matrix, one may include cell membrane receptor peptide ligands through covalent bonding to use the hydrogel as tissue engineering vs more hydrophobic like PLGA. However, a major shortcoming of hydrogels is their poor mechanical properties, which cause problems in their management [153]. Sterilization problems are also very tricky. The use of hydrogels in tissue engineering has both benefits and drawbacks, and the latter will have to be addressed before hydrogels can be considered useful and convenient in this emerging area [154].

9.9.4 Wound Healing

From the late 1970s or early 1980s hydrogels have been used in wound healing. As discussed above, a hydrogel is a crosslinked polymer that has capacity to hold and absorb plenty of water in its structural network. Hydrogels behave as damped wound dressing material and have the capacity to retain and absorb wound exudates along with external bodies like bacteria in their structural network. Furthermore, hydrogels are prepared to enhance the proliferation of fibroblast by minimizing fluid loss from the surface of the wound and keeping the wound away from foreign noxae indispensable for quick wound curing. Hydrogels make it easy to establish a microclimate for biosynthetic reactions on the surface of the wound that are essential for cellular activities. Proliferation of fibroblasts is demanded for entire epithelialization of the injury that initiate from the wound edges. Keratinocytes can transfer on the surface, as hydrogels help to keep the wound moist. Depending upon the behaviour of the polymer, hydrogels may be transparent and provide a soothing, cooling, or cushioning effect on wound surface. One of the major benefits of transparent hydrogels is that the wound can be monitored without removing the wound dressing. The phenomenon of angiogenesis can be initiated by a semi-occlusive hydrogel dressing which is started due to momentary hypoxia. Angiogenesis of the wound confirms the development of granulation tissue by sustaining an adequate supply of nutrients and oxygen to the surface of the wound. Hydrogel films attached to polymer films or fabric are usually fixed to the wound surface with bandages or adhesives [155].

9.9.5 Gene Delivery

Gene delivery has been developed for the treatment of many inherited diseases as it can correct the substandard genes that cause these inherited diseases. The deliverance of a suitable therapeutic gene, e.g. DNA, into cells that will regulate, revamp, or replace the imperfect gene from which the disease originates is the crucial stage for the gene therapy. However, DNA is a hydrophilic, negatively charged molecule, thus its delivery into the nucleus of the cells as required for it to pass through the negatively charged hydrophobic cell membrane is not practicable. Subsequently, gene delivery carriers, also known vehicles or vectors, have been developed. Naturally, viruses are gene carriers, and these were initially used for gene delivery [156, 157]. However, viruses have many drawbacks, the most important of which is the immune reaction that they can cause, therefore non-viral carriers have now been developed [158]. Many of these are polymer based, as polymers are safer and cheaper than viruses, and also easier to amend as compared to other gene delivery carriers like liposomes [159]. When polymer gene delivery carriers are used, the major steps are (i) polymer and DNA complexation, (ii) inclusion of polymer or DNA complex (polyplex) on the cells for a specific time period (the transfection time), (iii) elimination of the complex from the cells, and (iv) incubation (cells are left to incubate for a certain time until results are noticed). The complex formation phenomenon is normally executed at room temperature while the transfection and incubation periods are at body temperature, i.e. 37 °C, for the survival of cells. Interestingly, thermoresponsive polymers are used to improve the transfection efficiency by modulating the temperature during either the complex formation or the transfection or incubation time span [160].

9.10 Conclusions

Hydrogels are extensively found in daily products and their potential has not been exhaustively investigated so far. This chapter reviews the fabrication and composition of hydrogels along with their different properties, and the natural and synthetic polymers used for the development of hydrogels in the presence of different crosslinking agents. The major characteristics of hydrogels related to clinical, pharmaceutical, and biomedical applications are identified, particularly for the applications of hydrogels in contact lenses, oral drug delivery, wound healing, tissue engineering matrices, and gene delivery.

References

1 Croisier, F. and Jérôme, C. (2013). Chitosan-based biomaterials for tissue engineering. *Eur. Polym. J.* 49: 780–792.
2 Singh, N.K., Nquyen, Q., Kim, B.S., and Lee, D.S. (2015). Nanostructure controlled sustained delivery of human growth hormone using injectable, biodegradable, pH/temperature responsive nanobiohybrid hydrogel. *Nanoscale* 7: 3043–3054.
3 Klymenko, A., Nicolai, T., Benyahia, L. et al. (2014). Multiresponsive hydrogels formed by interpenetrated self-assembled polymer networks. *Macromolecules* 47: 8386–8393.
4 Joglekar, M. and Trewyn, B.G. (2013). Polymer-based stimuliresponsive nanosystems for biomedical applications. *Biotechnol. J.* 8: 931–945.
5 Ravichandran, R., Sundarrajan, S., Venugopal, J.R. et al. (2012). Advances in polymeric systems for tissue engineering and biomedical applications. *Macromol. Biosci.* 12: 286–311.
6 Álvarez-Lorenzo, C. and Concheiro, A. (2014). Smart drug delivery systems: from fundamentals to the clinic. *Chem. Commun.* 50: 7743–7765.
7 Buwalda, S.J., Boere, K.W.M., Dijkstra, P.J. et al. (2014). Hydrogels in a historical perspective: from simple networks to smart materials. *J. Control. Release* 190: 254–273.
8 Desai, M.S. and Lee, S.-W. (2015). Protein-based functional nanomaterial design for bioengineering applications. *WIREs Nanomed. Nanobiotechnol.* 7: 69–97.
9 Vermonden, T., Censi, R., and Hennink, W.E. (2012). Hydrogels for protein delivery. *Chem. Rev.* 112: 2853–2888.

10 John, J.V., Johnson, R.P., Heo, M.S. et al. (2015). Polymer-block-polypeptides and polymer-conjugated hybrid materials as stimuliresponsive nanocarriers for biomedical applications. *J. Biomed. Nanotechnol.* 11: 1–39.

11 Gübeli, R.J., Ehrbar, M., Fussenegger, M. et al. (2012). Synthesis and characterization of PEG-based drug responsive biohybrid hydrogels. *Macromol. Rapid Commun.* 33: 1280–1285.

12 Kutikov, A. and Song, J. (2015). Biodegradable PEG-based amphiphilic block copolymers for tissue engineering applications. *ACS Biomater Sci. Eng.* 1: 463–480.

13 Bouyer, E., Mekhloufi, G., Rosilio, V. et al. (2012). Proteins, polysaccharides and their complexes used as stabilizers for emulsions: alternatives to synthetic surfactants in the pharmaceutical field? *Int. J. Pharm.* 436: 359–378.

14 Singh, D., Han, S.S., and Shin, E.J. (2014). Polysaccharides as nanocarriers for therapeutic applications. *J. Biomed. Nanotechnol.* 10: 2149–2172.

15 Nitta, S.K. and Numata, K. (2013). Biopolymer-based nanoparticles for drug/gene delivery and tissue engineering. *Int. J. Mol. Sci.* 14: 1629–1654.

16 Grijalvo, S., Mayr, J., Eritja, R., and Díaz, D.D. (2016). Biodegradable liposome-encapsulated hydrogels for biomedical applications: a marriage of convenience. *Biomater. Sci.* 4: 555–574.

17 Wichterle, O. and Lim, D. (1960). Hydrophilic gels for biological use. *Nature* 185: 117–118.

18 Tsang, V.L. and Bhatia, S.N. (2004). Three-dimensional tissue fabrication. *Adv. Drug Deliv. Rev.* 56: 1635–1647.

19 Haque, M.A. and Gong, J.P. (2013). Multi-functions of hydrogel with bilayer-based lamellar structure. *React. Funct. Polym.* 73: 929–935.

20 Nie, J., Lu, W., Ma, J. et al. (2015). Orientation in multi-layer chitosan hydrogel: morphology, mechanism, and design principle. *Sci. Rep.* 5: 1–7.

21 Giri, T.K., Thakur, A., Alexander, A. et al. (2012). Modified chitosan hydrogels as drug delivery and tissue engineering systems: present status and applications. *Acta Pharm. Sin. B* 2: 439–449.

22 Seliktar, D. (2016). Designing cell-compatible hydrogels for biomedical applications. *Science* 336: 1124–1128.

23 Tomme, S.R.V., Storm, G., and Hennink, W.E. (2008). In situ gelling hydrogels for pharmaceutical and biomedical applications. *Int. J. Pharm.* 355: 1–18.

24 Lee, K.Y. and Mooney, D.J. (2001). Hydrogels for tissue engineering. *Chem. Rev.* 101: 1869–1880.

25 Van der Linden, H.J., Herber, S., Olthuis, W., and Bergveld, P. (2003). Stimulus-sensitive hydrogels and their applications in chemical (micro)analysis. *Analyst* 128: 325–331.

26 Jen, A.C., Wake, M.C., and Mikos, A.G. (1996). Review: hydrogels for cell immobilization. *Biotechnol. Bioeng.* 50: 357–364.

27 Bennett, S.L., Melanson, D.A., Torchiana, D.F. et al. (2003). Next-generation hydrogel films as tissue sealants and adhesion barriers. *J. Card. Surg.* 18: 494–499.

28 Ullah, F., Othman, M.B.H., Javed, F. et al. (2015). Classification, processing and application of hydrogels: a review. *Mater. Sci. Eng. C* 57: 414–433.

29 Ahmed, E.M. (2015). Hydrogel: preparation, characterization, and applications: a review. *J. Adv. Res.* 6: 105–121.

30 Awad, H.A., Wickham, M.Q., Leddy, H.A. et al. (2004). Chondrogenic differentiation of adipose-derived adult stem cells in agarose, alginate, and gelatin scaffolds. *Biomaterials* 25: 3211–3222.

31 McGuigan, A.P. and Sefton, M. (2007). Modular tissue engineering: fabrication of a gelatin-based construct. *Biomaterials* 1: 136–145.

32 Billiet, T., Vandenhaute, M., Schelfhout, J. et al. (2012). A review of trends and limitations in hydrogel-rapid prototyping for tissue engineering. *Biomaterials* 33: 6020–6041.

33 Lee, C.H., Singla, A., and Lee, Y. (2001). Biomedical applications of collagen. *Int. J. Pharm.* 221: 1–22.

34 Lee, C.R., Grodzinsky, A.J., and Spector, M. (2001). The effects of cross-linking of collagen-glycosaminoglycan scaffolds on compressive stiffness, chondrocyte-mediated contraction, proliferation, and biosynthesis. *Biomaterials* 22: 3145–3154.

35 Park, S.N., Park, J.C., Kim, H.O. et al. (2002). Characterization of porous collage/hyaluronic acid scaffold modified by 1-ethyl-3-(3-dimethylaminopropyl) carbodiimide cross-linking. *Biomaterials* 23: 1205–1212.

36 Schoof, H., Apel, J., Heschel, I., and Rau, G. (2001). Control of pore structure and size in freeze-dried collagen sponges. *J. Biomed. Mater. Res.* 58: 352–357.

37 Tan, W., Krishnaraj, R., and Desai, T.A. (2001). Evaluation of nanostructured composite collagen–chitosan matricies for tissue engineering. *Tissue Eng.* 7: 203–210.

38 Chen, G., Ushida, T., and Tateishi, T. (2001). Development of biodegradable porous scaffolds for tissue engineering. *Mater. Sci. Eng. C* 17: 63–69.

39 Huang, L., Naqapudi, K., Apkarian, R.P., and Chaikof, E.L. (2001). Engineered collagen—PEO nanofibers and fabrics. *J. Biomater. Sci. Polym. Ed.* 12: 979–993.

40 Drury, J.L. and Mooney, D.J. (2003). Hydrogels for tissue engineering: scaffold design variables and applications. *Biomaterials* 24: 4337–4351.

41 Tan, H., Huang, D., Lao, L., and Gao, C. (2009). RGD modified PLGA/gelatin microspheres as microcarriers for chondrocyte delivery. *J. Biomed. Mater. Res. B Appl. Biomater.* 91: 228–238.

42 Huang, Y., Onyeri, S., Siewe, M. et al. (2005). In vitro characterization of chitosan-gelatin scaffolds for tissue engineering. *Biomaterials* 26: 7616–7627.

43 Tan, H. and Marra, K.G. (2010). Injectable, biodegradable hydrogels for tissue engineering applications. *Materials* 3: 1746–1767.

44 Kleinman, H.K. and Martin, G.R. (2005). Matrigel: basement membrane matrix with biological activity. *Semin. Cancer Biol.* 15: 378–386.

45 Morritt, A.N., Bortolotto, S.K., Dilley, R.J. et al. (2007). Cardiac tissue engineering in an in vivo vascularized chamber. *Circulation* 115: 353–360.

46 Ponce, M.L. (2009). Tube formation: an in vitro matrigel angiogenesis assay. *Methods Mol. Biol.* 467: 183–188.

47 Zhu, J. and Marchant, R.E. (2011). Design properties of hydrogel tissue-engineering scaffolds. *Expert Rev. Med. Devices* 8: 607–626.

48 Kenny, P.A., Lee, G.Y., Myers, C.A. et al. (2007). The morphologies of breast cancer cell lines in three-dimensional assays correlate with their profiles of gene expression. *Mol. Oncol.* 1: 84–96.

49 Lee, G.Y., Kenny, P.A., Lee, E.H., and Bissell, M.J. (2007). Three-dimensional culture models of normal and malignant breast epithelial cells. *Nat. Methods* 4: 359–365.

50 Fessart, D., Begueret, H., and Delom, F. (2013). Three-dimensional culture model to distinguish normal from malignant human bronchial epithelial cells. *Eur. Respir. J.* 42: 1345–1356.

51 Santarius, M., Lee, C.H., and Anderson, R.A. (2006). Supervised membrane swimming: small G-protein lifeguards regulate PIPK signalling and monitor intracellular Ptdins(4,5)P2 pools. *Biochem. J.* 398: 1–13.

52 Schmitz, A.A., Govek, E.E., Böttner, B., and Van Aelst, L. (2000). Rho GTpases: signaling, migration, and invasion. *Exp. Cell Res.* 261: 1–12.

53 Tamura, M., Yanagawa, F., Sugiura, S. et al. (2015). Click-crosslinkable and photodegradable gelatin hydrogels for cytocompatible optical cell manipulation in natural environment. *Sci. Rep.* 5: 1–12.

54 Mano, J.F., Silva, G.A., Azevedo, H.S. et al. (2007). Natural origin biodegradable systems in tissue engineering and regenerative medicine: present status and some moving trends. *J. R. Soc. Interface* 4: 999–1030.

55 Oliveira, J.T. and Reis, R.L. (2011). Polysaccharide-based materials for cartilage tissue engineering applications. *J. Tissue Eng. Regen. Med.* 5: 421–436.

56 Zhao, W., Jin, X., Cong, Y. et al. (2013). Degradable natural polymer hydrogels for articular cartilage tissue engineering. *J. Chem. Technol. Biotechnol.* 88: 327–339.

57 Fraser, J.R., Laurent, T.C., and Laurent, U.B. (1997). Hyaluronan: its nature, distribution, functions and turnover. *J. Intern. Med.* 242: 27–33.

58 Toole, B.P. (2001). Hyaluronan in morphogenesis. *Semin. Cell Dev. Biol.* 12: 79–87.

59 Toole, B.P. (2004). Hyaluronan: from extracellular glue to pericellular cue. *Nat. Rev. Cancer* 4: 528–539.

60 Laurent, T.C. and Fraser, J.R. (1986). The properties and turnover of hyaluronan. *Ciba Found. Symp.* 124: 9–29.

61 Prestwich, G.D. (2008). Engineering a clinically-useful matrix for cell therapy. *Organogenesis* 4: 42–47.

62 Burdick, J.A. and Prestwich, G.D. (2011). Hyaluronic acid hydrogels for biomedical applications. *Adv. Mater.* 23: H41–H56.

63 Allison, D.D. and Grande-Allen, K.J. (2006). Review. Hyaluronan: a powerful tissue engineering tool. *Tissue Eng.* 12: 2131–2140.

64 Augst, A.D., Kong, H.J., and Mooney, D.J. (2006). Alginate hydrogels as biomaterials. *Macromol. Biosci.* 6: 623–633.

65 Baldwin, A.D. and Kiick, K.L. (2010). Polysaccharide-modified synthetic polymeric biomaterials. *Biopolymers* 94: 128–140.

66 Abbah, S.A., Lu, W.W., Chan, D. et al. (2008). Osteogenic behavior of alginate encapsulated bone marrow stromal cells: an in vitro study. *J. Mater. Sci. Mater. Med.* 19: 2113–2119.

67 Sundar, S., Kundu, J., and Kundu, S.C. (2010). Biopolymeric nanoparticles. *Sci. Technol. Adv. Mater.* 11: 1–13.

68 Grellier, M., Granja, P.L., Fricain, J.C. et al. (2009). The effect of the co-immobilization of human osteoprogenitors and endothelial cells within alginate microspheres on mineralization in a bone defect. *Biomaterials* 30: 3271–3278.

69 Freeman, I. and Smadar, C. (2009). The influence of the sequential delivery of angiogenic factors from affinity-binding alginate scaffolds on vascularization. *Biomaterials* 30: 2122–2131.

70 Wang, W. and Wang, A. (2010). Synthesis and swelling properties of pH-sensitive semi-IPN superabsorbent hydrogels based on sodium alginate-g-poly(sodium acrylate) and polyvinylpyrrolidone. *Carbohydr. Polym.* 80: 1028–1036.

71 Zhao, S., Cao, M., Li, H. et al. (2010). Synthesis and characterization of thermo-sensitive semi-IPN hydrogels based on poly- (ethylene glycol)-co-poly(epsilon-caprolactone) macromer, N-isopropylacrylamide, and sodium alginate. *Carbohydr. Res.* 245: 425–431.

72 Van Vlierberghe, S., Dubruel, P., and Schacht, E. (2011). Biopolymer-based hydrogels as scaffolds for tissue engineering applications: a review. *Biomacromolecules* 12: 1387–1408.

73 Sorlier, P., Denuzière, A., Viton, C., and Domard, A. (2001). Relation between the degree of acetylation and the electrostatic properties of chitin and chitosan. *Biomacromolecules* 2: 765–772.

74 Dong, Y.M., Qiu, W.B., Ruan, Y.H. et al. (2001). Influence of molecular weight on critical concentration of chitosan/formic acid liquid crystalline solution. *Polym. J.* 33: 387–389.

75 Patashnik, S., Rabinovich, L., and Golomb, G. (1997). Preparation and evaluation of chitosan microspheres containing biphosphonates. *J. Drug Target.* 4: 371–380.

76 Felt, O., Furrer, P., Mayer, J.M. et al. (1999). Topical use of chitosan in ophtalmology: tolerance assessment and evaluation of precorneal retention. *Int. J. Pharm.* 180: 185–193.

77 Song, J., Such, C.H., Park, Y.B. et al. (2001). A phase, I/IIa study on intra-articular injection of holmium-166-chitosan complex for the treatment of knee synovitis of rheumatoid arthritis. *Eur. J. Nucl. Med. Mol. Imaging* 28: 489–497.

78 Muzzarelli, R.A.A. (1997). Human enzymatic activities related to the therapeutic administration of chitin derivatives. *Cell Mol. Life Sci.* 53: 131–140.

79 Junginger, H.E. and Verhof, J.C. (1998). Macromolecules as safe penetration enhancers for hydrophilic drugs—a fiction? *Pharm. Sci. Technol. Today* 1: 370–376.

80 Hea, P., Davisa, S.S., and Illumab, L. (1998). In vitro evaluation of the mucoadhesive properties of chitosan microspheres. *Int. J. Pharm.* 166: 75–88.

81 Calvo, P., Vila-Jato, J.L., and Alonso, M.J. (1997). Evaluation of cationic polymer-coated nanocapsules as ocular drug carriers. *Int. J. Pharm.* 153: 41–50.

82 Ueno, H., Mori, T., and Fujinaga, T. (2001). Topical formulations and wound healing applications of chitosan. *Adv. Drug Deliv. Rev.* 52: 105–115.

83 Felt, O., Carrel, A., Baehni, P. et al. (2000). Chitosan as tear substitute: a wetting agent endowed with antimicrobial efficacy. *J. Ocul. Pharmacol. Ther.* 16: 261–270.

84 Liu, X.F., Guan, Y.L., Yang, D.Z. et al. (2001). Antibacterial action of chitosan and carboxymethylated chitosan. *J. Appl. Polym. Sci.* 79: 1324–1335.

85 Peter, M.G. (1995). Applications and environmental aspects of chitin and chitosan. *J. Macromol. Sci. Part A Pure Appl. Chem.* 32: 629–640.

86 Berger, J., Reist, M., Mayer, J.M. et al. (2004). Structure and interactions in covalently and ionically crosslinked chitosan hydrogels for biomedical applications. *Eur. J. Pharm. Biopharm.* 57: 19–34.

87 Shirakawa, M., Yamatoya, K., and Nishinari, K. (1998). Tailoring of xyloglucan properties using an enzyme. *Food Hydrocoll.* 12: 25–28.

88 Ruel-Gariépy, E. and Leroux, J.-C. (2004). In situ forming hydrogels-review of temperature-sensitive systems. *Eur. J. Pharm. Biopharm.* 58: 409–426.

89 Nisbet, D.R., Crompton, K.E., Hamilton, S.D. et al. (2006). Morphology and gelation of thermosensitive xyloglucan hydrogels. *Biophys. Chem.* 121: 14–20.

90 Klouda, L. and Mikos, A.G. (2008). Thermoresponsive hydrogels in biomedical applications. *Eur. J. Pharm. Biopharm.* 68: 34–45.

91 Zhang, X., Wu, D., and Chu, C.-C. (2004). Synthesis and characterization of partially biodegradable, temperature and pH sensitive Dex-MA/PNIPAAm hydrogels. *Biomaterials* 25: 4719–4730.

92 Huang, X. and Lowe, T.L. (2005). Biodegradable thermoresponsive hydrogels for aqueous encapsulation and controlled release of hydrophilic model drugs. *Biomacromolecules* 6: 2131–2139.

93 Rahfoth, B., Weisser, J., Sternkopf, F. et al. (1998). Transplantation of allograft chondrocytes embedded in agarose gel into cartilage defects of rabbits. *Osteoarthr. Cartil.* 6: 50–65.

94 Lima, E.G., Chao, P.-h.G., Ateshian, G.A. et al. (2008). The effect of devitalized trabecular bone on the formation of osteochondral tissue-engineered constructs. *Biomaterials* 29: 4292–4299.

95 Byers, B.A., Mauck, R., Chiang, I.E., and Tuan, R.S. (2008). Transient exposure to transforming growth factor beta 3 under serum-free conditions enhances the biomechanical and biochemical maturation of tissue-engineered cartilage. *Tissue Eng. A* 14: 1821–1834.

96 Bhatia, S.K. (2011). *Engineering Biomaterials for Regenerative Medicine: Novel Technologies for Clinical Applications*. New York: Springer.

97 Rabenstein, D.L. (2002). Heparin and heparan sulfate: structure and function. *Nat. Prod. Rep.* 19: 312–331.

98 Turnbull, J., Powell, A., and Guimond, S. (2001). Heparan sulfate: decoding a dynamic multifunctional cell regulator. *Trends Cell Biol.* 11: 75–82.

99 Greinacher, A. (2006). Heparin-induced thrombocytopenia: frequency and pathogenesis. *Pathophysiol. Haemost. Thromb.* 35: 37–45.

100 Baldwin, A.D., Robinson, K.G., Militar, J.L. et al. (2012). In situ crosslinkable heparin-containing poly(ethylene glycol) hydrogels for sustained anticoagulant release. *J. Biomed. Mater. Res. A* 100: 2106–2118.

101 Tae, G., Kim, Y.J., Choi, W.-I. et al. (2007). Formation of a novel heparin-based hydrogel in the presence of heparin-binding biomolecules. *Biomacromolecules* 8: 1979–1986.

102 Nie, T., Akins, R.E. Jr.,, and Kiick, K.L. (2009). Production of heparin-containing hydrogels for modulating cell responses. *Acta Biomater.* 5: 865–875.

103 Benoit, D.S.W., Collins, S.D., and Anseth, K.S. (2007). Multifunctional hydrogels that promote osteogenic human mesenchymal stem cell differentiation through stimulation and sequestering of bone morphogenic protein 2. *Adv. Funct. Mater.* 17: 2085–2093.

104 Kim, M., Kim, Y.-J., Gwon, K., and Tae, G. (2012). Modulation of cell adhesion of heparin-based hydrogel by efficient physisorption of adhesive proteins. *Macromol. Res.* 20: 271–276.

105 Freudenberg, U., Hermann, A., Welzel, P.B. et al. (2009). A star-PEG–heparin hydrogel platform to aid cell replacement therapies for neurodegenerative diseases. *Biomaterials* 30: 5049–5060.

106 Kharkar, P.M., Kiick, K.L., and Kloxin, A.M. (2013). Designing degradable hydrogels for orthogonal control of cell microenvironments. *Chem. Soc. Rev.* 42: 7335–7372.

107 Bromberg, L. (1998). Polyether-modified poly(acrylic acid): synthesis and applications. *Ind. Eng. Chem. Res.* 37: 4267–4274.

108 Craig, D.Q.M., Tamburic, S., Buckton, G., and Newton, J.M. (1994). An investigation into the structure and properties of carbopol® 934 gels using dielectric spectroscopy and oscillatory rheometry. *J. Control. Release* 30: 213–223.

109 Ayers, D., Cuthbertson, J.M., Schroyer, K., and Sullivan, S.M. (1996). Polyacrylic acid mediated ocular delivery of ribozymes. *J. Control. Release* 38: 167–175.

110 Kadajji, V.G. and Betageri, G.V. (2011). Water soluble polymers for pharmaceutical applications. *Polymers* 3: 1972–2009.

111 Gibas, I. and Janik, H. (2010). Review: synthetic polymer hydrogels for biomedical applications. *Chem. Chem. Technol.* 4: 297–304.

112 Ramires, P.A., Miccoli, M.A., Panzarini, E. et al. (2005). In vitro and in vivo biocompatibility evaluation of a polyalkylimide hydrogel for soft tissue augmentation. *J. Biomed. Mater. Res. B Appl. Biomater.* 72: 230–238.

113 Merrill, E.W., Salzman, E.W., Wan, S. et al. (1982). Platelet-compatible hydrophilic segmented polyurethanes from polyethylene glycols and cyclohexane diisocyanate. *Trans. Am. Soc. Artif. Intern. Organs* 28: 482–487.

114 Whitesides, G.M., Ostuni, E., Takayama, S. et al. (2001). Soft lithography in biology and biochemistry. *Annu. Rev. Biomed. Eng.* 3: 335–373.

115 West, J.L. and Hubbell, J.A. (1996). Separation of the arterial wall from blood contact using hydrogel barriers reduces intimal thickening after balloon injury in the rat: the roles of medial and luminal factors in arterial healing. *React. Polym.* 25: 13188–13193.

116 Hern, D.L. and Hubbell, J.A. (1998). Incorporation of adhesion peptides into non-adhesive hydrogels useful for tissue resurfacing. *J. Biomed. Mater. Res.* 39: 266–276.

117 Peppas, N.A., Hilt, J.Z., Khademhosseini, A., and Langer, R. (2006). Hydrogels in biology and medicine: from molecular principles to bionanotechnology. *Adv. Mater.* 18: 1345–1360.

118 Huh, K.M. and Bae, Y.H. (1999). Synthesis and characterization of poly(ethylene glycol)/poly(l-lactic acid) alternating multiblock copolymers. *Polymer* 40: 6147–6155.

119 Sawhney, A.S., Pathak, C.P., and Hubbell, J.A. (1993). Bioerodible hydrogels based on photopolymerized poly(ethylene glycol)-co-poly (α-hydroxy acid) diacrylate macromers. *Macromolecules* 26: 581–587.

120 Maitra, J. and Shukla, V.K. (2014). Cross-linking in hydrogels – a review. *Am. J. Polym. Sci.* 4: 25–31.

121 Chirani, N., Yahia, L.H., Gritsch, L. et al. (2015). History and applications of hydrogels. *J. Biomed. Sci.* 4: 1–23.

122 Sung, H.-W., Huang, D.-M., Chang, W.-H. et al. (1999). Evaluation of gelatin hydrogel crosslinked with various crosslinking agents as bioadhesives: *in vitro* study. *J. Biomed. Mater. Res.* 46: 520–530.

123 Weber, L.M., Lopez, C.G., and Anseth, K.S. (2009). Effects of PEG hydrogel crosslinking density on protein diffusion and encapsulated islet survival and function. *J. Biomed. Mater. Res. A* 90: 720–729.

124 Campoccia, D., Doherty, P., Radice, M. et al. (1998). Semisynthetic resorbable materials from hyaluronan esterification. *Biomaterials* 19: 2101–2127.

125 Prestwich, G.D., Marecak, D.M., Marecak, J.F. et al. (1998). Controlled chemical modification of hyaluronic acid. *J. Control. Release* 53: 93–103.

126 Hennink, W.E. and van Nostrum, C.F. (2002). Novel crosslinking methods to design hydrogels. *Adv. Drug Delivery Rev.* 54: 13–36.

127 Lim, F. and Sun, A.M. (1980). Microencapsulated islets as bioartificial pancreas. *Science* 210: 908–910.

128 Nakamae, K., Miyata, T., Jikihara, A., and Hoffman, A.S. (1994). Formation of poly(glucosyloxyethyl methacrylate)–concanavalin A complex and its glucose sensitivity. *J. Biomater. Sci. Polym. Ed.* 6: 79–90.

129 Morris, J.E., Hoffman, A.S., and Fisher, R.R. (1993). Affinity precipitation of proteins with polyligands. *Anal. Biochem.* 41: 991–997.

130 Hoffman, A.S. (2012). Hydrogels for biomedical applications. *Adv. Drug Deliv. Rev.* 64: 18–23.

131 Chandra, P., Dhaval, N., Upendra, D.P. et al. (2013). A conceptual overview on superporous hydrogel for controlled release drug delivery. *Int. J. Pharm. Integr. Life Sci.* 1: 1–13.

132 Hoare, T.R. and Kohane, D.S. (2008). Hydrogels in drug delivery: progress and challenges. *Polymer* 49: 1993–2007.

133 Berger, J., Reist, M., Mayer, J.M. et al. (2004). Structure and interactions in covalently and ionicallycrosslinked chitosan hydrogels for biomedical applications. *Eur. J. Pharm. Biopharm.* 57: 19–34.

134 Rosiak, J.M. and Yoshii, F. (1999). Hydrogels and their medical applications. *Nucl. Instrum. Methods Phys. Res., Sect. B* 151: 56–64.

135 Caló, E. and Khutoryanskiy, V.V. (2015). Biomedical applications of hydrogels: a review of patents and commercial products. *Eur. Polym. J.* 65: 252–267.

136 Dimida, S., Demitri, C., De Benedictis, V.M. et al. (2015). Genipin-cross-linked chitosan-based hydrogels: reaction kinetics and structure-related characteristics. *J. Appl. Polym. Sci.* 132.

137 Bemiller, J.N. (1967). Acid-catalyzed hydrolysis of glycosides. *Adv. Carbohydr. Chem.* 22: 25–108.

138 Holme, H.K., Foros, H., Pettersen, H. et al. (2001). Thermal depolymerization of chitosan chloride. *Carbohydr. Polym.* 46: 287–294.

139 Jin, J., Song, M., and Hourston, D.J. (2004). Novel chitosan-based films cross-linked by genipin with improved physical properties. *Biomacromolecules* 5: 162–168.

140 Sung, J.H., Hwang, M.R., Kim, J.O. et al. (2010). Gel characterisation and in vivo evaluation of minocycline-loaded wound dressing with enhanced wound healing using polyvinyl alcohol and chitosan. *Int. J. Pharm.* 392: 232–240.

141 Ono, K., Saito, Y., Yura, H. et al. (2000). Photocrosslinkable chitosan as a biological adhesive. *J. Biomed. Mater. Res.* 49: 289–295.

142 Tan, H., Chu, C.R., Payne, K.A., and Marra, K.G. (2009). Injectable in situ forming biodegradable chitosan-hyaluronic acid based hydrogels for cartilage tissue engineering. *Biomaterials* 30: 2499–2506.

143 Ranjha, N.M. and Khan, S. (2013). Chitosan/poly (vinyl alcohol) based hydrogels for biomedical applications: a review. *J. Pharm. Altern. Med.* 2: 30–41.

144 Lloyd, A.W., Faragher, R.G.A., and Denyer, S.P. (2002). Ocular biomaterials and implants. *Biomaterials* 22: 769–785.

145 Hu, X., Hao, L., Wang, H. et al. (2011). Hydrogel contact lens for extended delivery of ophthalmic drugs. *Int. J. Polym. Sci.* 2011: 1–9.

146 Venkatesh, S., Sizemore, S.P., and Byrne, M.E. (2007). Biomimetic hydrogels for enhanced loading and extended release of ocular therapeutics. *Biomaterials* 28: 717–724.

147 Xu, J., Li, X., and Sun, F. (2010). Cyclodextrin-containing hydrogels for contact lenses as a platform for drug incorporation and release. *Acta Biomater.* 6: 486–493.

148 Yang, J., Chen, J., Pan, D. et al. (2013). pH-sensitive interpenetrating network hydrogels based on chitosan derivatives and alginate for oral drug delivery. *Carbohydr. Polym.* 92: 719–725.

149 Mukhopadhyay, P., Sarkar, K., Bhattacharya, S. et al. (2014). pH sensitive N-succinyl chitosan grafted polyacrylamide hydrogel for oral insulin delivery. *Carbohydr. Polym.* 112: 627–637.

150 Remuñán-López, C., Portero, A., Vila-Jato, J.L., and Alonso, M.J. (1998). Design and evaluation of chitosan/ethylcellulose mucoadhesive bilayered devices for buccal drug delivery. *J. Control. Release* 55: 143–152.

151 Patel, V.R. and Amiji, M.M. (1996). Preparation and characterization of freezedried chitosan-poly (ethylene oxide) hydrogels for site-specific antibiotic delivery in the stomach. *Pharm. Res.* 13: 588–593.

152 Ahmadi, F., Oveisi, Z., Samani, S.M., and Amoozgar, Z. (2015). Chitosan based hydrogels: characteristics and pharmaceutical applications. *Res. Pharm. Sci.* 10: 1–16.

153 Hutmacher, D.W. (2001). Scaffold design and fabrication technologies for engineering tissues – state of the art and future perspectives. *J. Biomater. Sci. Polym. Ed.* 12: 107–124.

154 Hoffman, A.S. (2002). Hydrogels for biomedical applications. *Adv. Drug Deliv. Rev.* 54: 3–12.

155 Pal, K., Banthia, A.K., and Majumdar, D.K. (2009). Polymeric hydrogels: characterization and biomedical applications – a mini review. *Des. Monomers Polym.* 12: 197–220.

156 Merdan, T., Kopeček, J., and Kissel, T. (2002). Prospects for cationic polymers in gene and oligonucleotide therapy against cancer. *Adv. Drug Deliv. Rev.* 54: 715–758.

157 Felgner, P.L. (1997). Nonviral strategies for gene therapy. *Sci. Am.* 276: 102–106.

158 Crommelin, D.J.A., Storm, G., Jiskoot, W. et al. (2003). Nanotechnological approaches for the delivery of macromolecules. *J. Control. Release* 87: 81–88.

159 Kabanov, A.V. (1999). Taking polycation gene delivery systems from in vitro to in vivo. *Pharm. Sci. Technol. Today* 2: 365–372.

160 Ward, M.A. and Georgiou, T.K. (2011). Thermoresponsive polymers for biomedical applications. *Polymers* 3: 1215–1242.

10

Natural Aerogels as Thermal Insulators

Mohammadreza Saboktakin and Amin Saboktakin

Nanostructured Laboratory, NanoBMat Company, GmbH, Hamburg, Germany

Zhao et al. [1] have studied a new three-dimensional (3D) graphene-based aerogel material for construction. Graphene, two-dimensional (2D) one-atom thick carbon sheets with a similar benzene-ring structure, has attracted tremendous interest among researchers from different fields due to its outstanding physiochemical properties, such as a large specific surface area ($2630\,m^2\,g^{-1}$) [2], high-speed electron mobility ($20\,000\,cm^2$ Vs) [3], excellent thermal conductivities ($5000\,W\,m^{-1}\,K^{-1}$) [4], and well electrocatalytic activity [5]. In recent years, graphene and its derivatives have been used to make carbon-based materials, for instance one-dimensional (1D) tube-in-tube nanostructures [6], 2D layer-stacked films [7–9], and 3D graphene hydrogels [10–13] and aerogels [14–18]. On the other hand, graphene oxide (GO) has many hydrophobic basal plane and hydrophilic oxygen-containing groups, including epoxy, hydroxyl, carboxyl, and carbonyl groups, which allow GO to be functionalized through covalent and non-covalent approaches [19–22]. Thus, GO and its derivatives have become candidate materials for various technological fields, including optoelectronics, energy storage, catalysis, nanoelectronic devices, gas sensors, super capacitors, and drug-delivery systems [23–25]. Meanwhile, hydrophobic basal plane and hydrophilic oxygen-containing groups also provide the potential for the formation of graphene nanosheets (GNSs). It is generally recognized that various non-covalent forces (including hydrogen bonding, and ionic, amphiphilic, and π–π interactions) exist in graphene-based hydrogels. As we know, traditional hydrogels, because of their excellent properties, such as biocompatibility, rubber elasticity, equilibrium swelling, network structure characteristics, and environmental sensitivity, have been widely used in agriculture, biomedicine, and industry [26]. However, traditional hydrogels and aerogels have drawbacks, such as poor mechanical properties and limited functional properties [1]. In order to achieve the integration of nanostructural materials into macroscopic devices, a large amount of efforts has been undertaken to make GNS self-assembly and to improve traditional hydrogels. Since a pH-sensitive GO composite hydrogel was reported by Bai et al. [27], some graphene-based hydrogels have started to emerge, such as GO/conducting polymer composite hydrogels [11], macroscopic multifunctional graphene-based hydrogels and aerogels (GHAs) [13], temperature-sensitive GO nanocomposite hydrogels [28], and supramolecular hybrid hydrogels from graphene with block copolymers [29]. A graphene-based aerogel was obtained from a graphene hydrogel, which needed a

Bio Monomers for Green Polymeric Composite Materials, First Edition.
Edited by P.M. Visakh, Oguz Bayraktar and Gopalakrishnan Menon.
© 2020 John Wiley & Sons Ltd. Published 2020 by John Wiley & Sons Ltd.

drying process, such as freeze-drying [9, 10, 13, 14] or a supercritical CO_2 drying [14, 16]. This graphene-based aerogel, with interconnected 3D macroporous graphene network structures [9, 10, 13, 14], high electrical conductivity [14, 16], ultra-flyweight, and highly compressible properties [30], has also aroused tremendous interest among researchers.

Mohamad Ibrahim et al. [31] studied an aerogel-based coating for energy-efficient building envelopes. New thermal regulations (RT 2012) limit the primary energy consumption for new buildings to $50\,\text{kWh}\,\text{m}^{-2}$ per year and require better thermal comfort through limiting overheating during the summer season. Thus, it is crucial to find solutions to reduce energy consumption and enhance thermal comfort. Super-insulating materials (SIMs), such as silica aerogels and vacuum insulation, are promising techniques to be used in building envelopes to obtain the desired objectives. Renovation of existing buildings has high priority in many countries, including France, because these buildings represent such a high proportion of energy consumption and will be around for decades. Several studies [32–35] have shown that the most efficient way to curb energy consumption in the building sector (new and existing) is the reduction of the heat losses by improving the insulation of the building envelope (roof, floor, wall, and windows). A move beyond the current thermal performance of the building envelope is essential to realize the intended energy reduction in buildings. For retrofitting and even for new buildings in cities, the thickness of internal or external insulation layers becomes a major issue of concern. For systems such as domestic hot water a reduction in thickness is essential, therefore there is a growing interest in SIMs such as aerogels. Silica aerogels are silica-based dried gels that have very low weight and excellent thermal insulation performance. Specifically, they have high porosity (80–99.8%), low density, and low thermal conductivity (14 mW (m K)$^{-1}$) [36]. Silica aerogels are an innovative alternative to traditional insulation due to their high thermal performance, although the cost of this material remains high for cost-sensitive industries such as the building industry. Research is continuing to improve the insulation performance and to reduce the production costs of aerogels. The unique properties of aerogels offer many new applications in buildings [37]. The extraordinary low thermal conductivity of aerogels as well as their optical transparency allows them to be used for insulating building facades and insulating window panes. Two different types of silica aerogel-based insulating materials are used in the building sector: opaque silica aerogel-based materials and translucent insulation materials. Aerogel blankets/panels are beginning to be used to insulate walls, floors, attics, etc. Aspen Aerogels Inc. [38] has developed an insulation blanket based on silica aerogels called Spaceloft® [39], which has a thermal conductivity of 13.1 mW (m K)$^{-1}$. Spaceloft was used to convert an old mill house in Switzerland into an energy saving passive house. A case study was done in the United Kingdom where a number of governmental housing units were retrofitted by adding Spaceloft insulation layer at the interior wall surfaces [40]. A 44% reduction in the U-value, a 900 kWh yr^{-1} energy reduction, and a 400 kg yr^{-1} carbon emissions reduction were obtained. In another study, thermal performances and experimental tests were performed on walls and roofs using aerogel insulation [41]. Hot box measurements on wall assemblies containing aerogels showed that the R-value of wood-framed walls is improved by 9% and that of steel framed walls by 29%. Finite difference simulations performed on steel-framed wall assemblies using 0.6 cm thick aerogel strips showed that the aerogel can help to reduce the internal surface temperature differences between the center of the cavity and the stud location from 3.2 °C to only 0.4 °C. For roof structures, hot box measurements performed on the fastened metal roof insulated with aerogel strips showed an increase in the overall roof R-value of about 14%. Another type of silica aerogel-based material used in buildings is translucent insulation. These materials combine a low thermal conductivity with high transmittance of solar energy and daylight. Research has been conducted in the last decade on the development of highly insulating windows based on granular aerogel and monolithic aerogel. A new glazing element based on granular silica aerogel was developed at ZAE Bayern, Germany [42, 43]. The glazing consists of a 16 mm double-skin sheet made of polymethylmethacrylate filled with granular silica aerogels separated by two gaps filled with krypton or argon and glass panes installed at the ends. Three window systems have been developed. The first is a day-lighting system with total solar transmittance between 0.33 and 0.45, and a U value in the range 0.44–0.56 W m^{-2} K. The second system is a sun-protecting system developed by applying a lower emissivity (0.03) to the glass panes. A U-value of 0.37–0.47 W m^{-2} K, a visible transmittance in the range 0.19–0.38, and a solar transmittance between 0.17 and 0.23 were achieved. The third system is a solar collector formed by placing a heat exchanger between a layer of aerogels and the two glass panes. Aerogels are advanced materials almost like solid smoke, in which the aerogel resembles a hologram, appearing to be a projection rather than a solid object. They consist of more than 96% air. The remaining 4% is a wispy matrix of silicon dioxide [44]. Aerogels, consequently, are one of the lightest solids ever conceived. An aerogel is made by the sol-gel process. During this process, organic compounds containing silica undergo a chemical reaction, producing silicon oxide (SiO$_2$) [45]. This mixture is a liquid at the start of the reaction and becomes more and more viscous as the reaction proceeds. When the reaction is completed, the solution loses its fluidity and the whole reacting mixture turns into a gel. This gel consists of a 3D network of silicon oxide filled with the solvent [46]. During the special drying procedure, the solvent is extracted from the gel body leaving the silicon oxide network filled with air. This product is called an aerogel. Silica aerogels (SiO$_2$) are unique porous materials composed of more than 90% air and less than 10% solid silica in the form of highly crosslinked network structures, which results in low thermal conductivity and a large surface area (500–1000 m^2 g^{-1}), pore size (5–100 nm), and pore volume (1.5–4.5 cm^3 g^{-1}) [47–49]. The pore structure of aerogels is comparable to that of large pore mesostructures, therefore aerogels that have an open-pore structure can be readily adapted to polymer nanocomposites as reinforcing agents. An organic polybutadiene (PBD) rubber-based aerogel insulation material has been developed that will provide superior thermal insulation in architecture, exhibiting the flexibility, toughness, and durability typical of the parent polymer, yet with the low density and superior insulation properties associated with aerogels [50]. The rubbery behaviors of PBD rubber-CMS based aerogels are able to overcome the weak and brittle nature of conventional inorganic and organic aerogel insulation materials [8, 9]. Inorganic aerogels such as silica aerogels demonstrate many unusual and useful properties. There are several strategies to overcoming the drawbacks associated with the weakness and brittleness of silica aerogels. The development of a flexible fiber-reinforced silica aerogel composite blanket is one promising approach, providing a conveniently fielded from factor that is relatively robust toward handling in industrial environments compared to silica aerogel monoliths [51–55]. However, flexible silica aerogel composites still have a brittle,

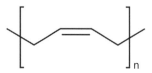

Figure 10.1 The chemical structure of 1,4-cis-polybutadiene.

dusty character that may be undesirable, or even intolerable, in certain applications [56–60]. Although crosslinked organic aerogels such as resorcinol-formaldehyde (RF), polyisocyanurate, and cellulose aerogels show very high impact strength, they are also very brittle, with little elongation [61, 62]. In the present study, we report the mechanical reinforcement of hyperbranched 1,4-cis-polybutadiene/CMS nanocomposites using a silica aerogel as the reinforcing agent [63]. The PBD-based rubber aerogel nanocomposite is a very flexible, no-dust, and hydrophobic organic compound that demonstrates the following ranges of typical properties such as density, shrinkage factor, and thermal conductivity [64–67]. The second component of the polymeric system is 1,4-cis-polybutadiene. For many applications, this compound is becoming increasingly popular to reinforce the concrete with small. Organic polymer aerogels are generally prepared from sol-gel polymerization followed by supercritical drying to remove liquid from the delicate gel structure without collapse or shrinkage. Organic polymer aerogels have been synthesized from resorcinol–formaldehyde, melamine-formaldehyde, phenolic-furfural, polyurethane (PU)-dichloromethane, cresol-formaldehyde, phloroglucinol-formaldehyde, epoxy-amine, diamine-aromatic dianhydride, 1,3-dimethoxybenzene or 1,3,5-trimethoxybenzene-formaldehyde, hydroxylated benzene derivatives-alkyl or aryl aldehydes, and aromatic dianhydrides-aromatic diamines or combined aromatic and aliphatic diamine systems (Figure 10.1).

Organic resorcinol-formaldehyde aerogels are produced from the base-catalyzed, aqueous polycondensation of resorcinol (1,3-dihydroxy benzene) with formaldehyde, which reacts further to form a crosslinked, dark red gel. Figure 10.2 outlines the resorcinol-formaldehyde reaction and depicts the formation of a crosslinked polymer network. Variables, such as reactant ratio, pH, and temperature, influence the crosslinking chemistry and growth process that take place prior to gelation. The size and number of resorcinol-formaldehyde clusters generated during the polymerization are controlled by the resorcinol/catalyst ratio in a formulation, thus the nanostructures can be controlled by the chemical reactions. Pekala et al. reported the synthesis of resorcinol-formaldehyde aerogels in detail. Sol-gel polymerization of resorcinol and formaldehyde with sodium carbonate as the base catalyst was performed at $358 + 3\,K$ for seven days, then the resorcinol-formaldehyde gels were washed by 0.125% trifluoroacetic acid to assist in the further condensation of hydroxymethyl groups remaining in the gels to form ether bridges between resorcinol molecules. Multiple exchanges with fresh acetone were used to remove the residual water. All acetone was then removed from the resorcinol-formaldehyde gels by CO_2 extraction. Dry resorcinol formaldehyde aerogels were finally obtained by the supercritical drying of resorcinol-formaldehyde gels with CO_2 at critical temperature and pressure of 304.1 K and 7.4 MPa. Resorcinol-formaldehyde aerogels are obtained in monolithic form, composed of the interconnected beads of diameter smaller than 10 nm and cell size less than 100 nm in an open-celled structure. An improved process for high-temperature production of organic aerogels was claimed in which the gel is directly extracted

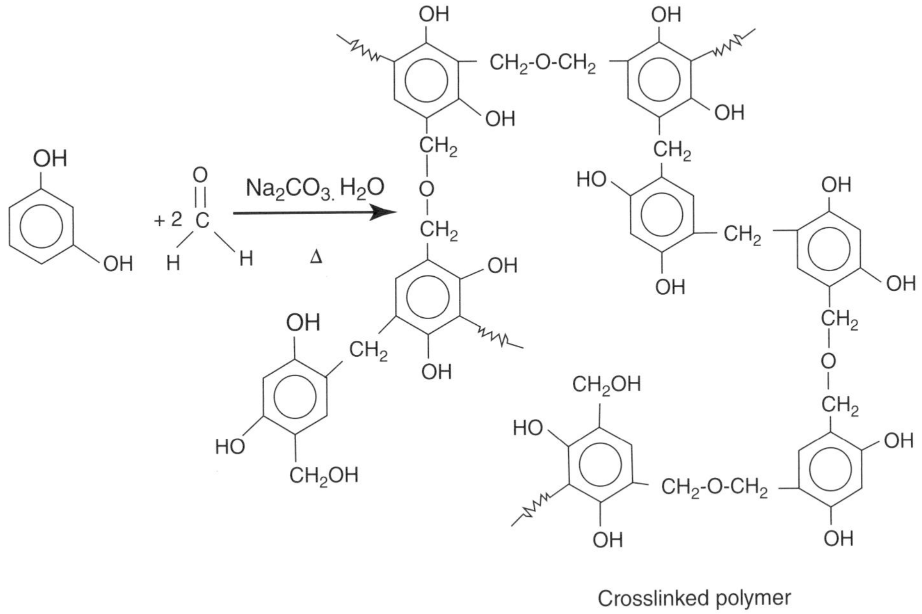

Figure 10.2 Schematic diagram of the sol-gel polymerization of resorcinol with formaldehyde.

supercritically without the need for exchange with a low-temperature solvent (i.e. CO_2) prior to the extraction of the fluid in pores [68].

Pekala et al. also synthesized melamine-formaldehyde aerogels from the aqueous, sol-gel polymerization of melamine with formaldehyde using a procedure analogous to the synthesis of resorcinol-formaldehyde aerogels [69]. Figure 10.3 shows a schematic

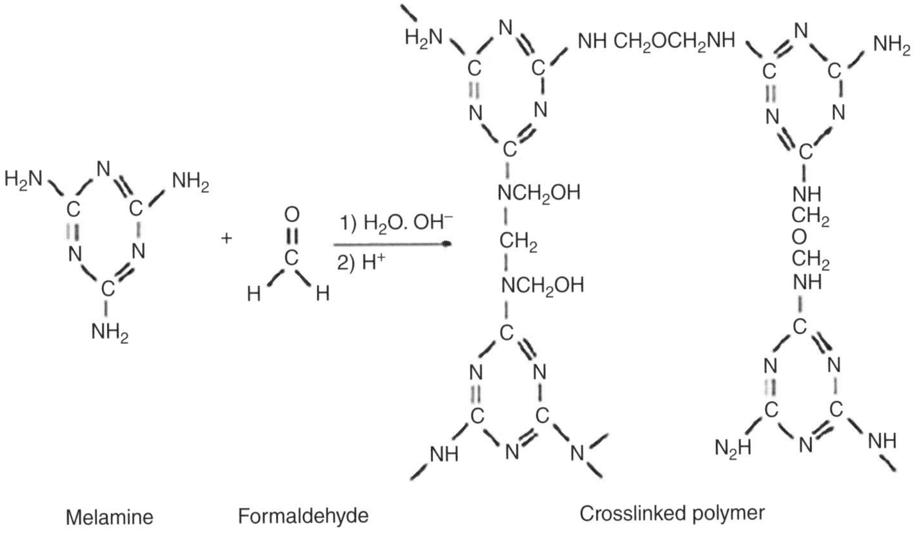

Melamine   Formaldehyde   Crosslinked polymer

Figure 10.3 Schematic diagram of the reaction of melamine with formaldehyde.

diagram for the reaction of melamine with formaldehyde. These aerogels have ultrafine cells of less than 50 nm diameter and continuous porosities. The aerogel density, transparency, and nanoporosity are controlled by varying the reactant concentration of the starting solution. Resorcinol–formaldehyde aerogels are easily produced with densities of 0.035–0.100 g cm^{-3}. Melamine-formaldehyde aerogels have densities of 0.1–0.8 g cm^{-3}. These aerogels are transparent. Resorcinol-formaldehyde aerogels are dark red, while melamine-formaldehyde aerogels are colorless. Resorcinol-formaldehyde aerogels with a high bulk density of 0.8–1.0 g cm^{-3} were recently prepared by the inverse phase suspension polycondensation of resorcinol-formaldehyde with sodium carbonate as catalyst in peanut oil, followed by drying the sol-gel spheres under alcohol supercritical drying conditions. It was easy with this fabrication method to avoid the accumulation of polymerization heat during gelation, and easy to remove the products from the reaction container.

In comparison to free radical copolymerization, step-growth thermosetting systems yield a more uniform network structure of gels than free radical chemistry. Mixing stoichiometric quantities of epoxy and amine (2 : 1 mol ratio, as each amine reacts with two epoxies) in the presence of tetrahydrofuran as the solvent at 333 K produces epoxy-amine gels. Since the volatile organic solvents occupy the pores of gels, drying the samples under supercritical conditions maintains the morphology of the pores. By changing the solvent content, materials with systematically tailored pore sizes up to 100 nm can be synthesized. Although aerogels are typically fabricated by subjecting a sol-gel precursor to critical-point drying (CPD) in order to remove background liquid without collapsing the network, supercritical drying is difficult to apply at an industrial scale because of its expensive and potentially dangerous character. Various methods to synthesize these materials at lower cost have been explored. Resorcinol-formaldehyde aerogels can be made using supercritical acetone. The displacement of acetone by CO_2 is no longer necessary, so the process is shortened compared to drying with supercritical CO_2. This technique was suggested for large-scale production for industrial purposes. Other methods such as subcritical drying and adding a surfactant to a polymerizable component were claimed to prepare porous polymer aerogels. An ambient-pressure drying process has been developed that causes moderate shrinkage of the organic gels. Tamon et al. [70] studied hot air drying, drying in vacuo, and freeze-drying with *t*-butanol for resorcinol-formaldehyde xerogels. The freeze-dried gels have the smallest shrinkage. It is applicable of freeze-drying to prepare resorcinol-formaldehyde cryogels. The molar ratio is one of the parameters, which can lead to a tailoring of the dried gel texture. With small angle X-ray scattering (SAXS), Czakkel et al. [71] reported that freeze-drying in *t*-butanol resulted in a high surface area (>2500 m^2 g^{-1}) in the gel, but the carbon cryogel was not structurally stable. Xerogels obtained by heating the resorcinol-formaldehyde hydrogel in an inert atmosphere had a compact structure and displayed low specific surface area (<900 m^2 g^{-1}). The conventional convective drying, with controlled air temperature, velocity, and humidity, was also tried to produce resorcinol-formaldehyde xerogels. In comparison with supercritical, vacuum or freeze-drying methods, this technique enjoys easy industrial implementation and is a non-expensive process. When synthesis conditions are adequate, it is possible to produce nanoporous resorcinol–formaldehyde xerogels by using convective drying to remove the solvent. First, 0.5 g of corn starch and 120 ml of 2-propanol were placed in a 500 ml vessel and stirred for 2 hours. Then 5 g of sodium hydroxide was added and reacted for 1 hour at 78–80 °C. After that, 10 g of chloroacetic acid was added to the vessel and the mixture was stirred for another 2 hours at 50 °C. The product was filtered and washed several times with ethanol, then dried under vacuum. The resulting CMS was crushed in a mortar [degree of substitution (DS) = 0.49]. A silica aerogel was prepared using sodium silicate as the starting material. The sodium silicate was diluted with deionized water [sodium silicate:deionized water (weight ratio = 1:3)] and 1.0 M aqueous HCl solution was used to modify the pH of the silica sol to 5. The obtained silica was then stirred for 1 minute, after which the sol suspension was aged until gelation occurred. To remove the sodium, the hydrogels were washed with deionized water three times, after which the silica hydrogel was collected. Next, butanol solution (pH adjusted to 2 by HCl solution) containing ETMS (5 wt% to the hydrogel) was added to the hydrogel. The mixture was then refluxed at 110 °C (~10 hours) until the pore water was exchanged with butanol. The water was removed using a Dean-Stark trap. Finally, the ETMS-modified alcogel was dried at 130 °C for 3 hours in a vacuum oven. To prepare the 1,4-cis-polybutadiene polymer nanocomposites, a pre-determined amount of silica aerogel was added to the butadiene monomer and mixed at 25 °C for 10 minutes. A stoichiometric amount of cobalt dialkylthiocarbamat catalyst $(2.0 \times 10^{-4}\,M)$ in toluene solvent was then added to the mixture and mixed at 25 °C for another 10 minutes. The resulting slurry was out-gassed under vacuum and transferred to a silicon mold. Pre-curing of nanocomposite was conducted at 25 °C for 2 hours to complete the crosslinking. The CMS was then mixed at 25 °C for 30 minutes. After drying at 37 °C for 48 hours, the mean diameter of the dried nanocomposite was determined by a sieving method using USP standard sieves. A solvent exchange step is required during the synthesis of silica aerogels to dry the hydrogels. Hexane with low surface tension has generally been used to exchange pore water even when a supercritical CO_2 drying tool is used to remove the organic solvent confined in the silica pores. A surface modification process is also necessary to induce hydrophobicity on the silica aerogel surface to avoid water adsorption. The hydrophobic aerogel can preserve its low thermal conductivity due to the absence of water in the pores. In this study, a silica aerogel was synthesized via simultaneous solvent exchange and surface modification in a butanol solution containing ETMS. Figure 10.4 shows the scanning

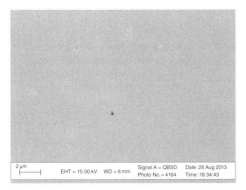 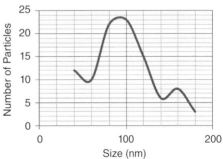

Figure 10.4 (a) Morphology of 1,4-cis-polybutadiene–CMS silica aerogel nanocomposites by emission scanning electron microscopy. (b) The size distribution of 1,4-cis-polybutadiene–CMS silica aerogel nanocomposites determined from the SEM picture.

Table 10.1 The average particle sizes of 1,4-cis-polybutadiene–CMS aerogel nanocomposites.

Variables	Values	Average particle size (nm)
Aerogel content (wt%)	0	25.79 ± 0.12
	1	28.34 ± 0.06
	3	32.76 ± 0.23
	5	45.00 ± 0.16
Polymerization time (h)	0.5	20.16 ± 0.12
	1	23.76 ± 0.08
	1.5	29.76 ± 0.20
	2	35.02 ± 0.16
Temperature (°C)	25	23.65 ± 0.16

electron microscopic (SEM) morphology of the silica aerogel. The synthesized silica aerogel exhibited a porous network structure with 5–100 nm pores.

The particle size of the aerogels was measured (Table 10.1). It can be seen that the average pore size of the silica aerogel is large enough to allow polybutadiene chains to readily fill the internal space of the mesostructured aerogel. Moreover, the structure of the aerogel was found to be open or 3D. Furthermore, modification of the silica surface with hydrophobic ETMS results in good compatibility with the 1,4-cis-polybutadiene chains. Tensile data describing the 1,4-cis-polybutadiene nanocomposites containing 1–5% silica aerogel were obtained from load-displacement plots of dogbone-shaped specimens. The stress-strain curves as a function of silica loading for the 1,4-cis-polybutadiene–CMS nanocomposites are provided in Figure 10.5.

The silica aerogel clearly provided composites with improved mechanical performance. The modulus, strength, elongation, and toughness of the 1,4-cis-polybutadiene–CMS silica aerogel nanocomposites prepared from the silica aerogel generally increased as the silica loading increased (Figures 10.6–10.9).

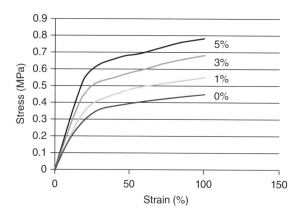

Figure 10.5 Stress-strain curves of 1,4-cis-polybutadiene–CMS silica aerogel nanocomposites containing different aerogel loadings.

Figure 10.6 Loading dependence of the tensile strength of 1,4-cis-polybutadiene–CMS silica aerogel nanocomposites.

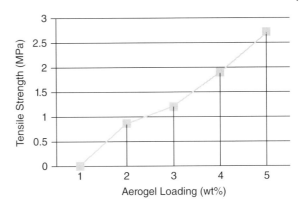

Figure 10.7 Loading dependence of the tensile modulus of 1,4-cis-polybutadiene–CMS silica aerogel nanocomposites.

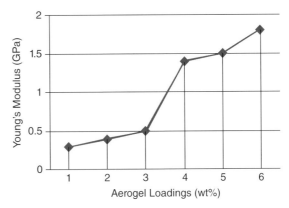

Figure 10.8 Loading dependence of the elongation at break of 1,4-cis-polybutadiene–CMS silica aerogel nanocomposites.

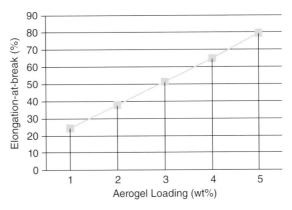

The stress-strain curves of 1,4-cis-polybutadiene–CMS–silica aerogel nanocomposites prepared from butadiene monomer, cobalt dialkylthiocarbamat catalyst, and silica aerogel as reinforcing agent are shown in Figure 10.10.

The tensile data for different silica loadings are summarized in Table 10.2.

The silica aerogel is an effective reinforcing agent for the 1,4-cis-polybutadiene and CMS matrix. For example, at an aerogel loading of 4–5 wt%, the tensile strength, modulus, and toughness of the polymer matrix were increased by 9, 15, 7%, respectively, but the elongation at break decreased by 11%. The mechanical analysis

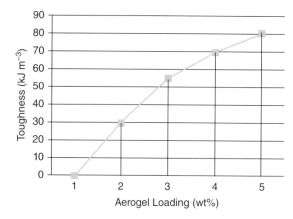

Figure 10.9 Loading dependence of the toughness of 1,4-cis-polybutadiene–CMS silica aerogel nanocomposites.

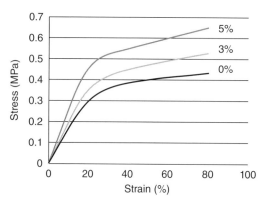

Figure 10.10 Comparison of stress-strain curves of 1,4-cis-polybutadiene–CMS silica aerogel nanocomposites containing different aerogel loadings.

Table 10.2 Tensile properties of 1,4-cis-polybutadiene–CMS silica aerogel nanocomposite.

Aerogel content (wt%)	Tensile modulus (MPa)	Tensile strength (MPa)	Toughness (kJ m⁻³)	Elongation (%)
0	1.52	0.50	240	42.3
1	1.65	0.92	484	72.9
3	1.87	1.26	896	75.3
5	3.56	1.97	1298	86.2

demonstrated that the silica aerogels improved the stiffness and yield stress of 1,4-cis-polybutadiene. Similar improvements in tensile modulus and strength have been observed for 1,4-cis-polybutadiene–mesoporous nanocomposites. The temperature dependence of the storage modulus (E') and tan δ of pure 1,4-cis-polybutadiene and 1,4-cis-polybutadiene–CMS–aerogel nanocomposites are shown in Figure 10.11.

The nanocomposites had a higher storage modulus than that of the matrix. The storage modulus of the nanocomposites increased with filler content. The E' values determined at 30 and 110 °C were compared for all samples and are presented in Table 10.3. The E' systematically increased with filler content in a fashion similar to that observed for the

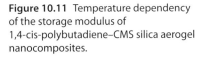

Figure 10.11 Temperature dependency of the storage modulus of 1,4-cis-polybutadiene–CMS silica aerogel nanocomposites.

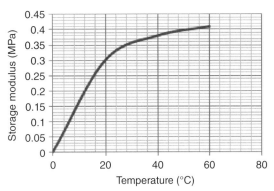

Table 10.3 Storage modulus and T_g values of 1,4-cis-polybutadiene–CMS silica aerogel nanocomposites.

Aerogel content (wt%)	Storage modulus		T_g (°C)
	30 °C	110 °C	
0	2042	12.65	104
3	2698	20.19	103
5	2845	23.14	103

tensile tests. Figure 10.11 also shows the temperature dependency of tan δ for the pure polymer and 1,4-cis-polybutadiene–CMS aerogel nanocomposites. The T_g values were determined at the maximum peaks of tan δ. As shown in Table 10.3, there was little or no shift in the T_g values at different silica loadings.

Carbon nanotube aerogels are novel materials. Aerogels are ultralight, highly porous materials typically fabricated by subjecting a wet-gel precursor to CPD or lyophilization (freeze-drying) in order to remove background liquid without collapsing the network. Microscopically, aerogels are composed of tenuous networks of clustered nanoparticles, and the materials often have unique properties, including very high strength-to-weight and surface-area-to-volume ratios. To date most aerogels are fabricated from silica [68] or pyrolized organic polymers [69, 70]. Practical interest in the former stems from their potential for ultralight structural media, radiation detectors, and thermal insulators [1], and in the latter from their potential for battery electrodes and supercapacitors [2]. Here we will investigate the properties of a new class of aerogels based on carbon nanotubes (CNTs). Small-diameter CNTs, such as single- and few-wall CNTs, are exciting candidates for electrically conducting aerogels. Individually, these nanotubes are extraordinarily stiff [4] and their electrical conductivity can be very large [71, 72]. Furthermore, ensembles of such nanotubes are useful aerogel precursors: they form electrically percolating networks at very low volume fractions [6] and elastic gels in concentrated suspensions through van der Waals interaction mediated crosslinking [73]. Here we report the creation of CNT aerogels from aqueous gel precursors by CPD and freeze-drying. CNT aerogels have been produced previously as intermediate phases during the process of drawing nanotube fibers [74] from a furnace and during the process of making sheets from multiwall CNT forests. By contrast, our aerogels were derived directly from CNT networks in suspension, and we could readily manipulate the network properties as a result. The flexibility afforded by this process enabled us to control CNT concentration, to utilize optimized CNT dispersion processes [75], to reinforce the networks with, for example, polyvinyl alcohol (PVA), and to infiltrate or backfill them with polymeric fluids. The CNT aerogels are synthesized by processing methodologies and their electrical and mechanical properties can be characterized. The CNT aerogels supported thousands of times their own weight after PVA reinforcement and, depending on processing conditions, their electrical conductivity ranged as high as c. 1 S cm^{-1}. Although our starting chemical vapor deposition (CVD) nanotube material contained single- and few-wall CNTs (the latter being predominantly double-wall CNTs, DWNTs), the dispersion and preparation processes employed here are directly applicable to pure single-wall carbon nanotubes (SWNTs). CNT aerogel electrical and structural properties are also expected to be similar to pure SWNT samples because the electrical [76] and tensile [77] properties of bulk SWNTs and DWNTs are comparable. Reinforcement of aerogels with PVA polymer [78–80] improved the strength and stability of the aerogel. In this case, the aerogels could support at least 8000 times their own weight. All of the aerogels (with or without PVA reinforcement) were highly porous, with pore sizes ranging from tens of nanometers up to 1 μm. Transmission electron microscopy (TEM) images like the one in Figure 10.12, prepared by shredding (pulling apart) unreinforced CNT aerogels above a TEM sample grid, suggest that nanotubes in the aerogels form a random filamentous network [81] with little bundling.

In the absence of bundling, the nanotube surface area is very large, a unique feature potentially useful for applications that require materials with large surface-area-to-volume ratios, including chemical sensors, reaction catalysts, and novel electrodes. The freeze-dried samples sometimes have a second porosity length scale of the order of tens to hundreds of micrometers as a result of ice crystals that form during the freezing process and displace nanotubes within the network. The densities of the CNT aerogels studied ranged from 10–30 mg ml^{-1} with no PVA reinforcement to 40–60 mg ml^{-1} for aerogels reinforced in a 1 wt% PVA bath. The PVA content in the aerogels could be estimated by subtracting the known mass of the CNTs in each sample, as determined from the sample volume and initial CNT gel concentration. The PVA content in the aerogel increased with increasing bath PVA concentration and also depended on the CNT concentration. PVA to CNT weight ratios ranged between 1 and 6 for all reinforced samples, with most samples ranging between 2 and 3 [82]. The electrical conductivity of

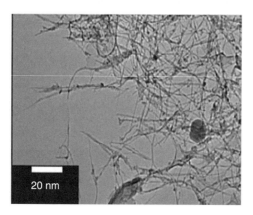

Figure 10.12 TEM image of an aerogel.

the aerogels was found to depend on several factors, including nanotube and PVA content, and the drying process (i.e. freeze-drying or CPD). The freeze-dried samples were consistently less conductive than the CPD samples. This difference can be attributed to disruptions of the nanotube network that occur during the freezing process, such as those described above. Samples created by CPD suffered significantly less mechanical distortion and had more reproducible conductivities. Figure 10.3 summarizes our results for a variety of CNT and PVA bath concentrations. In Figure 10.13, aerogel conductivity is plotted as function of CNT content, while the PVA bath concentration is constant at 1 wt%. In Figure 10.3b, the conductivity is plotted against PVA bath concentration, while the CNT content is fixed at $7.5 \, \text{mg ml}^{-1}$. As can be seen, a $7.5 \, \text{mg ml}^{-1}$ CNT aerogel with no PVA had a conductivity of nearly $1 \, \text{S cm}^{-1}$, significantly higher than typically obtained in solid CNT/polymer composites with comparable nanotube volume fractions of order 1% [83]. Although the addition of PVA into the bath reinforced the physical structure of the aerogel, the electrical conductivity of the reinforced aerogels was initially significantly reduced. After reinforcement in a 1 wt% PVA solution, for example, the conductivity dropped by five orders of magnitude to c. $10{-}5 \, \text{S cm}^{-1}$. This level of conductivity is more typical of solid polymer composites with comparable nanotube volume fractions of order 1%. For comparison, aligned CNT/PVA fibers, which are significantly denser (c. $1.3 \, \text{g cm}^{-3}$), have conductivities of c. $10 \, \text{S cm}^{-1}$.

Although the physical reinforcement provided by PVA is often desirable, the concurrent reduction in conductivity may present problems for certain applications. Interestingly, high-current pulses applied to a PVA-reinforced aerogel sample produced

Figure 10.13 Conductivities of as-prepared CNT aerogel samples.

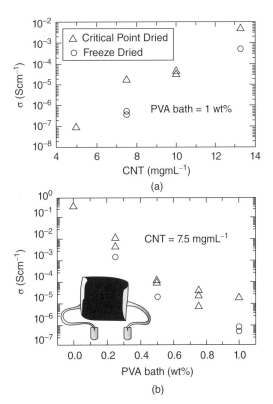

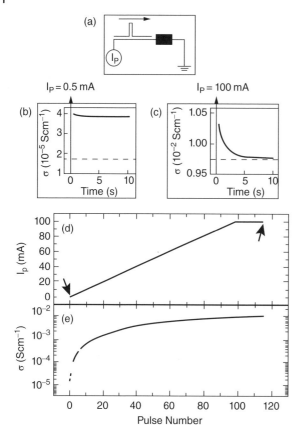

Figure 10.14 (a) Discrete current pulses applied across a sample. (b) A 15 ms, 0.5 mA current pulse was applied to a 13.3 mg ml^{-1} CNT CPD sample reinforced in a 0.5 wt% PVA bath.

stepwise, irreversible increases in network electrical conductivity. The net increase could be quite large, for example several orders of magnitude, as shown for a CPD sample in Figure 10.14. In a typical configuration, the sample was placed under high vacuum and 15 ms current pulses of increasing amplitude were applied across the sample as shown in Figure 10.14a. The current pulses were spaced at least 30 seconds apart and kept as short as possible to minimize bulk heating of the sample, which can cause significant distortion. The maximum current applied was 100 mA.

In summary, the carbon nanotube aerogels from wet-gel precursors differ significantly from the gelatin nanotube foams [84] and high-temperature aerogels [85–89] reported previously. The nanotube aerogels were strong and electrically conducting, offering potential improvement over current carbon aerogel technologies for applications such as sensors, actuators, electrodes, thermoelectric devices, etc., as well as silica-based aerogel applications. The aerogels could be significantly reinforced by small amounts of PVA, albeit at the cost of reduced conductivity; the network conductivity, however, could be at least partially restored by applying current pulses through the sample. Lastly, the aerogel structures could be backfilled with polymeric fluids, which, in turn, could be cured to create novel electrically conducting solid composites. Polymer/clay nanocomposites have been an area of extensive research for the past 15 years, and have resulted in a limited number of commercial products. These composite materials typically rely upon exfoliation of layered, smectic clays, such that individual clay platelets are dispersed within a polymeric phase, generating a network which can transfer physical loads, and can serve to create both polymer crystallization and confinement [90]. The requirement of exfoliation (typically an endothermic step) and the generally high level of incompatibility of silicates with organic polymers are significant technical hurdles that must be overcome in order to produce useful composites in this manner. Aerogels are typically inorganic materials composed primarily of air, with inorganic skeletal structures. Silica is the material most commonly converted to aerogels – these materials have been known since the 1930s, when Kistler coined the term "aerogel" [91, 92]. Since that time, silica aerogels have been utilized in applications including sensors, catalyst supports, thermal/acoustic insulation, and as collection media for particles found in space [93–96]. In each case, extremely low density was the driving force for use of aerogels; these materials suffer, however, from high manufacturing costs. The ability to convert layered clays into nanostructured aerogels was first described Mackenzie [97] and Call [98]; the resultant fibrous montmorillonite structures were described as possessing reasonable rigidity, but poor thermal stability at 110 °C for an extended time or when desiccated over phosphorus pentoxide. Similar processing of non-swelling clays, such as kaolin, only produced fine powders. The structural features of highly compressible "gel skeletons" produced by vacuum sublimation of frozen thixotropic clay gels in water or benzene were first described by Weiss [99], Hoffman [100], and Norrish [101]. Processing parameters for clay aerogels and definition of a structure in which clay particles are linked edge-to-face much like a "house of cards" owing to opposite surface and edge charges that exist in clays were reported by Van Olphen [102]. The effects of process parameters, such as clay concentration and freezing rates, upon the size and shape of resultant clay aerogels was investigated by Nakazawa [103]. Building on these prior studies, we recently reported a robust process for the production of organically modified clay aerogels [104]. Despite a modest number of studies spanning half a century, very little has been done in the past with these clay aerogels. The results of these composite studies, using clay aerogels as nanostructured fillers, are presented. Also presented are business startup activities associated with these materials (including the launch of AeroClay Inc. in 2006).

The aerogel blanket is a new material that has very low thermal conductivity that makes it a good candidate for insulating walls. In fact, nowadays aerogel blankets are used to improve the energy performance of existing walls. Using aerogel blankets as insulation material in a wall means replacing the existing insulation material. When a material in a building component is replaced in order to improve a specific performance, it is important to verify any effect on the other functional requirements of the wall, such as fire performance. The purpose of this chapter is to evaluate the fire performance of a non-load bearing external wall insulated by aerogel blankets. In order to achieve this aim, experiments and numerical simulations were carried out. First, two wall models using aerogel blankets as insulation materials were designed. The fire behavior of the designed walls was investigated and the models were modified according to the fire requirements. Finally, the moisture properties of the aerogel blanket were measured and compared to other common insulation materials. The results indicated that although aerogel blanket thermal properties are excellent, its fire reaction is not so advantageous. However, if aerogel blankets are placed in a safe wall position, the thickness of the wall can be reduced. Because of this thickness reduction, using aerogel blankets as insulation material could be an interesting alternative wherever space is an important issue. Aerogel blankets are an innovative alternative to traditional insulation materials in building and construction applications, but their cost is still high in the building industry. Some examples of aerogel blanket applications in building and construction are (Aspen Aerogels):

- roof insulation
- floor and balcony insulation
- external insulation
- internal insulation
- heat bridge treatments, e.g. located on the ledges of windows
- services, e.g. pipe insulation.

One manufacturer that develops this product is Aspen Aerogels Inc., whose aerogel blanket product for wall insulation is named Spaceloft. Spaceloft can be used to improve the insulation capacity of existing walls. Although blankets of 10 mm of thickness are currently available, it is also possible to have them in larger thickness or to apply them in several layers. Aspen Aerogels Inc. was consulted in order to discover how the material should be applied to the walls. In external insulation, the Spaceloft layer or layers are directly applied to the existing wall using mechanical and/or adhesive to attach the blanket to the wall (Figure 10.15).

Two reference walls were chosen to be studied. The first wall was a wood-based wall, which is a common wall material in Sweden. The other wall was a concrete-based wall because concrete is used in multi-family houses. The studied walls were non-load-bearing external walls of multi-family houses. The designed reference walls were modified by adding aerogel blankets, keeping constant the thermal performance of the walls. The calculations were performed by using the constant thermal conductivity

Figure 10.15 External insulation (Aspen Aerogels).

of the wall insulation materials at room temperature. An organic polybutadiene (PB) rubber-based aerogel insulation material has been developed that will provide superior thermal insulation and inherent radiation protection, exhibiting the flexibility, resiliency, toughness, and durability typical of the parent polymer, yet with the low density and superior insulation properties associated with the aerogels. The rubbery behaviors of the PBD rubber-based aerogels are able to overcome the weak and brittle nature of conventional inorganic and organic aerogel insulation materials. Additionally, with higher content of hydrogen in their structure, PBD rubber aerogels will also provide inherently better radiation protection than inorganic and carbon aerogels. Since PBD rubber aerogels also exhibit good hydrophobicity due to their hydrocarbon molecular structure, they will provide better performance reliability and durability as well as simpler, more economic, and environmentally friendly production over conventional silica or other inorganic-based aerogels, which require chemical treatment to make them hydrophobic. Inorganic aerogels such as silica aerogels demonstrate many unusual and useful properties. There are several strategies to overcoming the drawbacks associated with the weakness and brittleness of silica aerogels. Development of a flexible fiber-reinforced silica aerogel composite blanket is one promising approach, providing a conveniently fielded form factor that is relatively robust toward handling in industrial environments compared to silica aerogel monoliths. However, flexible silica aerogel composites still have a brittle, dusty character that may be undesirable, or even intolerable, in certain applications.

Although crosslinked organic aerogels such as resorcinol-formaldehyde, polyisocyanurate, and cellulose aerogels show very high impact strength, they are also very brittle, with little elongation (i.e. less rubbery). Also, silica and carbon aerogels are less efficient radiation shielding materials due to their lower content of hydrogen. The present invention relates to maleinized polybutadiene (or polybutadiene adducted with maleic anhydride)-based aerogel monoliths and composites, and the methods for preparation. Hereafter, they are collectively referred to as polybutadiene aerogels. Specifically, the polybutadiene aerogels of the present invention are prepared by mixing a maleinized polybutadiene resin, a hardener containing a maleic anhydride reactive group, and a catalyst in a suitable solvent, and maintaining the mixture in a quiescent state for a sufficient period of time to form a polymeric gel. After aging the gel at elevated temperatures for a period of time, a uniformly stronger wet based aerogel is then obtained by removing interstitial solvent by supercritical drying. The mesoporous maleinized polybutadiene-based aerogels contain an open-pore structure, which provides inherently hydrophobic, flexible, nearly unbreakable, less dusty aerogels with excellent thermal and physical properties. The materials can be used as thermal and acoustic insulation, radiation shielding, and vibration-damping materials. The organic PB-based rubber aerogels are very flexible, no-dust, and hydrophobic organics that demonstrate the following ranges of typical properties: densities of $0.08–0.255\,\text{g cm}^{-3}$, shrinkage factor (raerogel/rtarget) of 1.2–2.84, and thermal conductivity values of $20.0–35.0\,\text{mW (m-K)}^{-1}$. Based on Doshi studies, insulating coatings are prompting significant interest with increasing environmental and safety awareness and sensitivity to energy efficiency. Aerogels are the best thermal insulators in the world. Cabot's Nanogel® aerogel has a thermal conductivity of $12\,\text{mW (m-K)}^{-1}$ (compared to still air or PU foam at $26\,\text{mW (m-K)}^{-1}$) and is currently being used extensively in day-lighting and oil and gas applications. The performance of Nanogel aerogel is based on the unique nanostructure of the particles that is not impacted by size. The studies will discuss how the new line of Nanogel aerogel products can be used to create the next generation of insulative coatings and the properties of such coatings. The benefits of these coatings include energy efficiency and burn protection. Basic formulation principals and data from thermal tests on sample coatings are shared here. The data shows the superior nature of the material's thermal performance at work, reducing surface temperatures by 60–100 °C for 200 °C substrates. Nanogel aerogel-based thermally insulative coatings are complementary to IR reflective (low e) coatings and cases that show this synergy at work will be exemplified. For new buildings, HECK AERO enables lower exterior wall thicknesses, reducing thickness to a minimum. Thus thermal insulation is about 29 times more efficient than solid brick. When compared to concrete, thermal insulation actually is 105 times more efficient. Thermal insulation is expressed here in terms of heat loss coefficient (K), and thermal conductivity (l) at ambient conditions. Lower values for K and l mean higher insulation. As can be seen, to get the same insulation level, a traditional foam insulation layer needs to be up to five times thicker (50 mm) than an aerogel layer (10 mm). In addition, acrylic foam could burn and generate toxic and hazardous fumes in case of fire. TAASI's (inorganic) aerogels do not burn or produce hazardous fumes. The TAASI Corporation provides various types of aerogel products with different apparent densities and insulation properties (Tables 10.4 and 10.5).

Different routes for the production of the cellulose aerogels have been investigated. For the production of the wet-gel precursors cellulose is either directly dissolved in N-methylmorpholine-N-oxide (NMMO) or NaOH, or derivatives (acetate or carbamate) are used. These routes are described below, followed by the other steps necessary to produce aerogels from these precursors. It has been known for some time that cellulose can be dissolved in aqueous NaOH solutions in a rather narrow temperature and concentration range. CEMEF1™ is investigating this in the AeroCell project. Solutions of celluloses with the distribution percentage ranging from 170 to 500 were prepared at different concentrations at −6 °C and a fixed NaOH concentration of 7.6%. They were gelled by increasing the temperature and finally regenerated, mostly in water, but other solvents (alcohols and acids) were also tested.

NMMO is a solvent that can completely dissolve cellulose without prior derivatization. The region in the NMMO/water/cellulose phase diagram where complete dissolution is possible, is still rather small. However, a wider range of cellulosic materials can be dissolved compared to NaOH. Cellulose-NMMO solutions were prepared in the following way [105]. Pulp, NMMO (50–75%), and stabilizer were placed in a kneader. The pulp was soaked and afterwards the temperature was increased and the excess water was removed by applying low pressure. The kneader was kept under these conditions until all the excess water was removed. Normally during this time the pulp will be dissolved completely. The pressure was then raised and the solution was kneaded until the pulp was completely dissolved if necessary. A wide variety of cellulosic materials can be easily dissolved in NMMO. We were able to dissolve commercial paper pulp and unbleached pulp in NMMO without any difficulties. The DP of the used materials ranges from 180 to 6200. The addition of other substances (fibers, polymers, soot) to the solution is also possible as long as they fulfill some basic requirements (e. g. stable in NMMO solution). The sol-gel synthesis carried out by CEP2™ is based on the formation of urethane bonding through addition reactions between cellulose acetate

Table 10.4 Published examples.

Aerogel type	Apparent density, ρ (kg m^{-3})	g/cc	Heat loss coefficient, K (W m^{-2} K)	Insulation thickness, d (mm)	Thermal conductivity, λ (10^{-3} W (m K)$^{-1}$)	Remarks and references
Reference: acrylic foam (PMMA)			1.8	50 (two layers)	(90)	Geotzberger and Wittwer 1985
SiO$_2$ aerogel						ibid.
(a) air-filled						
(b) vacuum	232		2.0	10	13 (20)	
	0.232		1.1	10	(11)	
	232					
	0.232					
SiO$_2$	270	0.27	1.5	11	13 (16.5)	λ = 10–20 for
	200	0.20		15		ρ = 0.1–0.2,
	105		0.6	22	(13.2)	Buttner et al.,
	0.105				13	1985, Fricke
	109				16.3	et al., 1985,
	0.109		1.5	9	(13.5)	Nilsson et al.
	80					1985
	0.080					
	75					
	0.075					
SiO$_2$-C	70–90	0.07–0.09			14.6	Lee et al. 1995
SiO$_2$-TiO$_2$	260	0.26			25	Kuhn et al.,
	200	0.20			12 (at ambient temperature)	1995, Wang et al., 1995
					38 (at 800 K)	
SiO$_2$-TiO$_2$-FeO	200	0.20				

PMMA, poly methyl methacrylate.

and isocyanate [106], namely the reaction of cellulose acetate with a polyfunctional isocyanate [methylene diphenyl diisocyanate (MDI), functionality = 2.7]. For this purpose, we dissolved commercial cellulose acetate (from Aldrich, M_n = 50 000) with an acetyl substitution degree (DS) equal to 2.4 in acetone. In solution, the unreacted hydroxyl groups can further react with MDI to crosslink. Extra-dry acetone (water content <0.02%) was used to prevent any reactions between isocyanate groups and water. Gelation requires the use of a catalyst. Dibutyltin dilaurate was chosen for its adequate catalytic activity toward urethane bond formation. The associated sol route process was performed at room temperature under slow mechanical stirring. The cellulose acetate concentration (mass ratio cellulose acetate and solvent) was fixed to 10% and catalyst concentration (mass ratio of catalyst and reagents) to 5%. In addition we varied the mass ratio between cellulose acetate and isocyanate. During the syneresis stage, the gels were washed three times with pure acetone. AeroCell precursors

Table 10.5 Prediction of silica aerogel's thermal conductivity.

Apparent density, ρ		Thermal conductivity
(kg m^{-3})	(g ml^{-1})	
109	0.109	13
140	0.14	19
200	0.20	24
300	0.30	36
400	0.40	48
500	0.50	60
600	0.60	72

based on cellulose carbamate were prepared by FhGIAP3^{TM}. Cellulose carbamate was synthesized from activated dissolving pulp and urea at elevated temperatures. The resulting cellulose carbamate has a DP of 300–400 and a DS of 0.2–0.6. The cellulose carbamate was dissolved in caustic soda at an appropriate temperature and NaOH concentration. For precipitation and shaping two precipitation routes are available: chemical precipitation (acidic precipitation medium with additives) and thermal precipitation (via a suitable temperature regime). Cellulose carbamate solutions with 1, 3, 5, and 9% concentrations were prepared and tested with respect to the effects of the chemical and thermal precipitation. In most cases the liquid solutions were poured in a (cylindrical) mold and then gelled/solidified. There are some other shaping methods that produce cellulose aerogels of different sizes and forms. GeniaLab4 used JetCutter technology [107] to produce beads of size 400–1200 μm from the cellulose solutions. The bead generation is achieved by cutting a solid jet of fluid coming out of a nozzle by rotating cutting wires into cylindrical segments, which then form beads due to surface tension on their way to a hardening device. By dripping the molten cellulose-NMMO solution through a hole into the regeneration bath small spheres can be produced. The cellulose-NMMO solutions can also be cast into films or spun into fibers, which offers the possibility of continuous production. NASA's Glenn Research Center has developed a way to reinforce silica aerogels using a conformal coating of polymer to significantly enhance their strength and durability. These polymer-reinforced aerogels are lightweight, porous, and have low thermal conductivity, which makes them an ideal material for numerous automotive applications. Made from lightweight, mechanically strong materials, they can be used for acoustic and thermal insulation (e.g. engine firewall), vibration damping, or anywhere structural foam is currently used in vehicles.

References

1 Zhao, Z., Wang, X., Qiu, J. et al. (2014). Three-dimensional graphene-based hyderogels/aerogel materials. *Rev. Adv. Mater. Sci.* 36: 137–151.
2 Stoller, M.D., Park, S., Zhu, Y.-W. et al. (2008). Reduction of graphite oxide using alcohols. *Nano Lett.* 8: 3498.
3 Novoselov, K.S., Geim, A.K., and Morozov, S.V. (2004). Electric field effect in atomically thin carbon films. *Science* 306: 666.

4 Balandin, A.A., Ghosh, S., Bao, W. et al. (2008). Superior thermal conductivity of single-layer graphene. *Nano Lett.* 8: 902.

5 He, H. and Gao, C. (2011). Graphene nanosheets decorated with Pd, Pt, Au, and Ag nanoparticles: Synthesis, characterization, and catalysis applications. *Sci. China Chem.* 54: 397.

6 Zhu, Z.P., Su, D., Weinberg, G., and Schlogl, R. (2004). Supermolecular self-assembly of graphene sheets: Formation of tube-in-tube nanostructures. *Nano Lett.* 4: 2255.

7 Dikin, D.A., Stankovich, S., Zimney, E.J. et al. (2007). Preparation and characterization of graphene oxide paper. *Nature* 448: 457.

8 Chen, H.Q., Muller, M.B., Gilmore, K.J. et al. (2008). Mechanically strong, electrically conductive, and biocompatible graphene paper. *Adv. Mater.* 20 (18): 3557–3561.

9 Tang, L., Wang, Y., Li, Y. et al. (2009). Preparation, structure, and electrochemical properties of reduced graphene sheet films. *Adv. Funct. Mater.* 19: 2782.

10 Xu, Y.X., Sheng, K.X., Li, C., and Shi, G.Q. (2010). Strong, conductive, lightweight, neat graphene aerogel fibers with aligned pores. *ACS Nano* 4: 4324.

11 Tang, Z.H., Shen, S.L., Zhuang, J., and Wang, X. (2010). Noble-metal-promoted three-dimensional macroassembly of single-layered graphene oxide. *Angew. Chem.* 122: 4707.

12 Bai, H., Sheng, K.X., Zhang, P.F. et al. (2011). Graphene oxide/conducting polymer composite hydrogels. *J. Mater. Chem.* 21: 18653.

13 Zu, S.Z. and Han, B.H. (2009). Aqueous dispersion of graphene sheets stabilized by pluronic copolymers: Formation of supramolecular hydrogel. *J. Phys. Chem. C* 113: 13651.

14 Cong, H.P., Ren, X.C., Wang, P., and Yu, S.H. (2012). A binary functional substrate for enrichment and ultrasensitive SERS spectroscopic detection of folic acid using graphene oxide/Ag nanoparticle hybrids. *ACS Nano* 6: 2693.

15 Zhang, X.T., Sui, Z.Y., Xu, B. et al. (2011). Highly elastic graphene oxide–epoxy composite aerogels via simple freeze-drying and subsequent routine curing. *J. Mater. Chem.* 21: 6494.

16 Vickery, J.L., Patil, J., and Mann, S. (2009). Fabrication of graphene-polymer nanocomposites with higher order three-dimensional architectures. *Adv. Mater.* 21: 2180.

17 Worsley, M.A., Pauzauskie, P.J., Olson, T.Y. et al. (2010). Synthesis of graphene aerogel with high electrical conductivity. *J. Am. Chem. Soc.* 132: 14067.

18 Chen, M.X., Zhang, C.C., Li, X.C. et al. (2013). Topotactical conversion of carbon coated Fe-based electrodes on graphene aerogels for lithium ion storage. *J. Mater. Chem. A* 1: 2869.

19 Dreyer, D.R., Park, S., Bielawski, C.W., and Ruoff, R.S. (2010). The chemistry of graphene oxide. *Chem. Soc. Rev.* 39: 228.

20 Kim, J., Cote, L.J., Kim, F. et al. (2010). Graphene oxide sheets at interfaces. *J. Am. Chem. Soc.* 132: 8180.

21 Kim, F., Cote, L.J., and Huang, J. (2010). Graphene oxide: surface activity and two-dimensional assembly. *Adv. Mater.* 22: 1954.

22 Stankovich, S., Dikin, D.A., and Dommett, G.H.B. (2006). Graphene-based composite materials. *Nature* 442: 282.

23 Xu, Y.X. and Shi, G.Q. (2011). Assembly of chemically modified graphene: methods and applications. *J. Mater. Chem.* 21: 3311.

24 Blake, P., Brimicombe, P.D., Nair, R.R. et al. (2008). Graphene-based liquid crystal device. *Nano Lett* 8: 1704.

25 Eda, G. and Chhowalla, M. (2009). Graphene-based composite thin films for electronics. *Nano Lett.* 9: 814.

26 Slaughter, B.V., Khurshid, S.S., and Fisher, O.Z. (2009). Hydrogels in regenerative medicine. *Adv. Mater.* 21: 3307.

27 Bai, H., Li, C., Wang, X.L., and Shi, G.Q. (2010). A pH-sensitive graphene oxide composite hydrogel. *Chem. Commun.* 46: 2376.

28 Ma, X.M., Li, Y.H., Wang, W.C. et al. (2013). Fillers and reinforcements for advanced nanocomposites. *Eur. Polym. J.* 49: 389.

29 Liu, J.H., Chen, G.S., and Jiang, M. (2011). Electrochemically sensitive supra-crosslink and its corresponding hydrogel. *Macromolecules* 44: 7682.

30 M. Ibrahim, E. Wurtz, P. Achard, P.H. Biwole, Fostering energy efficiency in buildings through aerogel-based renders. Hal-Archive Ouvertes, 1–11.

31 Hu, H., Zhao, Z.B., Wan, W.B. et al. (2013). Ultralight and highly compressible graphene aerogels. *Adv. Mater.* 25: 2219.

32 International Energy Agency. Energy conservation in buildings and community systems, ECBCS news 2011 issue 54. www.ecbcs.org.

33 Enkvist P.A., Naucler T., and Rosander J. A cost curve for greenhouse gas reduction, The McKinsey Quarterly, 2007.

34 Energy Efficiency Watch, Final Report, 2013. Improving and implementing national energy efficiency strategies in the EU framework.

35 Verbeeck, G. and Hens, H. (2004). Energy savings in retrofitted dwellings: economically viable? *Energy & Build.* 37 (7): 747–754.

36 Soleimani, D.A. (2007). Silica aerogel: synthesis, properties and characterization. *J. Mater. Process. Technol.*

37 TAASI. http://www.taasi.com/apps.htm (2011, date of last access).

38 Aspen Aerogels Inc. http://www.aerogel.com (2014, date of last access).

39 Spaceloft® Safety Data Sheet, Retrieved 9 February 2011, from http://www.aerogel.com/products/pdf/Spaceloft_MSDS.pdf.

40 http://www.aerogel.com/markets/Case_Study_Interior_Wall_web.pdf (2012, date of last access).

41 Kosny J., Petrie T., Yarbrough D., Childs P., Syed A.M., and Blair C. Nano-scale insulation at work: thermal performance of thermally bridged wood and steel structures insulated with local aerogel insulation. Thermal performance of the exterior envelopes of buildings X, Proceedings of ASHRAE THERM X, Clearwater, FL, 2007.

42 Reim, M., Korner, W., Manara, J. et al. (2005). Silica aerogel granulate material for thermal insulation and daylighting. *Sol. Energy* 79: 131–139.

43 Reim, M., Beck, A., Korner, W. et al. (2002). Highly insulation aerogel glazing for solar energy use. *Sol. Energy* 72 (1): 21–29.

44 Fischer, F., Rigacci, A., Pirard, R., and Achard, P. (2006). Cellulose-based aerogels. *Polymer* 47 (22): 7636–7645.

45 Brinker, C.J. and Scherer, G.W. (1990). *Sol-Gel Science: The Physics and Chemistry of Sol-Gel Processing*. New York: Academic Press.

46 Hushing, N. and Schubert, U. (1998). Aerogels-airy materials: chemistry, structure, and properties. *Angew. Chem. Int. Ed.* 37 (1): 22–45.

47 Rao, A.P., Rao, A.V., and Pajonk, G.M. (2005). Hydrophobic and physical properties of the two step processed ambient pressure dried silica aerogels with various exchanging solvents. *J. Sol-Gel Sci. Technol.* 36 (3): 285–292.

48 Bhagat, S.D. and Rao, A.V. (2006). Surface chemical modification of TEOS based silica aerogels synthesized by two step (acid-base) sol-gel process. *Appl. Surf. Sci.* 252 (12): 4289–4297.

49 Haranath, D., Waagh, P.B., Pajonk, G.M., and Rao, A.V. (1997). Influence of sol-gel processing parameters on the ultrasonic sound velocities in silica aerogels. *Mater. Res. Bull.* 32 (8): 1079–1089.

50 Tan, C., Fung, M., and Newman, J.K. (2001). Organic aerogels with very high impact strength. *Adv. Mater.* 13 (9): 644–646.

51 Jin, W. and Brennan, J.D. (2002). Properties and applications of proteins capsulated within sol-gel derived materials. *Anal. Chim. Acta* 461: 1–36.

52 Zheng, J.K., Pang, J.B., Qiu, K.Y., and Wei, Y. (2000). Synthesis of mesoporous silica materials with hydroxyacetic acid derivateds as templates via a sol-gel process. *J. Inorg. Organomet. Polym.* 10: 103–113.

53 Smirnova, I., Suttruengwong, S., and Arlt, W. (2004). Feasibility study of hydrophilic and hydrophobic silica aerogels as drug delivery systems. *J. Non-Cryst. Solids* 350: 54–60.

54 Soleimani, D. and Abbasi, M.H. (2008). Silica aerogel; synthesis, properties and characterization. *J. Mater. Process. Technol.* I9 (9): 10–26.

55 Graham, A.L., Carison, C.A., and Edmiston, P.L. (2002). Development and characterization of molecularly imprinted sol-gel materials for the selective detection of DDT. *Anal. Chem.* 74: 458–467.

56 Radha, G. and Ashok, K. (2008). Molecular imprinting in sol-gel matrix. *Biotechnol. Adv.* 26: 533–547.

57 Dai, S., Shin, Y.S., Branes, C.E., and Toth, L.M. (1997). Enhancement of uranyl adsorption capacity and selectivity on silica sol-gel glasses via molecular imprinting. *Chem. Mater.* 9, 2521–5.

58 Shustak, G., Marx, S., Turyan, I., and Mandler, D. (2003). Application of sol-gel technology for electroanalytical sensing. *Electroanalysis* 15: 398–408.

59 Chaudhury, N.K., Gupta, R., and Gulia, S. (2007). Sol-gel technology for sensor applications. *Def. Sci. J.* 57: 241–253.

60 Schmidt, M. and Schwertfeger, F. (1998). Applications for silica aerogel products. *J. Non-Cryst. Solids* 225 (1): 364–368.

61 Gupta, R. and Kumar, A. (2008). Bioactive materials for biomedical applications using sol-gel technology. *Biomed. Mater.* 3: 034005.

62 Jin, H., Nishiyama, Y., Wada, M., and Kuga, S. (2004). Nanofibrillar cellulose aerogels, Colloids Surf. *A* 240 (3): 63–67.

63 Lin, J. and Wang, W. Novel low-polyimide/mesoporous silica composite films: preparation, microstructure, and properties. *Polymer* 48 (1): 318–329.

64 Potapo, Y., Borisov, Y., Panfilov, D. et al. (2004). Research of polymer concrete. Part VII: Strength of continuously reinforced polymer concrete with various kinds of fibers. *J. Sci. Israel Technolog. Advantages* 6 (3–4): 71–74.

65 Seal, A., Bose, N.R., and Dakui, S.K. (2001). Mechanical properties of glass polymer multilayer composite. *Bull. Mater. Sci.* 24 (2): 197–201.

66 Zhang, S., Ye, L., and Mai, Y.W.A. (2001). Study on polymer composite strengthening systems for concrete columns. *Appl. Compos. Mater.* 7 (3): 125–138.

67 Rana, A.K. and Jayachandran, K. (2000). Jute fiber for reinforced composites and its prospects. *Mol. Cryst. Liq. Cryst.* 353: 35–45.

68 Hrubesh, L.W. (1998). *J. Non-Cryst. Solids* 225: 335.

69 Pekala, R.W., Farmer, J.C., Alviso, C.T. et al. (1998). Carbon aerogels for electrochemical applications. *J. Non-Cryst. Solids* 225: 74.

70 Takuji Yamamoto, Shin R. Mukai, Akira Endo, Masaru Nakaiwa, Hajime Tamon, Interpretationof structure formation during the sol-gel transition of aresorcinol-formaldehyde solution by population balance, *Journal of Colloid and Interface Science*, Volume 264, Issue 2, 15 August2003, Pages 532–537.

71 Czakkel O, Nagy B, Geissler E, László K., Effect of molybdenum on the structure formation of resorcinol-formaldehyde hydrogelstudied by coherent x-ray scattering, *J ChemPhys.* 2012 Jun 21; 136(23):234907. doi: 10.1063/1.4729465.

72 Bryning, M.B., Islam, M.F., Kikkawa, J.M. and Yodh, A.G. (2005) Very Low Conductivity Threshold in Bulk Isotropic Single-WalledCarbon Nanotube-Epoxy Composites. *Advanced Materials*, 17, 1186–1191.

73 L.A. Hough, M.F. Islam, P.A. Janmey, A.G. Yodh, Viscoelasticity of Single Wall Carbon Nanotube Suspensions, *Phys. Rev. Lett.* 2004, 93, 168 102.

74 Vigolo, B., Coulon, C., Maugey, M. et al. (2005). An experimental approach to the percolation of sticky nanotubes. *Science* 309: 920.

75 Li, Y.-L., Kinloch, I.A., and Windle, A.H. (2004). Direct spinning of carbon nanotube fibers from chemical vapor deposition synthesis. *Science* 304: 276.

76 Zhang, M., Fang, S.L., Zakhidov, A.A. et al. (2005). Strong, transparent, multifunctional, carbon nanotube sheets. *Science* 309: 1215.

77 Islam, M.F., Rojas, E., Bergey, D.M. et al. (2003). High weight fraction surfactant solubilization of single-wall carbon nanotubes in water. *Nano Lett.* 3: 269.

78 Wei, J.Q., Zhu, H.W., Jiang, B. et al. (2003). Significantly enhanced thermoelectric properties of ultralong double-walled carbon nanotube bundle. *Carbon* 41: 2495.

79 Li, Y.J., Wang, K.L., Wei, J.Q. et al. (2005). Low voltage energy-saving double-walled carbon nanotube electric lamps. *Carbon* 43: 31.

80 Vigolo, B., Pénicaud, A., Coulon, C. et al. (2000). Macroscopic fibers and ribbons of oriented carbon nanotubes. *Science* 290: 1331.

81 Vigolo, B., Poulin, P., Lucas, M. et al. (2002). Improved structure and properties of single-wall carbon nanotube spun fibers. *Appl. Phys. Lett.* 81: 1210.

82 Poulin, P., Vigolo, B., and Launois, P. (2002). Hot-drawing of single and multiwall carbon nanotube fibers for high toughness and alignment. *Carbon* 40: 1741.

83 Zhou, W., Islam, M.F., Wang, H. et al. (2004). Small angle neutron scattering from single-wall carbon nanotube suspensions: evidence for isolated rigid rods and rod networks. *Chem. Phys. Lett.* 384: 185.

84 Yang, J., Tian, M., Jia, Q.-X. et al. (2007). Improved mechanical and functional properties of elastomer/graphite nanocomposites prepared by latex compounding. *Acta Materialia* 55: 6372–6382.

85 Bryning, M.B., Milkie, D.M., Islam, M.F. et al. (2005). Carbon nanotube aerogels. *Appl. Phys. Lett.* 87, 161 909.

86 Wang, J., Angnes, L., Tobias, H. et al. (1993). Capacitive deionization of NH_4ClO_4 solutions with carbon aerogel electrodes. *Anal. Chem.* 65: 2300.

87 Du, F.M., Guthy, C., Kashiwagi, T. et al. (2006). Polymer nanotube nanocomposites: synthesis, properties, and applications. *J. Polym. Sci.*, Part B 44: 1513.

88 Hjortstam, O., Isberg, P., Soderholm, S., and Dai, H. (2004). Applied Physics A: Materials Science & Processing. *Appl. Phys. A Mater. Sci. Process.* 78: 1175.

89 Nabeta, M. and Sano, M. (2005). Nanotube foam prepared by gelatin gel as a template. *Langmuir* 21: 1706.

90 Pinnavaia, T.J. and Beall, G.W. (2002). *Polymer-Clay Nanocomposites*. New York: John-Wiley.

91 Kistler, S.S. (1932). Colloidal silica: fundamentals and applications. *J. Phys. Chem.* 36: 52–64.

92 Kistler, S.S., Sol-Gel Processing and Applications, US Patent 2,093,454, 21 September 1937.

93 Ayers, M.R. and Hunt, A.J. (1998). Observation of the aggregation behavior of silica sol-gels. *J. Non-Cryst. Solids* 225: 343.

94 Morley, K.S., Licence, P., Marr, P.C. et al. (2004). Supercritical fluids: A route to palladium-aerogel nanocomposites. *J. Mater. Chem.* 14: 1212–1217.

95 Schwertfeger, F., Zimmerman A., Wonnerm J., Scholl, F., and Schmidt, M., Commissioner of Patents v. Farbwerke Hoechst Aktiengesellschaft Vormals Meister Lucius & Bruning, US Patent 6,143,400 (to Hoechst AG), 7 November 2000.

96 NASA/JPL website http://stardust.jpl.nasa.gov/index.html.

97 Mackenzie, R.C. (1952). Lipid-mobilizing activity during fasting. *Nature* 171: 681–683.

98 Call, F. (1953). *Houben-Weyl Methods of Organic Chemistry*, 4e: Science, Nature, vol. I/1, 172–126.

99 Weiss, A., Fahn, R., and Hofmann, U. (1952). Coordination assemblies from a Pd(II)-cornered square complex. *Naturwiss.* 39: 351–352.

100 Hofmann, U., Fahn, R., and Weiss, A. (1957). Novel procedures for determining the static and dynamic flow behavior of suspensions containing clay minerals. *Kolloid Z.* 151: 97–115.

101 Norrish, K. and Rausell-Colon, J.A. (1962). Foam-like polymer/clay aerogels which incorporate air bubbles. *Clay Miner. Bull.* 5: 9–16.

102 van Olphen, H. (1967). Tough polymer aerogels incorporating a conformal inorganic coating for low flammability and durable hydrophobicity. *Clay Miner.* 15: 423–435.

103 Nakazawa, H., Yamada, H., Fujita, T., and Ito, Y. (1987). Smectites in the montmorillonite-beidellite series. *Clay Sci.* 6: 269–276.

104 Somlai, L.S., Bandi, S.A., Mathias, L.J., and Schiraldi, D.A. (2006). pH tailoring electrical and mechanical behavior of polymer–clay–nanotube aerogels. *AICHE J.* 52 (3): 1–7.

105 Josef Innerlohinger, Hedda K. Weber, and Gregor Kraft, (2006) Aerocellulose, Aerocellulose: Aerogelsand Aerogel-like Materials made from Cellulose, *Lenzinger Berichte*, 86 137–143.

106 Fischer, F., Rigacci, A., Pirard, R. et al. Cellulose-based aerogels. *Polymer* in Press.

107 Prüße, U., Dalluhn, J., Breford, J., and Vorlop, K.-D. (2000). Production of spherical beads by JetCutting. *Chem. Eng. Technol.* 23: 1105–1110.

Index

Bio Monomers for Green Polymeric Composite Materials, First Edition.
Edited by P.M. Visakh, Oguz Bayraktar and Gopalakrishnan Menon.
© 2020 John Wiley & Sons Ltd. Published 2020 by John Wiley & Sons Ltd.